An Elsevier Civil Engineering Compendium

GEOTECHNOLOGY COMPENDIUM I

An Elsevier Civil Engineering Compendium

GEOTECHNOLOGY COMPENDIUM I

A collection of papers from the journals

Soil Dynamics and Earthquake Engineering
Computers and Geotechnics
International Journal of Rock Mechanics and Mining Sciences
Tunnelling and Underground Space Technology
Geotextiles and Geomembranes
Journal of Terramechanics

2000

2002
AMSTERDAM – BOSTON – LONDON – NEW YORK – OXFORD – PARIS
SAN DIEGO – SAN FRANCISCO – SINGAPORE – SYDNEY – TOKYO

ELSEVIER SCIENCE Ltd
The Boulevard, Langford Lane
Kidlington, Oxford OX5 1GB, UK

First edition 2002

Library of Congress Cataloging in Publication Data
A catalog record from the Library of Congress has been applied for.

British Library Cataloguing in Publication Data
A catalogue record from the British Library has been applied for.

ISBN: 0-08-044095-9

Transferred to digital printing 2005
Printed and bound by Antony Rowe Ltd, Eastbourne

PREFACE

This compendium is made up of a selection of the best and most representative papers from a group of Elsevier's geotechnology journals. Selections were made by the journals' Editorial teams.

The papers all appeared during 2000 and we hope they will provide a useful snapshot of the field for researchers, students and practitioners.

The journals from which papers have been selected and their Editors are:

Soil Dynamics and Earthquake Engineering *L. Finn, M. Erdik*
Computers and Geotechnics, *G. Pande, S. Pietruszczak*
International Journal of Rock Mechanics and Mining Sciences *J. Hudson*
Tunnelling and Underground Space Technology *R. Sterling, E. Broch, J. Zhao*
Geotextiles and Geomembranes *R. K. Rowe*
Journal of Terramechanics *G. Blaisdell, S. A. Shoop, P. W. Richmond*

Each paper appears in the same format as it was published in the journal; citations should be made using the original journal publication details.

It is intended that this compendium will be the first in a series of such collections. A compendium has also been published in the area of structural engineering-
(ISBN: 0-08-044038-X)

For full details of Elsevier's geotechnology publications, please visit the Elsevier website at www.elsevier.com or (where access is allowed) log on to ScienceDirect at www.sciencedirect.com.

James Sullivan
Elsevier Science Ltd
Langford Lane
Kidlington
Oxford OX5 1GB
United Kingdom

Email: j.sullivan@elsevier.co.uk

CONTENTS

Tunnelling and Underground Space Technology

Geotextiles and Geomembranes

Journal of Terramechanics

Papers from

Soil Dynamics and Earthquake Engineering

Energy dissipating restrainers for highway bridges

J.-M. Kim[a], M.Q. Feng[b,*], M. Shinozuka[c]

[a]*Department of Ocean Civil Engineering, Yosu National University, Yosu, Chonnam 550-749, South Korea*
[b]*Department of Civil and Environmental Engineering, University of California, Irvine, CA 92695-2175, USA*
[c]*Department of Civil Engineering, University of Southern California, Los Angeles, CA 90089-2531, USA*

Accepted 25 September 1999

Communicated by A. Cakmak

Abstract

Recent destructive earthquakes have demonstrated the vulnerability of highway bridges to collapse due to excessive movement beyond the available seat widths at expansion joints. This paper investigates the efficacy of using energy dissipating restrainers at expansion joints for preventing collapse of highway bridges in the event of a severe earthquake. The restrainer consists of a nonlinear viscous damper and an elastic spring connected in parallel or in series. Two-dimensional finite element analysis using bilinear hysteretic models for bridge substructure joints and nonlinear gap elements for expansion joints is performed on example bridges with one or two expansion joints. The analytical study demonstrates that the energy dissipating restrainers are effective in reducing the relative opening displacements and impact forces due to pounding at the expansion joints, without significantly increasing ductility demands in the bridge substructures. © 2000 Elsevier Science Ltd. All rights reserved.

Keywords: Bridge; Expansion joint; Energy dissipating; Restrainer; Earthquake

1. Introduction

Following the 1971 San Fernando earthquake, the California Department of Transportation (Caltrans) identified 1250 bridges as having vulnerable expansion joint hinges susceptible to collapse due to seismic response beyond the available seat widths [1,2]. To prevent unseating, these bridges have been retrofitted with restrainers made of steel rods or cables under Caltrans' phase I retrofit program. These restrainers are typically designed to work statically, rather than dissipate energy dynamically [3]. Post-earthquake evaluations from recent earthquakes have shown several cases where the restrainers failed. The partial collapse of the Gavin Canyon Undercrossing during the 1994 Northridge earthquake is one of the examples [4,5].

Use of energy dissipation devices as restrainers for expansion joints was proposed by one of the authors immediately following the1994 Northridge earthquake. In the previous studies performed by her and her associates, effort was made to demonstrate that linear viscous dampers are in principle effective in reducing the displacement at expansion joints [6–9]. They focused on the Gavin Canon Undercrossing and pounding effects at the expansion joints

were not considered in the analysis. In this study, nonlinear viscous dampers combined with elastic springs in parallel or in series are examined. Two general models representing typical Caltrans bridges built with expansion joints were considered for nonlinear response analysis [10]. The nonlinearities include yielding of columns, pounding of adjacent bridge frames at expansion joints, and nonlinear characteristics of the energy dissipating restrainers. The SAP2000/Nonlinear finite element computer code was used for the two-dimensional (longitudinal and vertical directions) seismic response analysis of the bridges involving nonlinearities [11].

2. Analytical models of bridges and energy dissipating restrainers

It is typical of California highway bridges with more than four spans to have expansion joints located nearly at inflection points (i.e. 1/4 to 1/5th of spans) to allow for thermal expansion. The bridge superstructures consist of reinforced or prestressed concrete box girders. Two typical Caltrans bridges with expansion joints are chosen in this study.

---

* Corresponding author.

Reprinted from *Soil Dynamics and Earthquake Engineering* **19 (1)**, 65-69 (2000)

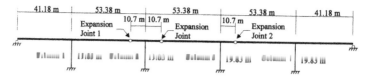

Fig. 1. Elevation of example bridges.

1. Model TY1H: five span bridge with one expansion joint.
2. Model TY2H: five span bridge with two expansion joints.

The geometry and boundary conditions of bridge models are shown in Fig. 1. The material and cross-sectional properties of the models are as follows: Young's modulus = 27.79 GPa, mass density = 2.40 mg/m^3, cross-section area and moment of inertia of the box girders are, respectively 6.936 m^2 and 4.787 m^4, and 4.670 m^2 and 1.735 m^4 for columns. These bridges have dominant horizontal vibration modes with periods of 0.5–1.0 s in each isolated frame separated by expansion joints.

The bridges are modeled with the SAP2000/Nonlinear finite element computer program. The nonlinearities involved in the bridge analytical model are depicted in Fig. 2(a). The plastic hinge formed in the bridge column is modeled as a hysteretic model. The expansion joint is constrained in the relative vertical movement, while allowing horizontal opening movement and rotation. The closure at the joint, however, is resisted by a linear impact spring with stiffness of 100 MN/m once the movement exceeds the initial gap width of 2.54 cm.

The energy dissipating restrainers examined in this study are nonlinear viscous dampers with the damping exponent of 0.5 combined with elastic springs in parallel or in series. The nonlinear viscous damper, whose force–velocity and force–displacement relationships are given in Fig. 2(b), is considered since it can prevent itself from generating excessively large force at large velocity and also dissipate larger energy at small velocity, compared with a linear viscous damper. The restrainers are installed at expansion joints between two adjacent bridge frames.

3. Results of nonlinear analysis

3.1. Input ground motions and restrainer parameter values

Four seismic ground motion, each with two components, are used as inputs for the two-dimensional simulation analysis. They are recorded during the El Centro earthquake (NS-component, UD-component, 1940), the Taft earthquake (N21E-component, UD-component, Lincoln School Tunnel, 1952), the Loma Prieta (EW-component, UD-component, Dumbarton Bridge, 1989) and the Northridge earthquake (NS-component, UD-component, Newhall, 1994). The horizontal components of the original accelerations were linearly scaled so that their PGAs are 0.70 g in accordance with the maximum PGA in the seismic design spectra used by Caltrans. The vertical components of these ground motions are scaled accordingly. These selected motions represent a variety of earthquakes with different durations and frequency components.

Twenty-one different values of spring constant k as well as damping coefficient c are chosen. For the restrainers with the damper and the spring connected in parallel, the parameter values are in the ranges $0 \le k \le 3.5 \times 10^4$ kN/m and $0 \le c \le 3.5 \times 10^4$ kN s/m, while those connected in series are, $0 \le k \le 10.0 \times 10^5$ kN/m and $0 \le c \le 3.5 \times 10^4$ kN s/m. Totally, 441 cases were analyzed for each model bridge and each earthquake input.

3.2. Relative displacements at expansion joints

Table 1 lists peak opening displacement at expansion

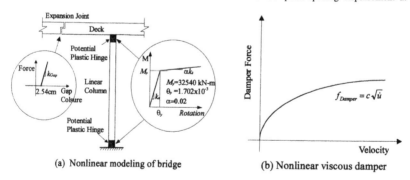

(a) Nonlinear modeling of bridge   (b) Nonlinear viscous damper

Fig. 2. Nonlinear models. (a) Nonlinear modeling of bridge; (b) Nonlinear viscous damper.

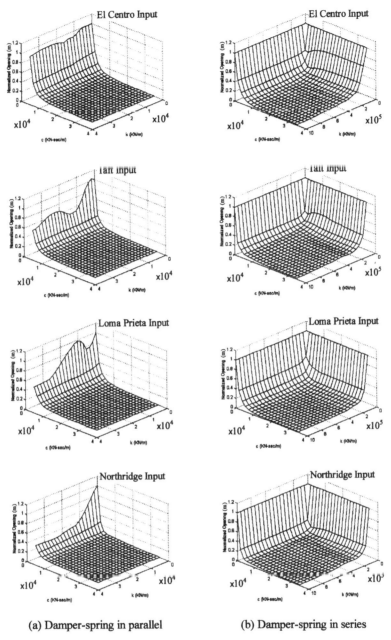

(a) Damper-spring in parallel   (b) Damper-spring in series

Fig. 3. Normalized peak opening displacements for bridge TY1H: (a) damper–spring in parallel; (b) damper–spring in series.

6

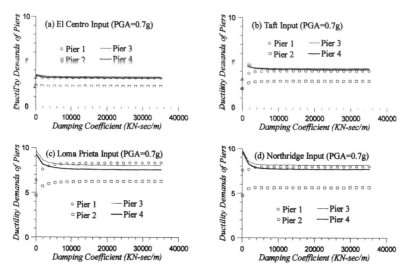

Fig. 4. Ductility demands in columns of bridge TY1H.

joints of the bridges without the energy dissipating restrainers installed under different earthquake ground motions. Normalized peak opening displacements at the expansion joints protected by the restrainers are plotted in Fig. 3 for bridge TY1H. The displacement values are normalized with respect to the peak openings without the restrainers.

For all the earthquake ground motions, the trends of the relative opening displacements at the expansion joints are quite similar for each type of restrainers. For the first type of restrainer, where the nonlinear damper and elastic spring are connected in parallel (with results shown in Fig. 3(a)), the effect of spring stiffness k on the peak relative displacement is not as pronounced as the effect of damping coefficient c. Particularly, when c is large the effect of elastic spring on the relative displacement is almost diminished. This suggests that the nonlinear viscous dampers are far more effective than the elastic springs (which are, in fact, the current steel cable/rod restrainers used in seismic retrofit of California bridges) in reducing the relative displacement at the expansion joint. Also, the nonlinear viscous damper alone can serve the purpose of the restrainer. For the second type of restrainer where the damper and the spring are connected in series (with results shown in Fig. 3(b)), the

larger the damping coefficient c and the spring stiffness k, the more effective the restrainer becomes.

3.3. Ductility demands in bridge substructures and impact forces

The effect of energy dissipating restrainers at the expansion joints on the bridge substructure performance is also examined. The ductility demands in the four columns of bridges TY1H are calculated and plotted in Fig. 4 as functions of the damping coefficient c under the four earthquakes. In this case, the elastic spring in the restrainer is eliminated, as it does not significantly contribute to the control of the opening displacement at the expansion joints as shown in Fig. 3(a). The ductility demand is defined as the ratio of plastic hinge rotation (θ) at the bottom of column to the yield rotation of the nonlinear spring (θ_y). Although the restrainer increases the ductility demand in columns 1 and 2 of bridge TY1H, it actually decreases the ductility demand in the rest of the columns. The ductility demands in all the columns approach a constant regardless of an increase of the damping coefficient, since the viscous dampers work to equalize the absolute horizontal displacement of the super-structures. More importantly, the maximum ductility

Table 1
Peak opening displacements at expansion joints without restrainers(the peak ground acceleration of scaled motion for the inputs is 0.70 g)

Bridges		El Centro Input (cm)	Taft Input (cm)	Loma Prieta Input (cm)	Northridge Input (cm)
TY1H		7.6	14.0	17.9	18.4
TY2H	Joint 1	29.0	21.8	40.4	49.4
	Joint 2	32.0	19.7	42.6	30.2

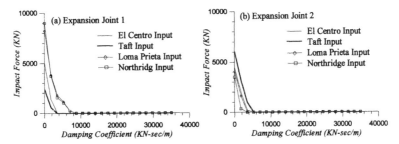

Fig. 5. Impact forces at expansion joints of bridge TY2H.

demand, among all the four columns, is associated with column 3 under the Loma Prieta earthquake when there is no restrainer installed ($c = 0$). Therefore, for this bridge with an identical cross-section for all the columns, the energy dissipating restrainer does not increase the maximum ductility demand in the columns. Similar observations can be made for bridge TY2H.

Fig. 5 plots the impact forces at the expansion joints 1 and 2 due to pounding of adjacent bridge frames of bridge TY2H under different earthquake ground motions, as functions of damping coefficient c and spring constant k. Again, elastic springs are not used for the restrainers in this analysis. The impact forces decrease substantially with increase of the damping in the restrainers. The same observations can be made for bridge TY1H. Therefore, the energy dissipating restrainers can significantly reduce the impact forces due to pounding, thus protecting the expansion joints from damage during severe earthquakes.

4. Conclusion

The efficacy of using energy dissipating devices consisting of nonlinear viscous dampers and elastic springs as restrainers for expansion joints of highway bridges is examined in this study through finite-element seismic response analysis. The nonlinearities of viscous dampers, plastic hinges of bridge substructures and pounding at the expansion joints are considered in the analysis. The results indicate that the nonlinear viscous dampers are significantly effective both in limiting the relative opening displacements and in reducing pounding forces at the expansion joints, without significantly increasing ductility demands in bridge substructures. The elastic springs, which are basically the current restrainers used in seismic retrofit of older bridges in California and elsewhere, are far less effective than the nonlinear viscous damper. Therefore, use of nonlinear viscous dampers at expansion joints is recommended. For older bridges with insufficient seat widths the dampers can be installed as restrainers for seismic retrofit purposes, while for newer bridges the dampers can be used to improve

seismic performance including reducing the negative effects of pounding at the expansion joints.

Acknowledgements

This study is supported by the National Center for Earthquake Research under the auspices of Federal Highway Administration (Contract No. DTFH61-92-C-00106) and the National Science Foundation (Grant No. CMS-9501796)

References

[1] Saiidi M, Maragakis E, Feng S. An evaluation of the current Caltrans seismic restrainer design method. Technical Report of Center for Civil Engineering Earthquake Research, No. CCEER92–8, 1992.

[2] Saiidi M, Maragakis E, Abdel-Ghaffer S, Feng S, O'Connor D. Response of bridge hinge restrainers during earthquakes–field performance, analysis and design, Technical Report of Center for Civil Engineering Earthquake Research, No. CCEER93–6, 1993.

[3] Caltrans, Bridge design specifications manual. California department of transportation, 1990.

[4] Buckle I, editor. The Northridge, California Earthquake of January 17, 1994: performance of Highway bridges (Technical Report NCEER-94-0008), 1994 National Center for Earthquake Engineering Research.

[5] Caltrans, Northridge Post Earthquake Investigation Report. California Department of Transportation - Division of Structures, 1994.

[6] Feng MQ, Chen W. Dampers as bridge hinge restrainers. Proceedings of the National Science Conference on Bridges and Highways, 1995 San Diego, CA.

[7] Klosek JT, Deodatis G, Shinozuka M. The Gavin canyon undercrossing: failure analysis under seismic loading and a retrofitting proposal using spring-damper systems. Proceedings of the National Science Conference on Bridges and Highways, 1995 San Diego, CA.

[8] Shinozuka M, Kim JM, Purasinghe R. Use of visco-elastic dampers at expansion joints to suppress seismic vibration of bridges. Proceedings of the National Seismic Conference on Bridges and Highways, Sacramento, CA 1997:487–98.

[9] Feng MQ, Kim JM, Shinozuka M, Purasinghe R. Visco-elastic dampers at expansion joints for seismic protection of bridges. Journal of Bridge Engineering 2000. To be published in February.

[10] Sultan M, Kawashima K. Comparison of the Seismic Design of Highway Bridges in California and in Japan. Recent selected publications of Earthquake Engineering Division, PWRI, Japan, 1994 Technical Memorandum of PWRI No. 3276.

[11] Computers and Structures, Inc., SAP2000/Nonlinear Users Manual, Berkeley, CA., 1990.

Seismic stability analysis of reinforced slopes

E. Ausilio, E. Conte, G. Dente[*]

Dipartimento di Difesa del Suolo, Università della Calabria, 87036 Rende, Cosenza, Italy

Accepted 27 January 2000

Abstract

In this paper, the seismic stability of slopes reinforced with geosynthetics is analysed within the framework of the pseudo-static approach. Calculations are conducted by applying the kinematic theorem of limit analysis. Different failure modes are considered, and for each analytical expressions are derived that enable one to readily calculate the reinforcement force required to prevent failure and the yield acceleration of slopes subjected to earthquake loading. Several results are presented in order to illustrate the influence of seismic forces on slope stability. Moreover, a suitable procedure based on the assessment of earthquake-induced permanent displacement is proposed for the design of reinforced slopes in seismically active areas. © 2000 Elsevier Science Ltd. All rights reserved.

Keywords: Seismic stability; Reinforced slope; Yield acceleration; Permanent displacement

1. Introduction

Construction of earth structures reinforced with geosynthetics has expanded extensively in the last twenty years, also in seismically active areas. Although severe damage was not observed during recent strong earthquakes [1–3], the performance of these structures under seismic loading is not fully known. This is due mainly to the almost absolute lack of well-documented case histories with specific regard to the properties of soil and reinforcement, earthquake characteristics and measurement of deformations earth structures undergo. Several studies have been, on the contrary, conducted on reduced-scale models using shaking table tests [4,5]. However, due to modelling limitations the results found are not often very meaningful when seeking to deduce the seismic performance of full-scale prototype structures from that of reduced-scale physical models.

To date, the theoretical approach is the most widely used to analyse seismic stability of reinforced earth structures. In the last few years, several analytical methods have been formulated and many parametric studies have been carried out to show the importance of the main design parameters [6–8]. The finite element method is certainly the most comprehensive approach to analyse the performance of soil structures subjected to seismic loading. However, its use usually requires high numerical costs and accurate

* Corresponding author. Tel.: + 39-0984-493086; fax: + 39-0984-493086.
  E-mail address: gdente@unical.it (G. Dente).

measurements of the properties of the component materials, which are often difficult to achieve. In addition, further difficulties arise with modelling failure in frictional materials [9]. This makes use of the finite element method not very attractive for current applications. Finite difference techniques have also been employed for seismic analysis of reinforced structures [6,10].

The majority of the methods used by practitioners are based on the pseudo-static approach where the effect of earthquake on a potential failure soil mass is represented in an approximate manner by a static force acting in the horizontal direction. The stability of the soil structure under this force is expressed by a safety factor that is usually defined as the ratio of the resisting force to the destabilising force. Failure occurs when the safety factor drops below one. The most common technique used for design is the limit equilibrium method. Many studies were conducted using this method, where the failure surface was assumed to have differing geometry [7,11–13]. In order to solve a limit state problem, limit analysis is also an effective methodology. Michalowski [8] has recently applied the kinematic theorem of limit analysis to calculate the reinforcement strength and length necessary to prevent collapse of earth structures, under the assumption that the reinforcement strength is uniformly distributed through the slope height or linearly increasing with depth. The solution proposed by Michalowski [8] should be used when a large number of reinforcement layers is installed. In the above studies, design charts have been presented to determine the reinforcement requirements for simple slopes.

Reprinted from *Soil Dynamics and Earthquake Engineering* **19** (3), 159-172 (2000)

(a)

(b)

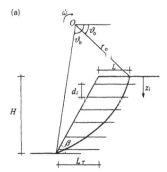

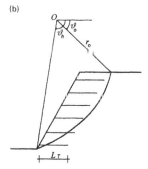

Fig. 1. Rotational slope failure mechanism: (a) log-spiral failure surface involving the reinforced zone; and (b) log-spiral failure surface extending within and beyond the reinforced zone.

As known, design based on pseudo-static analysis is generally considered conservative, since even when the safety factor drops below one the soil structure could experience only a finite displacement rather than a complete failure. A design procedure based on a tolerable displacement against sliding was proposed by Ling et al. [7]. They employed Newmark's sliding block method to evaluate permanent displacement of reinforced slopes during earthquakes. More recently, the same approach has been extended by Ling and Leshchinsky [14] to include the vertical component of ground acceleration.

In the present study, the kinematic theorem of limit analysis is applied to calculate both the forces required to ensure the stability and the yield acceleration for slopes reinforced with a discrete number of geosynthetic layers. Different possible failure mechanisms are considered, and for each of them analytical expressions that allow the above design parameters to be determined are derived. Comparisons of results obtained in this study to those existing in literature are shown, and the most significant differences are indicated. Results are also presented in order to illustrate the effect of seismic force on stability of reinforced earth structures. Finally, a suitable procedure based on the assessment of earthquake-induced permanent displacement is proposed for the design of reinforced slopes subjected to earthquakes.

2. Method of analysis

The kinematic theorem of limit analysis is applied here to analyse the stability of reinforced slopes under seismic loading. This theorem state that a slope will collapse if the rate of work done by external loads and body forces exceeds the energy dissipation rate for any assumed kinematically admissible failure mechanism. Applicability of the theorem requires that soil will be deformed plastically according to the normality rule associated with the Coulomb yield condition.

Following the pseudo-static approach, the effect of earth-

quake on a potential failure soil mass is represented by a force acting horizontally at the centre of gravity, which is calculated as the product of a seismic intensity coefficient and the weight of the potential sliding mass. An appropriate value of the seismic coefficient should be selected to account for possible acceleration amplification that is not implicitly considered in the analysis. The effects of pore pressure build-up and change of soil strength due to earthquake shaking are ignored. The analysis concerns slopes of homogeneous cohesionless soils, where the reinforcement layers are finite in number and have the same length. The reinforcement provides forces acting in the horizontal direction that are given by the tensile strength or pull-out resistance of the layers. As is usually assumed in the case of geosynthetics, resistance to shear, bending and compression is ignored. Under these assumptions, the rate of external work is due to soil weight and inertia force induced by earthquake and the only contribution to energy dissipation is that provided by the reinforcement.

The possible failure modes considered in this work are illustrated in Figs. 1–3. Specifically, Fig. 1 shows a rotational mechanism involving a log-spiral failure surface passing through the toe of the slope which may extend within (Fig. 1a) and also beyond (Fig. 1b) the reinforced zone. Moreover, the simple failure mechanism of Fig. 2 is also analysed, where the soil mass translates as a rigid body along a planar surface. Finally, in the mechanism schematised in Fig. 3, the reinforced soil mass slides over the

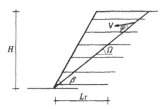

Fig. 2. Translational slope failure mechanism.

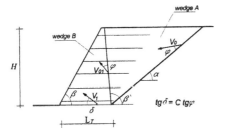

Fig. 3. Direct sliding mechanism.

bottom layer. This latter failure mode is known as the direct sliding mechanism. The slope is unstable when one of the above considered failure mechanisms occurs.

2.1. Log-spiral failure mechanism

In this failure mode, the reinforced soil mass above the failure surface rotates as a rigid body about the centre of rotation with angular velocity $\dot{\omega}$. The log-spiral failure surface (Fig. 1) is described by the equation

$$r = r_0 e^{(\vartheta - \vartheta_0)tg\varphi} \tag{1}$$

where r_0 = radius of the log-spiral with respect to angle θ_0; and φ = angle of soil shearing resistance. The assumed mechanism is completely specified by three variables: the height of the slope H, and the angles θ_0 and θ_h. Following Chang et al. [15], the rate of work due to soil weight and inertial force takes the form

$$\dot{W} = \gamma r_0^3 \dot{\omega}[f_1 - f_2 - f_3] + k_h \gamma r_0^3 \dot{\omega}[f_4 - f_5 - f_6] \tag{2}$$

where k_h = seismic coefficient; γ = soil unit weight; and the functions f_1–f_6 depend on the angles θ_0, θ_h, φ and β. Expressions for f_1–f_6 can be found in several works [8,15,16], for the sake of completeness they are also reported in the Appendix A of this paper. Eq. (2) provides the rate of work done by the body forces for a simple slope such as that shown in Fig. 1. For slopes of general shape the calculation for $\dot{W}$ may be carried out numerically. Moreover, when the slope is subjected to surcharge loads, the rate of work done by these loads and their inertia forces has to be added to $\dot{W}$ [15].

The energy dissipation rate during rotational failure due to reinforcement can be written as

$$\dot{D} = \dot{\omega} r_0 \sum_{i=1}^{n} T_i \left(\sin \theta_0 + \frac{z_i}{r_0} \right) \tag{3}$$

where z_i = depth of layer i measured downwards from the top of the slope; n = number of the reinforcement layers; T_i = force of the ith layer per unit width. Specifically, T_i = min$[T_u, T_p]$, where T_u = tensile strength of the reinforcement, and T_p = pull-out force which is usually expressed as

$$T_p = 2\gamma z_i l_i \mu tg\varphi \tag{4}$$

with l_i = anchorage length of ith layer, and μ = soil-reinforcement pull-out coefficient that ranges typically from 0.7 to 1.0.

As suggested by Ling et al. [7], expressing the total reinforcement force in a normalised form as

$$K = \frac{\sum_{i=1}^{n} T_i}{(1/2)\gamma H^2} \tag{5}$$

the value of T_i for each layer may be calculated in an approximate manner by the following expression:

$$T_i = K\gamma z_i d_i \tag{6}$$

with d_i = tributary area of layer i. Moreover, it can be shown that

$$\frac{H}{r_0} = a \tag{7}$$

where $a = \sin \theta_h\, e^{(\theta_h - \theta_0)tg\varphi} - \sin \theta_0$.

Substituting these latter relations into Eq. (3) gives

$$\dot{D} = \dot{\omega} r_0 K \gamma \sum_{i=1}^{n} \left[z_i d_i \left(\sin \theta_0 + \frac{z_i a}{H} \right) \right]. \tag{8}$$

Finally, equating the rate of energy dissipation to the rate of work of soil weight and seismic force leads to the expression for the normalised required force

$$K = \frac{H^2}{a^2} \frac{(f_1 - f_2 - f_3) + k_h(f_4 - f_5 - f_6)}{\sum_{i=1}^{n} [z_i d_i(\sin \theta_0 + (z_i a/H))]}. \tag{9}$$

When the reinforcement layers are uniformly spaced,

$$z_i = i\frac{H}{n} \quad \text{and} \quad d_i = \frac{H}{n} \tag{10}$$

consequently, Eq. (9) becomes

$$K = \frac{n^2}{a^2} \frac{(f_1 - f_2 - f_3) + k_h(f_4 - f_5 - f_6)}{\sum_{i=1}^{n} [i(\sin \theta_0 + (ia/n))]} \tag{11}$$

Eq. (9) or, alternatively, Eq. (11) provides a lower-bound solution for the reinforcement force necessary to prevent slope failure. In order to find the best estimation of K, an optimisation procedure needs to be used to maximise K with respect to θ_0 and θ_h. Once these angles are found, the geometry of the critical failure surface is completely defined. Moreover, one can also calculate length L, which expresses the distance between the failure surface at the top of the slope and the edge of the slope (Fig. 1). It is

$$\frac{L}{H} = \frac{1}{a} \left[\frac{\sin (\theta_h - \theta_0)}{\sin \theta_h} - a\frac{\sin (\theta_h + \beta)}{\sin \theta_h \sin \beta} \right]. \tag{12}$$

As will be shown later, L is a useful parameter to establish the reinforcement length to resist failure.

The kinematic theorem can be applied again to give the

upper-bound solution for the yield acceleration factor of the log-spiral failure mechanism:

$$k_y = \frac{K \sum_{i=1}^{n} [z_i d_i(\sin \theta_o + (z_i a/H))] - \dfrac{H^2}{a^2}(f_1 - f_2 - f_3)}{(H^2/a^2)(f_4 - f_5 - f_6)}$$

(13)

which, when the layers are uniformly spaced, becomes

$$k_y = \frac{K(a^2/n^2) \sum_{i=1}^{n} [i(\sin \theta_o + (ia/n))] - (f_1 - f_2 - f_3)}{(f_4 - f_5 - f_6)}.$$

(14)

The yield acceleration is defined, as the horizontal ground acceleration in the downhill direction required to bring the safety factor with respect to slope failure to one. However, it is assumed that the yield acceleration in the uphill direction is sufficiently large for failure in that direction not to occur. The critical seismic coefficient is obtained by minimising k_y with respect to θ_o and θ_h. This means taking the first derivatives of k_y and equating them to zero, i.e.

$$\begin{cases} \dfrac{\partial k_y}{\partial \theta_o} = 0 \\ \dfrac{\partial k_y}{\partial \theta_h} = 0 \end{cases}$$

(15)

Solving these equations and substituting the values of θ_o and θ_h thus obtained into the expression for k_y, the least upper-bound value for the yield acceleration factor is calculated. This critical value of k_y will be indicated in the following text as k_c.

It should be noted that the above expressions are derived under the assumption that the reinforced layers are firmly anchored in the soil. However, as will be shown later, when the seismic force increases the critical failure expands into the backfill, consequently some reinforcement layers located at the top of the slope could be pulled out if the anchorage length is not sufficient to sustain the required force (Fig. 1b). In such circumstance, it may be assumed that only the bottom layers contribute to ensure global stability of the slope by means of their tensile resistance [7]. Therefore, the expression for k_y is

$$k_y = \frac{\dfrac{T_u}{\gamma} \sum_{i=1}^{m} (\sin \theta_o + (z_i a/H)) - (H^2/a^2)(f_1 - f_2 - f_3)}{(H^2/a^2)(f_4 - f_5 - f_6)}$$

(16)

where m is the number of the reinforcement layers, located at the bottom, that are necessary to ensure slope stability. The maximum value of m can be evaluated by the relation

$$m T_u = \sum_{i=1}^{n} T_i$$

(17)

or equivalently in terms of K as

$$m = \frac{1}{2} K \gamma \frac{H^2}{T_u}.$$

(18)

2.2. Plane failure mechanism

The reinforced soil mass translates as a rigid body with velocity V (Fig. 2). The mechanism is specified by two variables: height H, and angle Ω that the failure plane makes with the horizontal. The rate of external work is due again to the rate of work done by soil weight and inertial force:

$$\dot{W} = GV \sin(\Omega - \varphi) + k_h GV \cos(\Omega - \varphi)$$

(19)

where G indicates the weight of the soil wedge, which is given by

$$G = \frac{1}{2} \gamma H^2 \frac{\sin(\beta - \Omega)}{\sin \Omega \sin \beta}$$

(20)

the energy dissipation rate due to the reinforcement is

$$\dot{D} = V \cos(\Omega - \varphi) \sum_{i=1}^{n} T_i$$

(21)

that owing to Eq. (5) becomes

$$\dot{D} = \frac{1}{2} V \cos(\Omega - \varphi) K \gamma H^2$$

(22)

Equating the rate of work due to internal forces to that of external forces yields

$$K = \frac{\sin(\beta - \Omega)}{\sin \Omega \sin \beta} [tg(\Omega - \varphi) + k_h]$$

(23)

which attains a maximum value when $(\delta K / \delta \Omega) = 0$. The expression for Ω that satisfies this condition is

$$\Omega = tg^{-1} \left\{ \frac{-k_h tg\varphi tg\beta + tg^2\varphi tg\beta + \sqrt{-tg\beta(tg^2\varphi + 1)(tg\varphi tg\beta + 1)(k_h - tg\varphi)}}{1 + tg^2\varphi + k_h tg^2\varphi tg\beta + tg\varphi tg\beta} \right\}$$

(24)

and the expression for the distance L between the failure surface and the edge of the slope is

$$\frac{L}{H} = \frac{\sin(\beta - \Omega)}{\sin \beta \sin \Omega}.$$

(25)

Similarly, the upper bound expression for the yield acceleration factor in the downhill direction can be obtained:

$$k_y = K \frac{\sin \Omega \sin \beta}{\sin(\beta - \Omega)} - tg(\Omega - \varphi)$$

(26)

it attains a minimum value at the wedge angle

$$\Omega = tg^{-1} \left\{ \frac{tg\beta[-Ktg\beta tg\varphi - tg^2\varphi - 1 + \sqrt{K(tg^2\varphi + 1)(1 + tg\beta tg\varphi)^2}]}{Ktg^2\beta tg^2\varphi - tg^2\varphi - 1} \right\}.$$

(27)

It should be noted that both Eqs. (24) and (27) become singular when $\beta = 90°$. However, this drawback can be

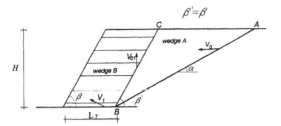

Fig. 4. Direct sliding mechanism, where energy dissipation due to reinforcement is zero (case $\beta' = \beta$).

easily overcome by using a value of β very close to $90°$.

2.3. Direct sliding mechanism

In direct sliding analysis, the soil mass is treated as a rigid body that translates outward over the bottom layer of the reinforcement when ground acceleration exceeds yield acceleration against sliding. The mechanism consists of two wedges (wedge A and wedge B) which are separated by a plane failure surface (Fig. 3). This mechanism is geometrically specified by embankment height H, total reinforcement length L_T (including anchorage length), and angles (α and β') that the planar failure surfaces make with the horizontal.

In order to find the expression for the yield acceleration factor, two situations are analysed. In the first case when $\beta' = \beta$ (Fig. 4), soil wedge A is assumed to slide with velocity V_o against the reinforced soil mass indicated as

wedge B. The two wedges are separated by a planar surface that is parallel to the slope face. Since energy dissipation is zero, the energy balance equation yields

$$G_A V_o \sin(\alpha - \varphi) + k_y G_A V_o \cos(\alpha - \varphi)$$

$$- G_B V_1 \sin \delta + k_y G_B V_1 \cos \delta = 0 \qquad (28)$$

where G_A and G_B indicate the weight of wedges A and B, respectively, and δ is the soil-reinforcement friction angle at the base of wedge B. It can be assumed that $tg\delta = C tg\varphi$, with C = reinforcement-soil interface coefficient. The magnitude of velocity V_1 is expressed as a function of V_o by the relation

$$V_1 = V_o \frac{\sin(\beta + 2\varphi - \alpha)}{\sin(\beta + \delta + \varphi)}. \qquad (29)$$

Substituting Eq. (29) into Eq. (28) and rearranging the terms leads to the following expression for k_y:

$$k_y = \frac{G_B \dfrac{\sin(\beta + 2\varphi - \alpha)}{\sin(\beta + \delta + \varphi)} \sin \delta - G_A \sin(\alpha - \varphi)}{G_A \cos(\alpha - \varphi) + G_B \dfrac{\sin(\beta + 2\varphi - \alpha)}{\sin(\beta + \delta + \varphi)} \cos \delta}. \qquad (30)$$

The yield acceleration factor attains the minimum value k_c, when the condition $(\partial k_y / \partial \alpha) = 0$ is satisfied.

In order to analyse the other situation, that is when $\beta' > \beta$ (Fig. 5), the energy dissipation rate along BC has to be included on the right-hand side of Eq. (28). It is

$$\dot{D} = V_1 \cos \delta \sum_{i=1}^{p} T_i \qquad (31)$$

where p = number of reinforcement layers intersected by the plane BC. As a consequence, the expression for k_y becomes

$$k_y = \frac{G_B \dfrac{\sin(\beta' + 2\varphi - \alpha)}{\sin(\beta' + \delta + \varphi)} \sin \delta - G_A \sin(\alpha - \varphi) + \dfrac{\sin(\beta' + 2\varphi - \alpha)}{\sin(\beta' + \delta + \varphi)} \cos \delta \sum_{i=1}^{p} T_i}{G_A \cos(\alpha - \varphi) + G_B \dfrac{\sin(\beta' + 2\varphi - \alpha)}{\sin(\beta' + \delta + \varphi)} \cos \delta} \qquad (32)$$

The critical value for k_y can be found by minimising Eq. (32) with respect to α and β'.

2.4. Comparisons to other solutions

It is instructive to compare the values of K and k_c calculated by means of the expressions derived in the previous sections to those obtained from other solutions publishing in the literature in order to show the possible differences. Several plots have been recently published by Ling et al. [7] and by Michalowski [8] to illustrate the effect of seismic forces on the stability of reinforcement slopes. Specifically, Ling et al. [7] conducted a parametric study using a procedure based on a pseudo-static limit equilibrium analysis. Results in terms of reinforcement force and length necessary to ensure stability were presented for a slope 5 m high having different inclination angles and soil

14

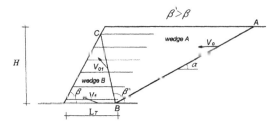

Fig. 5. Direct sliding mechanism, where energy dissipation due to reinforcement along the planar surface BC is included (case $\beta' > \beta$).

properties, and being reinforced by 20 equally spaced layers. An expression was also derived by the same authors, which allows the yield acceleration against direct sliding to be calculated. Regarding this, Ling et al. [7] considered a failure mechanism defined by two soil wedges (the retained wedge and the reinforced wedge) that are separated by an ideal vertical plane. The effect on k_c due to reinforcement forces was not included in the analysis. In turn, Michalowski [8] applied the kinematic theorem of limit analysis to evaluate the amount of reinforcement required to prevent slope collapse assuming that reinforcement strength is uniformly distributed or linearly increasing with depth. The solution found by Michalowski [8] is independent of the number of layers installed, and neither does it depend on their real location. Design charts were presented by the authors for different intensities of seismic force.

A comparison of the results obtained in this study to those presented by Ling et al. [7] and Michalowski [8] is given in Figs. 6 and 7, for two different slope angles ($\beta = 60$ and 90°). As can be seen from these figures, the equations derived in this work provide values for K that are in close agreement with those read from the graphs by Ling et al. [7]. The differences between the results are in fact less than 5%. Similarly, the results agree very well with those obtained from Michalowski [8] for slopes with a triangular distribution of reinforcement forces. However, significant differ-

ences can be noted in the case of slopes with a uniform distribution of forces, when a greater amount of reinforcement is generally required.

Figs. 8 and 9 compare the values of the critical acceleration factor against direct sliding as derived from Eqs. (30) and (32) with those calculated using the expression proposed by Ling et al. [7]. In the calculations it is assumed: $C = 0.8$, $n = 10$, $L = 5$ m, $H = 5$ m and $T_u = 20$ kN/m. As can be seen from Fig. 8, for the slope with $\beta = 60°$ the results are rather different due to differences in the geometry of the sliding mechanism considered. When φ is low the values of k_c calculated by means of the equations derived in this study are significantly larger than those presented by Ling et al. [7]. This is because the former take into account the effect of the reinforcement between the sliding wedges. However, as φ increases the differences between the results decrease, and when $\varphi > 40°$ the values of k_c calculated in this study are smaller than those derived from the expression proposed by Ling et al. [7]. The results are essentially equal when $\beta = 90°$ (Fig. 9), since in this case the failure mechanisms are the same in both analyses.

3. Results

In this section, a series of calculations is carried out to

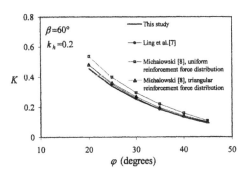

Fig. 6. Comparison of required reinforcement force K, for a slope with $\beta = 60°$.

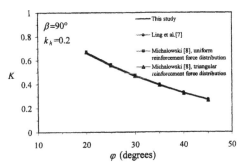

Fig. 7. Comparison of required reinforcement force K, for a slope with $\beta = 90°$.

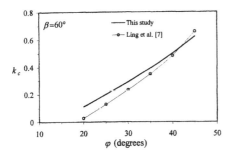

Fig. 8. Comparison of critical acceleration factor k_c, for a slope with $\beta =$ 60°.

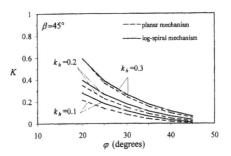

Fig. 10. Required force versus φ at different seismic coefficients for a slope with $\beta = 45$°.

illustrate the effect of inertial force on the stability of a reinforced slope 5 m in height. Both the rotational log-spiral failure mechanism and the planar translational mechanism are considered. It is assumed that the soil is cohesionless with unit weight $\gamma = 18$ kN/m³ and angle of soil shearing resistance φ ranging from 20 to 45°. Moreover, three different slope angles are considered ($\beta = 45$, 60 and 75°). The reinforcement consists of 10 equally spaced layers.

Figs. 10–12 show the required normalised force K obtained using Eqs. (11) and (23), for different values of the seismic coefficient ($k_h = 0.1$, 0.2 and 0.3). Specifically, the results derived from Eq. (11) are indicated by an unbroken line, whereas those calculated by Eq. (23) are marked by a dashed line. In practice, these figures can be used to compare the values of K obtained when the rotational or the translational failure mechanism occurs. As can be expected, K increases with decreasing φ and with increasing k_h. The increase in K is greater in the case of the steeper slope ($\beta = 75$°). As can be noted, the planar failure mechanism leads to values of K that are fairly close to those obtained when the log-spiral failure mechanism is considered. The slight differences between the results are sometimes more and other times less. This is because, when the log-spiral failure mode is assumed to occur, the required forces are affected

by the number of installed reinforcement layers whereas they are not dependent on n for the translational mechanism. The influence of n on K is illustrated in Fig. 13, where a slope with $\beta = 60$° is considered as an example. As can be seen, when $n = 5$ the values of K for the log-spiral mechanism are lower than those for the translational mechanism. However, as n increases the values of K increases, and when $n = 20$ the rotational mechanism becomes more critical than the translational mechanism. In any case, the influence of n on K appears not to be remarkable. This implies that for practical purposes the force required to prevent slope failure under seismic loading may be calculated using the relatively simpler Eq. (23) rather than Eq. (11).

Similar remarks can be made about the yield acceleration factor. From Figs. 14–16, it is seen that k_c increases as φ or K increases. Moreover, in the cases examined (when $n = 10$) the yield acceleration factor for the log-spiral failure mechanism is generally higher than that obtained for the plane failure mechanism. However, the differences between the results are again not remarkable, especially when the slope is gentle (Fig. 14).

Figs. 17 and 18 show the critical log-spiral failure surfaces determined for a slope with $\beta = 60$° and $\beta = 75$°, respectively, as the seismic coefficient increases from

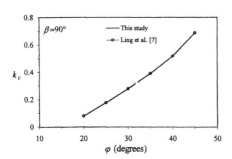

Fig. 9. Comparison of critical acceleration factor k_c, for a slope with $\beta = 90$°.

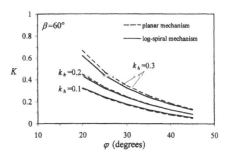

Fig. 11. Required force versus φ at different seismic coefficients for a slope with $\beta = 60$°.

16

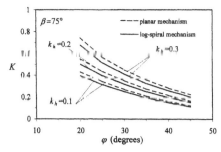

Fig. 12. Required force versus φ at different seismic coefficients for a slope with $\beta = 75°$.

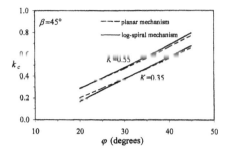

Fig. 14. Critical acceleration factor versus φ at different seismic coefficients for a slope with $\beta = 45°$.

0.1 to 0.3. The value of φ is assumed to be 30°. The results indicate that, as k_h varies from 0.1 to 0.3, the volume of the critical sliding mass expands into the backfill, particularly at the top of the slope. This implies that distance L between the failure surface and the edge of the slope may be used, in practice, to establish the length of the reinforcement, especially in the case of steep slopes and high values of the seismic intensity. Moreover, when the slope angle is nearly vertical and k_h is large, the log-spiral failure surface approaches a planar surface geometry (Fig. 18).

Referring to the cases considered in this section, the ratio (L/H) is plotted versus φ in Figs. 19–21. As can be noted, this ratio increases as φ decreases or k_h increases. Moreover, the results show that the planar failure mechanism leads to values of L/H that are rather conservative (i.e. longer reinforcement lengths) with respect to those calculated when the log-spiral mechanism is assumed, especially when the slope is gentle and φ is low. For a steep slope, however, the differences between the results are very slight (Fig. 21). Finally, Fig. 22 compares several values of L/H calculated using Eq. (25) for the case $\beta = 60°$ and $k_h = 0.2$ to those read from the graphs presented by Ling et al. [7] and Michalowski [8]. Although differences among the results can be noted, Eq. (25) gives results that generally fall within the range of the values for L. Therefore, this equation may come

in handy for a preliminary evaluation of the reinforcement length in the current applications.

4. Assessment of permanent displacement

The calculations carried out using the pseudo-static approach indicate that both force and length of the reinforcement increase considerably with an increase in the seismic force. Consequently, for large values of the seismic coefficient, the design of a reinforced soil structure could prove very expensive or even impracticable. In these circumstances, it is more reasonable to reduce the amount of the reinforcement and consequently accept that the structure is affected by permanent displacements during earthquakes. Due to the transient nature of ground motion, the slope could in fact experience only a finite displacement rather than a complete failure.

The calculations for permanent displacement are usually conducted using the sliding block method originally proposed by Newmark [17]. According to this method, the potential failure soil mass is treated as a rigid block on an inclined plane, which moves in the downhill direction whenever ground acceleration exceeds yield acceleration of the slope. Given a design

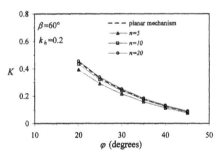

Fig. 13. Effect of n on K.

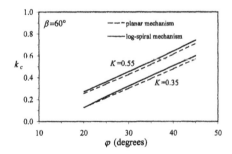

Fig. 15. Critical acceleration factor versus φ at different seismic coefficients for a slope with $\beta = 60°$.

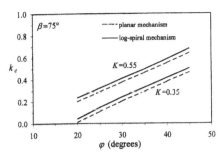

Fig. 16. Critical acceleration factor versus φ at different seismic coefficients for a slope with β = 75°.

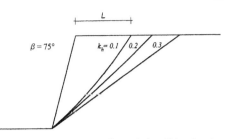

Fig. 18. Critical failure surfaces at different seismic coefficients for a slope with β = 75°.

accelerogram, the earthquake-induced displacement can be obtained by integrating twice the equation of motion, that is

$$\ddot{x} = [k_\mathrm{h}(t) - k_\mathrm{c}]g\frac{\cos(\varphi - \theta)}{\cos\varphi} \quad (33)$$

where $\ddot{x}$ = acceleration of the sliding block relative to the slip surface; $k_\mathrm{h}(t)g$ = ground acceleration time-history; $k_\mathrm{c}g$ = yield acceleration that is usually assumed to be constant with time; g = gravity acceleration; and θ = angle that the inclined plane makes with the horizontal. In the case of the rotational failure mechanism, it is more appropriate to express the equation of motion in terms of the angular rotation of the failure mass relative to the stable soil. The following equation can be derived [18]:

$$\ddot{\omega} = [k_\mathrm{h}(t) - k_\mathrm{c}]g\frac{R_\mathrm{gy}}{R_\mathrm{g}^2} \quad (34)$$

where $\ddot{\omega}$ = angular acceleration of the rotating mass; R_g = distance of the centre of gravity of the sliding mass from the centre of rotation; and R_gy = vertical component of R_g. When ω is small, the displacement of a point at the failure surface is given by the product of ω and the radius r. However, for practical purposes, Eq. (33) may be used instead of Eq. (34) to calculate in an approximate manner the permanent displacement due to the rotational mechanism, evaluating the inclination angle θ from the vector

diagram of the forces acting on the sliding soil mass under limit condition as shown in Fig. 23.

Newmark's approach is used extensively to carry out assessment of slope stability in seismically active areas. Based on earthquake acceleration records, many relationships between the expected permanent displacement and characteristic seismic parameters have been proposed for predicting permanent displacement of earth structures subjected to seismic loading. These relationships are very suitable to be used in current applications. Probabilistic methods were also developed. They provide an estimate of the expected permanent displacement with a given degree of confidence level or probability of exceedance [20–23]. A review of existing sliding block methods has been presented recently by Cai and Bathurst [19]. From the results shown by Cai and Bathurst [19], it appears that the following regression equation:

$$\log U = 0.9 + \log\left[\left(1 - \frac{k_\mathrm{c}}{k_\mathrm{h}}\right)^{2.53}\left(\frac{k_\mathrm{c}}{k_\mathrm{h}}\right)^{-1.09}\right] + 0.3t \quad (35)$$

that was proposed by Ambraseys and Menu [21], gives an adequate upper bound estimate of permanent displacement when a 95% confidence level is selected. In Eq. (35), U indicates the horizontal component of the permanent displacement in cm, and t is a function of the confidence level selected, which can be obtained from a normal distribution

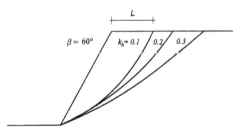

Fig. 17. Critical failure surfaces at different seismic coefficients for a slope with β = 60°.

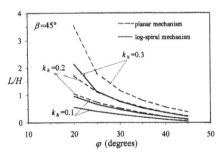

Fig. 19. L/H versus φ at different seismic coefficients for a slope with β = 45°.

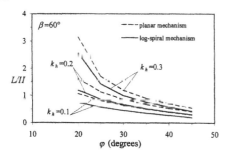

Fig. 20. L/H versus φ at different seismic coefficients for a slope with $\beta = 60°$.

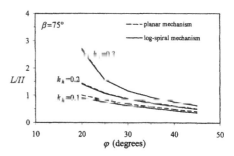

Fig. 21. L/H versus φ at different seismic coefficients for a slope with $\beta = 75°$.

table. Cai and Bathurst [24] used a similar approach to investigate the displacement response of reinforced retaining walls. They demonstrated reasonably good agreement between deterministic Newmark analyses and methods of the type used in this paper.

Comparisons of the permanent displacement calculated by means of Eq. (35) with field performance of several reinforced earth walls during recent strong earthquakes are illustrated in Table 1. All the data available about the geometry of the structures, the material properties and earthquake characteristics can be found in the paper by Ling et al. [7], who previously analysed the seismic performance of such reinforced earth structures. The same authors used surface cracking as an indication that permanent displacement had occurred. Only in three situations, was there evidence of cracking after the earthquake. In the other cases, no evidence of permanent deformation was observed. In the same way as Ling et al. [7], the calculations have been conducted here neglecting the facing contribution to seismic performance of the structures. For the sake of simplicity, the total required force to resist failure is determined using Eq. (23). The values of K are shown in Table 1 together with the

yield acceleration against direct sliding and the permanent displacement obtained by Eq. (35) assuming confidence levels of 50 and 95%. The measured crack width and the values of K, k_c and U calculated by Ling et al. [7] using different methodologies are also indicated in the same table. As can be seen, the results obtained in the present study are in good agreement with those presented by Ling et al. [7], in spite of the simplicity of the expressions used for calculation.

5. A design procedure based on the assessment of permanent displacement

In the following, a simple procedure for the design of reinforced slopes under seismic loading is presented. The main input data for the slope are height H, inclination angle β, angle of soil shearing resistance φ, and tensile strength of the reinforcement T_u. It is assumed that failure occurs according to the most critical failure mechanism, which is the mechanism with the smallest value of the yield acceleration factor. For a slope that has to be designed to resist an

Table 1
Data relative to several reinforced earth structures (modified by Ling et al. [7])

	Amagasaki wall Ref. [2]	Gould wall Ref. [13]	Valencia wall Ref. [13]	Valencia-top wall Ref. [13]	Seiken walls Ref. [25,26]	Seiken walls Ref. [25,26]
H (m)	4.7	4.6	6.4	2.2	5.5	5.5
L_T (m)	2.5	3.6	5.5	1.8	5.0	5.0
β (degrees)	90	86.4	86.4	86.4	78.7	78.7
φ (°)	35	33	33	33	37	37
γ (kN/m³)	20	20	20	20	18	18
T_u (kN/m³)	38	36	36	36	20	20
k_h	0.27	0.30	0.50	0.50	0.326	0.216
K Ling et al. [7]	0.452	0.482	0.773	0.744	0.375	0.286
K Eq. (23)	0.451	0.481	0.749	0.750	0.376	0.287
k_c Ling et al. [7]	0.310	0.312	0.324	0.318	0.427	0.427
k_c Eqs. (30) and (32)	0.306	0.310	0.322	0.316	0.415	0.415
Crack width (cm)	–	0.6	0.6	5	–	–
U Ling et al. [7] (cm)	–	–	16	6	–	–
U Eq. (35) (cm)	–	–	0.9–2.9	1–3.3	–	–

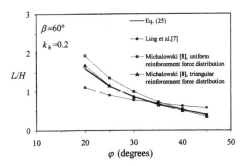

Fig. 22. Comparison of L/H for different values of φ.

Table 2
Required forces and anchorage length for the design example

Layer no.	$k_h = 0.35$		$k_h = 0.25$	
	$K = 0.407$	$L/H = 1.06$	$K = 0.314$	$L/H = 0.82$
	T_i (KN/m)	l_i (m)	T_i (KN/m)	l_i (m)
1	1.83	0.18	1.41	0.14
2	3.66	0.18	2.82	0.14
3	5.49	0.18	4.24	0.14
4	7.32	0.18	5.65	0.14
5	9.15	0.18	7.06	0.14
6	10.98	0.18	8.47	0.14
7	12.82	0.18	9.89	0.14
8	14.65	0.18	11.30	0.14
9	16.48	0.18	12.71	0.14
10	18.31	0.18	14.17	0.14
11	20.15	0.18	15.54	0.14
12	21.98	0.18	16.95	0.14

assigned peak ground acceleration $k_h g$, the normalised total force K required to prevent failure may be calculated by Eq. (11) or more easily by Eq. (23). Consequently, the values of T_i for each layer are derived from Eq. (6). Moreover, the total reinforcement length that is assumed to be the same for all the layers is

$$L_T = L + \max[l_i] \qquad (36)$$

where L = reinforcement length as derived from Eq. (12) or Eq. (25), and owing to Eqs. (4) and (6)

$$l_i = \frac{K d_i}{2 \mu t g \varphi} \qquad (37)$$

with i ranging from one to the total number of the layers, n.

If the length to resist failure is considered to be very long and/or the required reinforcement force is high, so that the number of layers is impracticable or even in some layers T_i

exceeds tensile strength, a reduced seismic coefficient is used and the calculations are repeated until suitable values for both K and L_T are obtained. The next step of the procedure consists of calculating the critical acceleration factor against direct sliding by means of Eqs. (30) and (32), and comparing it to that of the other failure mechanisms. The smallest value of k_c is used to evaluate the permanent displacement of the structure, in accordance with one of the existing methods based on Newmark's sliding block theory [19–24]. The main steps in the design procedure are illustrated by the flow chart in Fig. 24.

To show how the procedure may be used to design reinforced earth structures, a slope with $H = 6$ m and $\beta = 75°$ is considered as an example. In addition, it is assumed: $\varphi = 35°$, $\mu = 0.8$, $C = 0.8$, $T_u = 18$ kN/m and $k_h = 0.35$. The

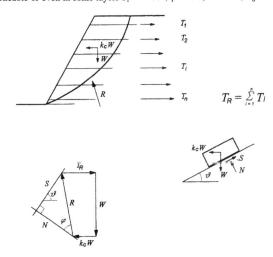

$$T_R = \sum_{i=1}^{n} T_i$$

Fig. 23. Calculation of the inclination angle θ.

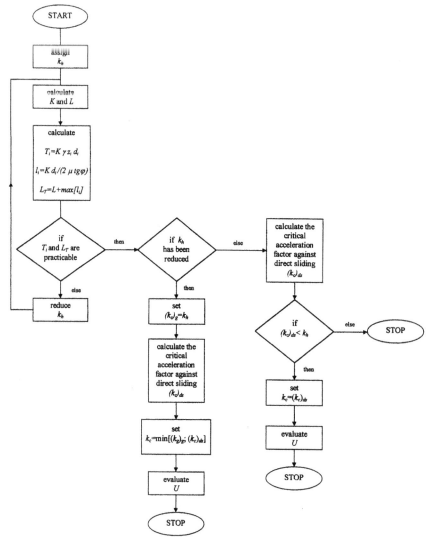

Fig. 24. Flow chart of the design procedure.

number of equally spaced reinforcement layers is set equal to 12. For the sake of simplicity, Eqs. (23) and (25) are used, respectively, to calculate the normalised total force K and length L/H required to ensure slope stability; these equations give $K = 0.407$ and $L/H = 1.06$. The force T_i and the anchorage length of each layer l_i may be evaluated, respectively, by using Eqs. (6) and (35). The results are shown in Table 2. As a consequence from Eq. (36), the total reinfor-

cement length is $L_T = 1.06 \times 6 + 0.18 = 6.5$ m, which is considered to be long. In addition, as can be noted from Table 2, for some layers $T_i > T_u$. Reducing k_h from 0.35 to 0.25, and repeating the process lead to values of K and L_T that are 0.314 and 5.1 m, respectively. Moreover, as shown in Table 2, all the forces T_i are smaller than T_u. However, we have to accept that permanent displacements occur during an earthquake with peak acceleration equal to $0.35g$. To evaluate displacement magnitude, the yield acceleration factor against direct sliding needs to be calculated. From Eqs. (30) and (32), one obtains values of k_c equal to 0.36 and 0.62, respectively, which are both higher than 0.25. Consequently, the most critical value of k_c is 0.25. Introducing this latter value into expression (35), for a 95% confidence level, gives

$$\log U = 0.9 + \log\left[\left(1 - \frac{0.25}{0.35}\right)^{2.53}\left(\frac{0.25}{0.35}\right)^{-1.09}\right] + 0.493$$

$$= 0.176$$

therefore, the calculated displacement is $U = 1.5$ cm, which may be considered acceptable.

6. Concluding remarks

The expressions derived in this paper by applying the kinematic theorem of limit analysis can be conveniently used to calculate a lower bound for the reinforcement force, K, required to prevent slope failure under seismic loading, or to obtain an upper bound for the critical acceleration factor, k_c, which is a key parameter to evaluate earthquake-induced permanent displacement using Newmark's sliding block method.

Although the rotational failure mechanism is generally more appropriate than the translational one, the results show that the latter failure mode leads to values of K and k_c that differ by only a few per cent from those calculated assuming a rotational mechanism. For the translational mechanism, where a planar failure surface is assumed to occur, closed-form solutions are derived that are very useful in current applications because of their simplicity.

The calculations carried out by the pseudo-static approach indicate that both reinforcement force and length increase significantly with an increase in seismic force. Consequently, for large values of the seismic coefficient, the design of reinforced soil structures could prove to be very expensive or even impracticable. In these circumstances, it is more reasonable to reduce the amount of the reinforcement and accept that the structure may experience permanent displacement during earthquakes. For this purpose, a suitable procedure based on the assessment of earthquake-induced permanent displacement using Newmark's sliding block method is proposed for design.

Appendix A

$$f_1 = \frac{\{(3tg\varphi\cos\theta_h + \sin\theta_h)\exp[3(\theta_h - \theta_o)tg\varphi] - 3tg\varphi\cos\theta_o - \sin\theta_o\}}{3(1 + 9tg^2\varphi)}$$

$$f_2 = \frac{1}{6}\frac{L}{r_o}\left(2\cos\theta_o - \frac{L}{r_o}\right)\sin\theta_o$$

$$f_3 = \frac{1}{6}\exp[(\theta_h - \theta_o)tg\varphi]\left[\sin(\theta_h - \theta_o) - \frac{L}{r_o}\sin\theta_h\right]$$

$$\times\left\{\cos\theta_o - \frac{L}{r_o} + \cos\theta_h\exp[(\theta_h - \theta_o)tg\varphi]\right\}$$

$$f_4 = \frac{1}{3(1 + 9tg^2\varphi)}\{(3tg\varphi\sin\theta_h - \cos\theta_h)\exp[3(\theta_h - \theta_o)tg\varphi]$$

$$- 3tg\varphi\sin\theta_o + \cos\theta_o\}$$

$$f_5 = \frac{1}{3}\frac{L}{r_o}\sin^2\theta_o$$

$$f_6 = \frac{1}{6}\exp[(\theta_h - \theta_o)tg\varphi]\left[\sin(\theta_h - \theta_o) - \frac{L}{r_o}\sin\theta_h\right]$$

$$\times\{\sin\theta_o + \sin\theta_h\exp[(\theta_h - \theta_o)tg\varphi]\}$$

References

[1] Collin JG, Chouery-Curtis VE, Berg RR. Field observations of reinforced soil structures under seismic loading. In: Proceedings of the International Symposium on Earth Reinforcement, Rotterdam, 1992. p. 223–28.

[2] Tatsuoka F, Tateyama M, Koseki J. Behavior of geogrid-reinforced soil retaining walls during the Great Hanshin-Awaji Earthquake. In: Proceedings of the First International Symposium on Earthquake Geotechnical Engineering, Tokyo, 1995. p. 55–60.

[3] Tatsuoka F, Tateyama M, Koseki J. Performance of soil retaining walls for railway embankments. Soils and Foundations (special issue) 1996;1:311–24.

[4] Koga, Y., Washida, S. Earthquake resistant design method of geotextile. In: Proceedings of the International Symposium on Earth Reinforcement, Rotterdam, 1992. p. 255–59.

[5] Yamanouchi T, Fukuda N. Design and observation of steep reinforced embankments. In: Proceedings of the Third International Conference on Case Histories in Geotechnical Engineering, St. Louis, MO, 1993. p. 1361–78.

[6] Bathurst RJ, Alfaro MC. Review of seismic design, analysis and performance of geosynthetic reinforced walls, slopes and embankments, Invited Keynote paper/lecture. In: Proceedings of the Third International Symposium on Earth Reinforcement, Kyushu, Japan, 1996. p. 887–918.

[7] Ling HI, Leshchinsky D, Perry EB. Seismic design and performance of geosynthetic-reinforced soil structures. Geotechnique 1997; 47(5):933–52.

[8] Michalowski RL. Soil reinforcement for seismic design of geotechnical structures. Computers and Geotechnics 1998;23:1–17.

[9] Zienkiewicz OC, Taylor RL. The finite element method. New York: McGraw-Hill, 1991.

[10] Bathurst RJ, Hatami K. Seismic response analysis of a geosynthetic reinforced soil retaining wall. Geosynthetics International 1998;5 (1/2):127–66.

[11] Richardson GN, Lee KL. Seismic design of reinforced earth walls. Journal of Geotechnical Engineering, ASCE 1975;101(2):167–88.

[12] Bonaparte R, Schmertmann GR, Williams ND. Seismic design of slopes reinforced with geogrids and geotextiles. In: Proceedings of the Third International Conference on Geotextiles, Vienna, 1986. p. 273–78.

[13] Bathurst RJ, Cai Z. Pseudo-static seismic analysis of geosynthetic reinforced segmental retaining walls. Geosynthetics International 1995;2(5):789–832.

[14] Ling HI, Leshchinsky D. Effects of vertical acceleration on seismic

design of geosynthetic-reinforced soil structures. Geotechnique 1998;48(3):347–73.

[15] Chang C, Chen WF, Yao JTP. Seismic displacements in slopes by limit analysis. Journal of Geotechnical Engineering, ASCE 1984;110(7):860–74.

[16] Crespellani T, Madai C, Vannucchi G. Earthquake destructiveness potential factor and slope stability. Geotechnique 1998;48(3):411–9.

[17] Newmark NM. Effects of earthquakes on dams and embankments. Geotechnique 1965;15(2):139–59.

[18] Ling HI, Leshchinsky D. Seismic performance of simple slopes. Soils and Foundations 1995;35(2):85–94.

[19] Cai Z, Bathurst RJ. Deterministic sliding block methods for estimating seismic displacements of earth structures. Soil Dynamics and Earthquake Engineering 1996;15:255–68.

[20] Lin JS, Whitman RV. Earthquake induced displacements of sliding blocks. Journal of Geotechnical Engineering, ASCE 1986; 112(1):44–59.

[21] Ambraseys NN, Menu JM. Earthquake-induced ground displacements. Earthquake Engineering and Structural Dynamics 1988;16(7):985–1006.

[22] Yegian MK, Marciano EA, Ghahraman VG. Earthquake-induced permanent deformations: probabilistic approach. Journal of Geotechnical Engineering, ASCE 1991;117(1):1158–67.

[23] Conte E, Rizzo G. A probabilistic analysis of earthquake-induced permanent displacements in slopes. Italian Geotechnical Journal 1996;1:62–70 (in Italian).

[24] Cai Z, Bathurst RJ. Seismic-induced permanent displacement of geosynthetic reinforced segmental retaining walls. Canadian Geotechnical Journal 1996;31:937–55.

[25] Ling HI, Tatsuoka F, Sato T, Tamura Y, Iwasaki K. Long-term behaviour of the geotextile-reinforced clay test embankment. In: Proceedings of the 24th National Convention of Japanese Society of Soil Mechanics and Foundation Engineering, Tokyo, 1986. p. 33–6. (in Japanese).

[26] Tatsuoka F, Murata O, Tateyama M, Nakamura K, Tamura Y, Ling HI, Iwasaki K, Yamauchi H. Reinforcing steep slopes with a nonwoven geotextile. In: Proceedings of the International Reinforced Soil Conference, Glasgow, 1991. p. 141–46.

Simulation of liquefaction beneath an impermeable surface layer

N. Yoshida[a], W.D.L. Finn[b],*

[a]*Sato Kogyo Company Ltd, Tokyo, Japan*
[b]*Anabuki Chair of Foundation Geodynamics, Kagawa University, Hayashi-cho, 2217-20 Shinmachi, Takamatsu 761-0396, Japan*

Accepted 6 May 2000

Abstract

A joint element is proposed, which can simulate the three phases of behaviour of an impermeable layer over a liquefied sand layer. The analysis tracks the post-liquefaction reconsolidation of the sand, the simultaneous development of a water film between the layers and the settlements resulting from the subsequent drainage of the water film. The element is incorporated in a finite element program, which can be used to simulate the behaviour of layered systems. The effectiveness of the program is demonstrated by simulation of the performance of a model soil deposit of two layers in a centrifuge test. © 2000 Elsevier Science Ltd. All rights reserved.

Keywords: Liquefaction; Reconsolidation; Joint element; Interface behaviour

1. Introduction

Lateral spreading of the ground and flow failures of embankments are two of the major hazards associated with liquefaction of saturated sands. In these cases, the resistance to movement relies substantially on the residual or steady state strength of the liquefied sand. However, in one instance, this strength may not be available soon after liquefaction. When the liquefied sand is covered by an impermeable layer, water will accumulate below it as the sand reconsolidates after liquefaction. Until the impermeable layer cracks and allows drainage of the accumulated water, there is no shearing resistance under the impermeable layer.

If the impermeable layer is sloping, a flow failure may occur. This type of failure is of serious concern in the case of earth dams, where often a relatively impermeable embankment surrounds a zone of liquefiable sand. This hazard has led to model studies using centrifuge tests to provide data for understanding and modelling such phenomena [1]. The detailed 2D analysis of this phenomenon from initial seismic excitation to the final post-liquefaction state is beyond our present capacity.

In this paper, a beginning is made towards modelling the behaviour of a relatively impermeable surface layer over a liquefied sand layer. A method of analysis is developed for the 1D case of a horizontal layered deposit. Liu and Dobry [2] tested such a soil deposit in a centrifuge test in which accel-erations, porewater pressures and settlements were monitored continuously. Data from this test provides a clear picture of what happens at the interface between the layers following liquefaction and the means of the data for validating proposed models for simulating the response of the system.

The basis of the proposed method is a joint element that simulates interface behaviour and that can be incorporated in a finite element code for seismic response analysis. The objectives of this paper are to describe the theory of such an element and demonstrate its capability to model interface behaviour after liquefaction. In what follows, first the joint element is described and the post-liquefaction properties needed for analysis are developed. Then, the behaviour of a model two-layer deposit in a seismic centrifuge test conducted by Liu and Dobry [2] is simulated. The simulation study shows that the joint element is capable of modelling adequately the main features of the effective stress seismic response of a horizontal impermeable layer over a liquefied sand layer.

2. Theory of joint element

2.1. Equilibrium equation

The equations defining the behaviour of the joint element as it accumulates free water in the interface between layers are developed for the case of vertical displacements. This limits the applicability to horizontally layered deposits. The joint element is shown in the open position in Fig. 1.

---
* Corresponding author.
E-mail address: finn@eng.kagawa-u.ac.jp (W.D.L. Finn).

Reprinted from *Soil Dynamics and Earthquake Engineering* 19 (5), 333-338 (2000)

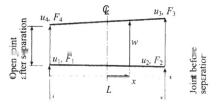

Fig. 1. Geometry of joint element.

Relative displacement between the top and bottom faces of the joint, $w = u_t - u_b$, is given by Eq. (1)

$$w = \{B\}\{u\} \tag{1}$$

Here u_t and u_b are the displacements at the top and bottom faces of the joint element. $\{B\}$, the interpolation function, and $\{u\}$, the nodal displacement vector, are given by

$$\{B\} = \left\{ -\frac{1}{2} + \frac{x}{L} \quad -\frac{1}{2} - \frac{x}{L} \quad \frac{1}{2} + \frac{x}{L} \quad \frac{1}{2} - \frac{x}{L} \right\} \tag{2}$$

$$\{u\} = \{ u_1 \quad u_2 \quad u_3 \quad u_4 \}^T \tag{3}$$

Here x is the co-ordinate along the joint, L is the length of the joint and u_1 to u_4 are the nodal displacements shown in Fig. 1 [3].

The constitutive equation for the joint for relative vertical displacements only is

$$\sigma' = -Sw \tag{4}$$

Here σ' is the effective normal *compressive* stress between the faces of the joint, and S is a parameter connecting effective stress and displacements normal to the joint. It has a large positive value when the joint closes and is zero when the joint opens. Equilibrium between the nodal forces on the joint and the element total stresses σ is expressed by

$$\{F\} = \int_{-L/2}^{L/2} -\{B\}^\sigma \, dx \tag{5}$$

When porewater pressures p develop in the joint, the total stress σ in the joint element is,

$$\sigma = \sigma' + p \tag{6}$$

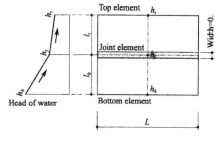

Fig. 2. Porewater pressure conditions around the joint element.

Substitution of Eqs. (1), (4) and (6) into Eq. (5) gives the equilibrium equation of the joint element as

$$\{F\} = [K]\{u\} - \{K_p\}p \tag{7}$$

where $[K]$ is the joint stiffness matrix given by

$$[K] = \frac{SL}{6} \begin{bmatrix} 2 & 1 & -1 & -2 \\ 1 & 2 & -2 & -1 \\ -1 & -2 & 2 & 1 \\ -2 & -1 & 1 & 2 \end{bmatrix} \tag{8}$$

and $\{K_p\}$ is a porewater pressure matrix given by

$$\{K_p\} = \int_{-L/2}^{L/2} \{B\}^T dx = \frac{L}{2} \{ -1 \quad -1 \quad 1 \quad 1 \}^T \tag{9}$$

2.2. Continuity equation for joint

The volume of water, W, flowing out of the joint element shown in Fig. 2 during the time dt is given by

$$dW = L\left(k_t \frac{h_t - h_j}{l_t} + k_b \frac{h_b - h_j}{l_b} \right)dt \tag{10}$$

where h_j denotes the head of water in the joint element, k_t and k_b are the permeabilities of the soil elements contacting the top and bottom faces of the joint, l_t and l_b are the distances to the far faces of these elements as shown in Fig. 2, and h_t and h_b, the associated water heads at these faces. By neglecting the velocity head, h_j is expressed as

$$h_j = \frac{p}{\rho_f g} - \frac{\{X\}^T\{b\}}{g} \tag{11}$$

where p denotes porewater pressure in the joint, ρ_f denotes mass density of porewater, g denotes the acceleration due to gravity, $\{X\}$ denotes the position vector, and $\{b\}$ denotes the gravity force vector.

Volume contraction of the joint element per unit length during the time increment dt during consolidation is given by $-\dot{w}dt$ or $-\{B\}\{\dot{u}\} \, dt$, where $\dot{u}$ is the velocity of flow. Integrating along the joint gives the total volume contraction as $-\{K_p\}^T\{\dot{u}\} \, dt$. The continuity equation is obtained by equating this volume change to the outflow water given by Eq. (10), while taking Eq. (11) into account

$$\{K_p\}^T\{\dot{u}\} = -\dot{W}$$
$$= \alpha(p - \rho_f\{X\}^T\{b\}) - \sum_i \alpha_i(p_i - \rho_f\{X_i\}^T\{b\}) \tag{12}$$

In this equation

$$\alpha_i = \frac{Lk_i}{\rho_f g l_i}, \qquad \alpha = \sum_i \alpha_i \tag{13}$$

where i indicates the top or bottom element forming the joint. Summation is conducted for both sides of the joint

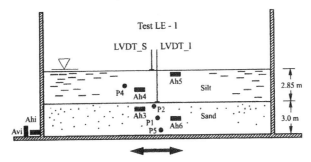

Fig. 3. Centrifuge model test at prototype scale showing the locations of horizontal accelerometers (A_h) and porewater pressure transducers (P) (after Ref. [2]).

element. The incremental forms of Eqs. (7) and (12), used in subsequent analyses, are given by Eqs. (14) and (16), respectively

$$[K]\{du\} - \{K_p\}dp = \{dF\} \tag{14}$$

where

$$dp = p(t + dt) - p(t) \tag{15}$$

and

$$\{K_p\}^T\{du\} - \alpha p\, dt + \sum \alpha_i p_i\, dt$$
$$= [-\alpha\rho_f\{X\}^T\{b\} + \sum \alpha_i\rho_f\{X_i\}^T\{b\}]dt \tag{16}$$

The Liu and Dobry [2] test, which will be analysed later, has a 1D displacement field. For this case, Eqs. (14) and (16) may be reduced to

$$S\, dw + dp = 0 \tag{17}$$

and

$$dw - \alpha p\, dt + (\alpha_t p_t + \alpha_b p_b)\, dt$$
$$= -\alpha z\rho_f g dt + \rho_f g(\alpha_t z_t + \alpha_b z_b)dt \tag{18}$$

per unit length of joint. In Eq. (18), z, z_t and z_b are co-ordinates. For the post-liquefaction case, when the joint is open, the change in external load is assumed to be zero. Eliminating the relative displacement $dw = du_t - du_b$ from Eqs. (17) and (18) gives

$$dp = -S\alpha p dt$$
$$+ S(\alpha_t p_t + \alpha_b p_b)\, dt + S\alpha z\rho_f g dt - S\rho_f g(\alpha_t z_t + \alpha_b z_b)dt \tag{19}$$

3. Modelling of volume change characteristics of sand after liquefaction

In order to predict the separation of the joint element or the thickness of the water film, the post-liquefaction volume change characteristics of the sand during dissipation of

porewater pressure must be known. Since the occurrence of liquefaction destroys the initial structure of the sand skeleton, the properties before the occurrence of liquefaction are not applicable in post-liquefaction analysis.

Inadomaru et al. [4] conducted cyclic triaxial tests on Toyoura sand in which the volume changes were measured during the dissipation of excess porewater pressure and plotted against the effective mean normal stress. Analysis of these data suggests that the relationship between effective stress and volumetric strain during reconsolidation after liquefaction may be described adequately by

$$\sigma'_m = D(e^{\epsilon_v/c} - 1) \tag{20}$$

where D and c are parameters that define the shapes of the reconsolidation curves, and ϵ_v is the recovered volumetric strain at σ'_m. The (-1) term ensures that the recoverable volumetric strain, ϵ_v, can be initialised at zero for the immediate post-liquefaction condition when $\sigma'_m = 0$. For the test data on Toyoura sand [4], the following expression for c is obtained

$$c = 0.0007 + 0.053\epsilon_{vo} \tag{21}$$

where ϵ_{vo} is the total recovered strain. When c is known, the constant D can be found from

$$D = \frac{\sigma'_{mo}}{e^{\epsilon_{vo}/c} - 1} \tag{22}$$

The 1D compressive modulus for reconsolidation, E_r, is obtained by differentiating Eq. (20) with respect to the volumetric strain, giving

$$\frac{d\sigma'_m}{d\epsilon_v} = \frac{\sigma'_m + D}{c} = E_r \tag{23}$$

4. Numerical example

4.1. Modelled centrifuge test

Liu and Dobry [2] conducted a centrifuge test at $50g$ acceleration on the model of two-layered level ground

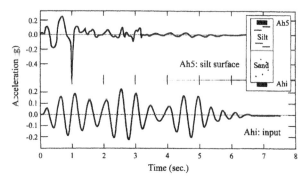

Fig. 4. Acceleration time histories of input and surface motions (after Ref. [2]).

shown in Fig. 3 at prototype scale. A 2.85 m silt layer overlies a 3 m sand layer and the water table is just above the surface of the silt. The locations of accelerometers, porewater pressure gauges, and LVDTs are also shown in Fig. 3. Unit weights of silt and sand are 18.9 and 19.2 kN/m^3, respectively. The relative density of the sand D_r is 40%. Permeability is estimated to be 1×10^{-5} cm/s for silt and 2.13×10^{-3} cm/s for sand on the basis of laboratory permeability tests. The centrifuge was operated at an acceleration of $50g$ so that the corresponding prototype permeabilities are 5×10^{-6} cm/s and 1.065×10^{-5} cm/s. This test will be used to verify the capability of the analysis described above.

The input accelerations have a peak value of about $0.2g$ and a duration of about 6 s at prototype scale as shown in Fig. 4. The surface acceleration record shows a significant drop in amplitude after about 1 s of shaking indicating that liquefaction has occurred.

Fig. 5 shows the distribution of excess porewater pressures with depth at different times based on readings from the four porewater pressure transducers. Liquefaction is judged to occur before 5 s, which is consistent with the

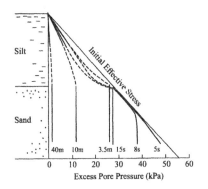

Fig. 5. Dissipation of excess porewater pressure during drainage (after Ref. [2]).

form of the acceleration record. Note that the porewater pressures remain essentially constant from 15 s to 4.5 min after liquefaction. Time histories of porewater pressure and settlement are presented in Section 4.2, which describes the simulation of the test.

4.2. Simulation of centrifuge test

The simulation was conducted using the 1D program DESRA-2C [5] in which the joint element was incorporated. The modified program is designated DESRA-2CJ [6].

The simulation begins immediately after liquefaction has occurred in the sand. There are three distinct phases in the post-liquefaction behaviour of the centrifuge model as shown schematically in Fig. 6. In phase 1, the sand reconsolidates after liquefaction, leaving a film of water between the silt and the sand layers. In the short time required for this, little drainage takes place through the silt. In phase 2, the interlayer water film begins to drain through the silt and the silt layer begins to settle until finally it contacts the sand layer. During this phase, the porewater pressure in sand remains constant at the value in the joint. Phase 3 now begins in which the sand layer undergoes additional reconsolidation under the weight of the silt layer, until finally an equilibrium state is achieved.

The form of Eq. (20) and the expression of c in Eq. (21) are assumed to be valid for the test sand in the absence of any information on the post-liquefaction properties of that sand. The initial volumetric strain ϵ_{vo} for the reconsolidation analysis of the sand is calculated from the final settlement of the silt layer. This is considered a good approximation. The assumption is necessary because the LVDT to measure the sand settlement did not function properly during the test. The values of c and D in Eq. (20) are determined as explained earlier.

The reconsolidation analysis can now be conducted on the basis of Eqs. (19) and (23), beginning at the instant of liquefaction. All simulations are conducted at prototype scale. The permeabilities based on the laboratory tests

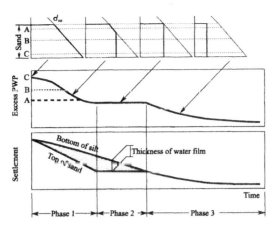

Fig. 6. Schematic representation of the three phases in the post-liquefaction behaviour of the layered system. A, B and C denote the locations where excess porewater pressure was measured. The relative durations of the various phases have been distorted to clearly show the short-duration early phases.

were used in the first trial simulation. The computed time-history of porewater pressure dissipation at the mid-height of the sand layer is shown in Fig. 7(a) for a duration of 50 s, in Fig. 7(b) for a duration of 10 min and in Fig. 7(c) for a duration of 60 min. The effective permeability of the sand in the centrifuge test after liquefaction must be greater than the $k = 1.065 \times 10^{-3}$ m/s measured on specimens of unliquefied sand because the computed time for maximum opening of the joint based on this permeability is about 25 s instead of the test value of about 15 s. A permeability of

2.13×10^{-3} m/s gives a good simulation of the test results as shown in Fig. 7(a). Such an increase in post-liquefaction permeability has also been noted by Arulanandan and Sybico [7].

The next phase is the drainage of the water layer through the silt. The test data indicate that the joint completely closes after about 4 min. For drainage through the silt a permeability of 8×10^{-5} m/s is necessary for the silt to make contact with the sand layer at the right time. This is 16 times greater than the measured permeability. The apparent increase in permeability may be due to cracks, which develop when the silt layer floats on the water or leakage of the contact between the silt and side of the centrifuge box.

For phase 3, in which the silt and sand are again in contact, the permeability of the silt must be reduced to 2×10^{-5} m/s for up to 10 min and then reduced to 5×10^{-6} m/s to get a good simulation. This reduction of permeability is consistent with the closing of cracks in the silt after contact with the sand.

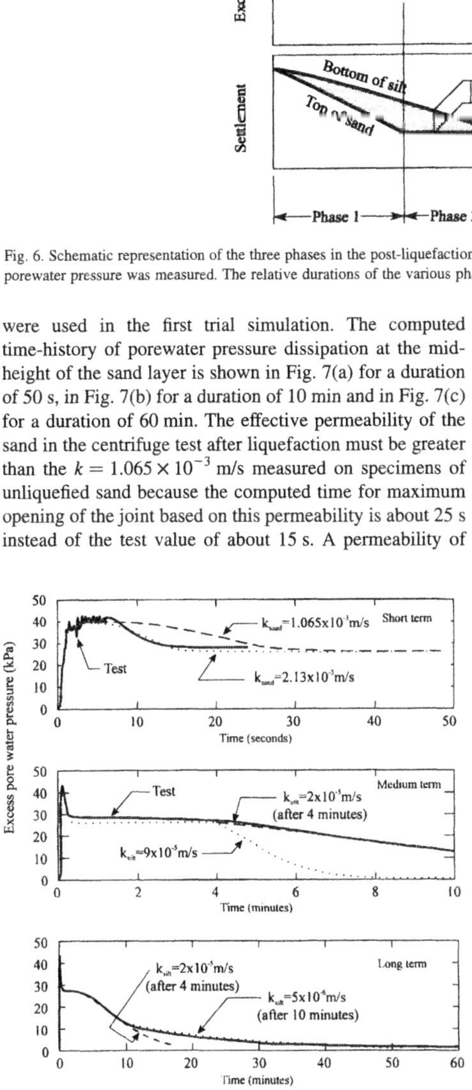

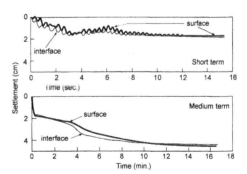

Fig. 7. Time-histories of porewater pressures over the short (0–50 s), medium (1–10 min), and long (0–60 min) terms.

Fig. 8. Measured settlements in centrifuge test.

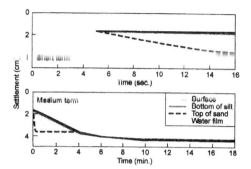

Fig. 9. Computer settlements in centrifuge test.

Fig. 8 shows measured settlements for both the short (0–16 s) and medium (0–18 min) terms. The settlements of the surface and the interface are nearly identical because the interface LVDT did not measure the sand settlement but remained attached to the bottom of the silt layer [2]. This means that there are no intermediate data available on the development on the thickness of the water film until the water film begins to dissipate through the silt. The abrupt occurrence of about 2 cm in the first 3–4 s is not compatible with the mechanics of the test model. It is most likely to be an initial adjustment in the upper layer and/or the LVDT. Fig. 9 shows the development of the water film and the contemporaneous settlement of the surface of the underlying sand layer in the short term, 0–16 s. The modelling starts at about 4 s, after the initial adjustments to the system noted above. In the medium term (also in Fig. 9) the water film begins to dissipate and the silt layer begins to approach the top of the sand layer. Contact is established at about 4 min, which seems to be a little later than that indicated by the interface curve in Fig. 8. Thereafter, as the porewater pressures dissipate, the effective stresses on the sand layer increase and additional settlements occur. Because of the problems of measuring the settlements accurately, reliable comparisons between measured and computed settlements as a function of time is not possible. However, comparisons between Figs. 8 and 9 suggest that the proposed method of analysis captures the evolving pattern of settlements rather well.

5. Conclusions

A joint element is presented for the analysis of what happens when a layer of sand liquefies beneath a relatively impermeable surface layer. The key responses that need to be modelled are the development and subsequent drainage of a film of water between the two layers as the sand layer reconsolidates after liquefaction. The ability of the method to simulate the response of such a two-layer system in a Liu and Dobry centrifuge test [2] is encouraging.

The major impediment to a satisfactory validation of the method of analysis is a lack of data on the post-liquefaction properties of sand and some uncertainty about the actual performance of the model in the centrifuge test.

The joint element has been incorporated in the dynamic effective stress program DESRA-2CJ [6]. In theory, this allows the entire sequence of behaviour from the initiation of shaking to the final settlement to be followed. The model is being extended to 2D analysis.

Acknowledgements

The joint element project was supported by Sato Kogyo Company Ltd, Tokyo, Japan, and by a grant to the second author from the Natural Science and Engineering Research Council of Canada.

References

[1] Arulanandan K, Seed HB, Yogachandran C, Muralectharan KK, Seed RB, Kabiramany K. Centrifuge study on volume changes and dynamic stability of earth dams. ASCE Journal of Geotechnical Engineering 1993;119(11):1717–31.

[2] Liu L, Dobry R. Centrifuge earthquake modelling of liquefaction and its effect on shallow foundations. Department of Civil and Environmental Engineering, RPI, New York, 1993.

[3] Goodman RE, Taylor RL. A model for the mechanics of jointed rock. Journal of Soil Mechanics, ASCE 1968;94(SM3):637–59.

[4] Inadomaru K, Tsujino S, Yoshida N. Fundamental study on the residual settlement of ground after liquefaction [in Japanese]. In: Proceedings of the 49th Annual Conference of the Japan Society of Civil Engineering, vol III, 1994. p. 498–9.

[5] Finn, Liam WD, Lee MKW, Yoshida N. DESRA-2C, dynamic effective stress response analysis of layered deposits incorporating soil yielding. Department of Civil Engineering, University of British Columbia, Vancouver, BC, Canada, 1991.

[6] Finn, Liam WD, Yoshida N, Lee MKW. DESRA-2CJ, dynamic effective stress response analysis of layered deposits incorporating soil yielding. Department of Civil Engineering, University of British Columbia, Vancouver, BC, Canada, 1996.

[7] Arulanandan K, Sybico J. Post-liquefaction settlement of sand mechanism and in situ evaluation. In: Proceedings of the Fourth Japan–US Workshop on Earthquake Resistant Design of Lifeline Facilities and Countermeasures for Soil Liquefaction, Honolulu, Hawaii. Technical Report NCEER-92-0019, 1992. p. 239–47.

Earthquake resistant construction techniques and materials on Byzantine monuments in Kiev

A. Moropoulou[a,*], A.S. Cakmak[b], N. Lohvyn[c,1]

[a]*Department of Materials Science and Engineering, Faculty of Chemical Engineering, National Technical University of Athens, Zografou Campus, 9, Iroon Polytechniou St., 15780 Athens, Greece*
[b]*Department of Civil Engineering and Operations Research, School of Engineering and Applied Science, Princeton University, Olden St, Princeton, NJ 08544, USA*
[c]*Povitroflotskyi prospeckt, 15-16, 252049, Kiev 49, Ukraine*

Accepted 17 May 2000

Abstract

The study of the behaviour of historic buildings that have suffered from earthquakes has become a valuable tool for the understanding of earthquake resistant construction techniques and materials. Byzantine monuments of the 11–13th century in Kiev have been studied to provide insights into their effective dynamic properties facing severe earthquake history in the area. The recessed brickworks according to the "concealed course" construction technique of the St. Sophia Cathedral (11th century), the Church of St. Michael in the Vydubytskyi Monastery (11th century), the Tithe Church of the Assumption of the Virgin (10th century) and the Cathedral of Assumption of the Virgin (11th century) in the Monastery of the Caves (Pecherskyi monastery) in Kiev were studied and the material properties of bricks and lime mortars with ceramic fill were investigated (mechanical strength tests, mineralogical, chemical and microstructural analysis). The results show major similarities with those of the Byzantine monuments in Istanbul (Theodosian Walls and Hagia Sophia — 6–11th century construction phases), giving evidence of earthquake resistant construction techniques and materials allowing for continuous stresses and strains. Hence, didactics on proper restoration techniques and materials are deduced aiming at their present safety in the face of future earthquakes. © 2000 Elsevier Science Ltd. All rights reserved.

Keywords: Earthquake resistant materials; Earthquake resistant construction techniques; Earthquake protection of monuments; Byzantine monuments in Kiev

1. Introduction

The study of the behaviour of historic buildings that have suffered from earthquakes has become a valuable tool for the understanding of earthquake resistant construction techniques and materials.

Structural preservation of historic buildings in seismic areas has evolved to become one of the important and relatively new issues in earthquake engineering. It encompasses the identification of the existing structural system and materials used [1] in the construction, including zones of previous repair [2], weakness, cracking and other structural discontinuities, linear and non-linear dynamic analysis, ambient vibration testing, soil and foundation investigations and the strong instrumentation of the monument [3]. On the basis of all these experimental and analytical investigations, alternatives of structural interventions towards the improvement of its structural worthiness can be formulated.

Structural studies to determine the earthquake worthiness of Hagia Sophia in Istanbul have proved that the monument's static and dynamic behaviour depends very strongly on the mechanical, chemical and microstructural properties of the masonry mortars and bricks. The results show a decrease of 5–10% in the natural frequencies, as the amplitude of accelerations increase and returns to their initial values, due to the non-linear nature of the masonry [4]. The analysis of the historic mortars has indicated that the amorphous C S H gel formation [5] between the crystalline phases of the calcite and the dispersed ceramic fragments allows for energy absorption by the structure during an earthquake, without affecting the materials' properties irreversibly, while the compatibility of the mortars to the original building units allows for continuous stresses and strains.

In the present work Byzantine monuments of the 11–13th century in Kiev are studied to provide insights into their

* Corresponding author. Fax: +30-1-772-3215.
E-mail addresses: amoropul@central.ntua.gr (A. Moropoulou), ahmet@princeton.edu (A.S. Cakmak).
1 Visiting Fellow, School of Engineering, Princeton University, NJ, USA.

Fig. 1. The church of St. Michae of Vydubytskyi Monastery in Kiev, 1070s: (1) view from south; (2) (next to 1) western facade; and (3) view of the stair tower from north-west; (4a,b) fragments of the recessed brickwork of the western facade (see 3:2, printed K).

effective dynamic properties by which they responded to the severe earthquake which occurred in the area in 1230 AD. The recessed brickworks according to the "concealed course" construction technique of the St. Sophia Cathedral (11th century), the Church of St. Michael in the Vydubyts-kyi Monastery (11th century), the Tithe Church of the Assumption of the Virgin (10th century) and the Cathedral of Assumption of the Virgin (11th century) in the Monastery of the Caves (Pecherskyi Monastery) in Kiev are studied and the material properties of bricks and lime mortars with cera-mic fill are investigated and compared to the dynamically effective ones of the already studied Byzantine monuments in Istanbul. Emphasis is given to the integrated study of the St. Michael Church in Kiev due to its proven structural worthiness throughout the seismic history of the area.

The church of St. Michael was built in the 1070s in the Monastery in Vydubytskyi, at the vicinity of ancient Kiev. The eastern part of the ancient church was destroyed several centuries ago and its western part is still preserved. In the 18th century a new eastern part of the church and a Baroque dome over it were built. The eastern part of the church was the same as in other Kyivan churches of that period [6–8].

It is interesting to note that both, narthex and the apses had no bonding in brickwork and foundation strips with the main cross-domed volume of the church. The absence of bonding between these parts of the church is the efficient building method, which allowed different parts of the build-ing to settle without splits.

The foundation strips of the church of St. Michael were put up from rubble on lime mortar with the addition of sand. The width of the strips is 1.7 m and their height 1.8 m. Longitudinal wooden beams were put into the lower parts of the founda-tions. The foundation strips were covered with a layer of mortar on which the contour of the walls was then drawn.

The width of the walls of the church is about 1.1 m. The walls and vaults of the ancient edifice are built in the "recessed" brickwork on lime mortar with ceramic fill (Fig. 1). The dimensions of the brick on average are 3.6–4.2 cm × 25–29 cm × 33–37 cm.

The outer surface of the church was originally covered with plaster of lime mortar with ceramic fill. The inner surface of the church was adorned with frescoes, its floors were covered with coloured glazed tiles and slate slabs [9].

In 1230 AD a strong earthquake of magnitude 6 on the

Fig. 2. (1–3) The recessed brickwork in the Church of the Assumption of the Virgin in the Monastery of the Caves, Kiev, 1070s. (4) The brickwork of the Church of St. Michael of the Vydubytskyi Monastery in Kiev, 1070s.

Richter scale occurred in the Kiev. A few churches collapsed, and other showed vertical cracks. St. Michael of Kiev showed no important earthquake damages. In the Church of the Assumption of the Virgin in the Monastery of the Caves vertical cracks and consequent demolitions occurred (Fig. 2(1–3)), while in Snt Sofia cracks were noticed on apses (Fig. 3(1,2)) and some Dome walls collapsed. This earthquake resistant behaviour of the St. Michael Church becomes very important for the understanding of earthquake resistant construction techniques and building materials on Byzantine monuments.

2. Construction techniques and materials

From the end of the 10th century through the beginning of the 12th century, Kyivan churches were built in "recessed" brickwork (Figs. 2–4), in which every second course of bricks was slightly deepened and covered with mortar of more than 5-cm thickness (Fig. 5a), following the "concealed course" construction technique (Fig. 5b). The type of the brickwork used in the earliest Kyivan buildings dated from the middle of the 10th century remains unidentifiable because of their fragmentary state.

The faces of a wall were built in recessed brickwork, while the core of a wall was usually made of rubble and spoiled bricks and poured over with lime- and ceramic mortar (Fig. 5a and b) which in studies regarding other Byzantine monuments have been found to infer to the masonry durability and high capacity in energy absorption [10]. After the building was finished the surface of the walls was covered with plaster. The plaster was made of lime mortar with carefully sieved crushed ceramic mass. The remains of the original ancient plaster can still be seen on the facades of the cathedral of St. Sophia (Fig. 3(1,2)) and the church of St. Michael in Vydubytskyi Monastery (Fig. 2(4)). The brickwork fragments with the original outer plaster were found during the archaeological excavations of the so-called Tithe church of the Assumption of the Virgin (10th century) and of the Cathedral of the Assumption of the Virgin (11th century) in the Monastery of the Caves (Pecherskyi Monastery) in Kiev (Fig. 2(1–3)).

32

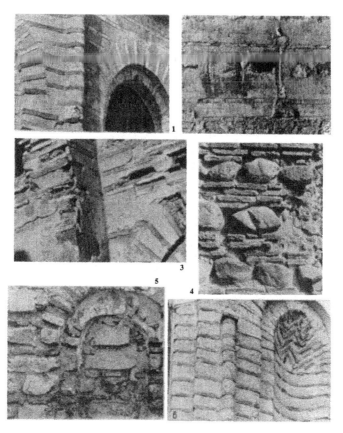

Fig. 3. The recessed brickwork in the old Kiev and Rus' (state of Kiev) monument of the 11–12th century: (1,2) the Cathedral of St. Sophia in Kiev, 1017–1030; (3,4) the Golden Gate, 11th century, Kiev; (5) the Church of St. George in Novgorod, 12th century; (6) the Cathedral of the Saviour in Chernikiv, 1030s.

To prevent deformations in brickwork and to strengthen the construction longitudinal wooden beams were placed in the middle of the walls (Fig. 5a). These beams formed a kind of a frame situated on a different level: at the bedding of walls, in the abutment of arches and arched openings of doors and windows, and in the basement of drums and domes. These wooden ties also served an antiseismic purpose, as has been discovered in historic buildings in Greece [11].

Large and comparatively thin bricks were used in ancient Kyivan buildings (Fig. 6(9–19)). Its dimensions were from around 26–36 cm and its thickness was 2.5–3 at the end of the 10–11th century, and in the second half of the 11th century to the beginning of the 12th century it was around 4–4.5-cm thick. For the manufacture of brick, different local types of clay and loam were used. The colour of baked bricks varied from pale pink and ochre to red and brown

(Fig. 7). The bricks were laid up on lime mortar with ceramic fill (Fig. 6(1–7)). The fill was made of slightly baked (<750°C) and then crushed sieved clay mass, currently called "tsemianka" (Fig. 8). The baking temperature of the bricks was 1000–1200°C.

3. Material properties

Various tests were performed to determine the elastic moduli and *mechanical properties* of the materials, such as the tensile strength.

Dynamic elastic moduli have been estimated at various brick and mortar locations on apses and dome ribs ranging as follows: brick, $E_b = 1$–3 GPa; mortar, $E_m = 0.6$–0.7 GPa; composite, $E_{bm} = 2.2$–2.5 GPa. The relevant

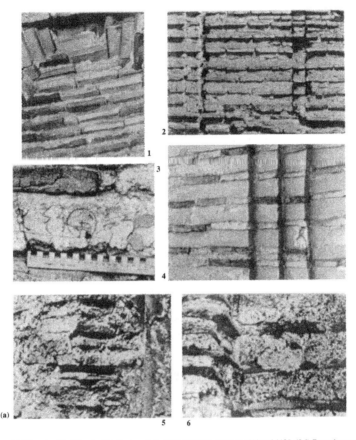

Fig. 4. (1–4) The recessed brickwork of the Church of the Saviour in Berestovo in Kyiv, between 1113 and 1125. (5,6) Byzantine monuments: Zeirek in Constantinople; Church in Antigoni.

estimated dynamic elastic moduli for Hagia Sophia [5] are: brick, $E_b = 3.1$ GPa; mortar, $E_m = 0.66$ GPa; composite, $E_{bm} = 1.83$ GPa.

Tensile strength tests performed on mortar samples show values between 1.25 and 2 MPa as compared to 0.7 and 1.5 MPa for Hagia Sophia.

The considerable mechanical strength along with the thickness of mortar joints, 1.2 times the brick thickness, allows for the consideration of these composites as early examples of reinforced concrete [12].

4. Mineralogical analysis

4.1. Optical microscopy

Thin sections of the mortar samples were examined under a polarizing Zeiss microscope (Fig. 9). The microphotographs taken allow for the petrographical–mineralogical characterization of the mortar constituents, as well as microscopic observations of the different mineral phases in the matrix.

Light coloured ceramic fragments and powder, as well as small rounded quartz grains are embedded in a finely crystallized matrix, well adhered. The calcitic matrix is well and homogeneously processed without any fossils and presents the coherence of hydraulic mortars. The cracks observed to trespass the ceramic fragments and the matrix are filled with secondary crystallized calcitic material indicating a self-healing effect to the earthquake damage.

4.2. X-ray diffraction analysis (XRD)

The XRD analysis of finely pulverized samples was

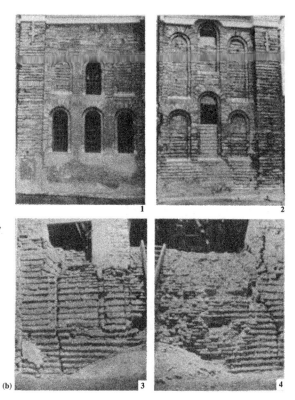

Fig. 4. (*continued*)

performed on a Siemens D-500 X-ray diffractometer, based on an automatic adjustment and analysis system with a Diffract-EVA quality analysis software, in order to identify the mineral constituents of the mortars (Fig. 10). To facilitate the direct observation of various spectra, a diffraction interval between 20-5 and 20-60, with a step of 0.02 was used.

The XRD spectrum of the matrix of the crushed brick sample shows abundance in calcite, enough quartz, some dolomite and feldspars and the presence of calcium silicate hydrate (CSH) and hematite (Hm) traces. This, compared to the results of a mortar sample from the Byzantine Walls of Istanbul differs only in the presence of Aragonite, due most probably to different climatic conditions governing the crystallization of $CaCO_3$ [13]. The spectra are indeed coinciding, showing the identity of either hydraulic lime or pozzolana admixed with aerial lime, rich in CSHs all over the spectra. The diversification of the spectra regarding CSH and Hm indicates the different provenance of the pozzolana or the clay used for the ceramic material correspondingly.

5. Chemical analysis

5.1. Thermo gravimetric analyses (TG/DTG) and differential thermal analyses (DTA)

Thermal analyses were performed by a Mettler TG 50, thermo balance, thermal analyser system. The weight loss of a mass of 20–50 mg sample was monitored through a heating cycle from 30 to 100°C. This process reveals thermal transformations such as dehydration, dehydroxidation, oxidation, and decomposition. In addition, crystalline transitions were observed. Differential thermal analyses (DTA) were performed on a Perking Elmer Thermo analyser DTA 1700, in order to elucidate the nature of the thermal transformations indicated by the TG–DTG plots.

Fig. 11 reports the values of percent of structurally bound water vs. the CO_2/H_2O ratio referred to the binder fraction (<63 μm), and shows an inverse relationship between them. The Kyivan sample values correspond to the zone of the highest structurally bound water values, above 5%,

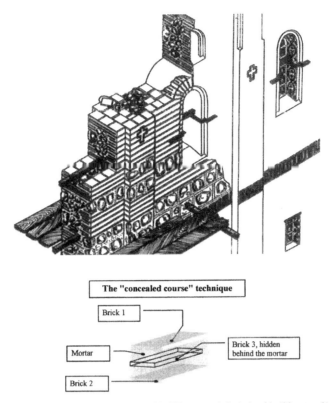

Fig. 5. (a) The structure of the recessed brickwork in Kyivan monuments of the 11th century to the beginning of the 12th century. (b) Schematic presentation of the "concealed course" construction technique.

with reference to the curve of similar Byzantine mortar samples from Hagia Sophia and Theodosian walls in Istanbul [14]. Taking into account that a small percentage of the aggregate is included in the binder fraction (<63 μm) as observed by optical microscopy and from the percentages of CO_2 and H_2O binder, the binder/aggregate ratio results from 1:2 to 1:4 in volume regarding structural mortars in Byzantine monuments in Istanbul [5]. The intermediate values in CO_2 regarding the Kyivan samples indicate their correspondence to a binder/aggregate ratio of around 1:3 in volume. This coincides with the guidelines of ancient technology regarding crushed brick lime mortars as presented by Vitruvius [15].

The percentage content of CO_2 in these fractions should be put in relation with the lime penetration into the ceramic structure, with the raw materials used for the production of the ceramic, and with the temperature at which it was baked. Finally, part of the CO_2 could be derived from the carbonation process of the calcium silico-aluminate hydrates

contained in the hardened mortar, that react slowly with CO_2, thus forming $CaCO_3$, whereas silica and alumina separate in amorphous state [16]. The constitutional water could be derived from the lime/ceramic interaction, which is much favoured if the baking temperature is around 700–750°C with raw materials of kaolinite–montmorillonite composition [17].

6. Microstructural analysis

Porosimetry studies were performed to determine the microstructural characteristics of the mortar and particularly in the length scale from 38 to 100 μm which are critical for determining the physico-chemical behaviour of cementitious materials, using a Carlo Erba Porosimeter 4000. Results are shown in Table 1 and Fig. 12.

The pore size distribution gives indications of the mortar production procedures and may predict under certain

Fig. 6. The samples of mortar and brick of the old Rus' monuments of the 11–12th century. *Mortars:* (1–4) the Church of St. Michael of the Vydubytskyi Monastery in Kiev, 1070s; (5,6,8) the Cathedral of St. Saviour in Chernihiv, 1030s; (7) the Church of the Assumption of the Virgin in the Monastery of the Caves (Pecherskyi Monastery) in Kiev, 1070s. *Bricks:* (9) the Church of Sts Borys and Hlib in Chernikiv, 12th century; (10–12) the Church of St. Michael of the Vydubytskyi Monastery in Kiev, 1070s; (13,14) the Church of the Assumption of the Virgin in the Monastery of the Caves (Pecherskyi Monastery) in Kiev, 1070s; (15,16) the Church of Sts Borys and Hlib in Vyshgorod, 11–12th century; (17–19) the Golden Gate, Kiev, 11th century.

conditions, the physicochemical state and mechanical strength of the mortar [18]. The pore size distributions are presented in Fig. 12 showing pore radii between 0.1 and 10 μm, as in the case of the Hagia Sophia mortars [5]. However, the diversified ration of small/large radii, as compared to the Hagia Sophia mortars, gives evidence of different aggregates granulometry and displays higher physico-chemical resistance to polluted and marine atmosphere [19]. Moreover, mortars containing ceramic materials show an increase of strength and modulus of elasticity during the first year of application and a slight decrease thereafter [20]. As shown in Table 1, the Kyivan samples are more porous and lighter than the ones from Hagia Sophia, and present higher mechanical strength and dynamic moduli of elasticity. These must be attributed to the lighter ceramic fragments due to their higher firing temperature (1000–1200°C) as compared to the ones from

Hagia Sophia (>850°C), as well as to the different granulometry of the brick fragments admixed. It is noticeable that the brick powder used in Kyivan mortars preparation arises from ceramics of very low firing temperature, hence providing higher contents of amorphous silicates to the formation of the hydraulic compounds in the mortar matrix. The presence of the gel phase considers a matrix formation of an advanced hydraulic composite, which allows for greater energy absorption and explains the good performance of the historic composites in resisting earthquakes [1].

7. Conclusions

Hence, the study of the behaviour of historic buildings that have suffered from earthquakes can become a valuable

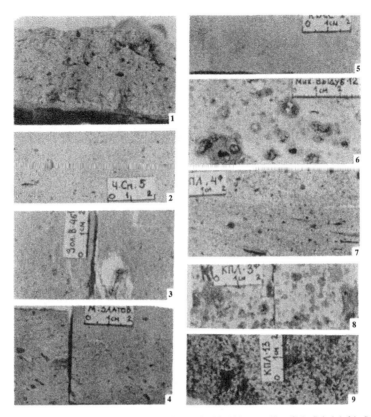

Fig. 7. The brick of monuments of the 10–12th century: (1) brick from the rotunda of the 10th century Kiev; (2) the Cathedral of the Saviour in Chernikiv, Tithe, 1030s; (3) the Golden Gate, Kiev, 11th century; (4) the Church of St. Michael of Golden Domes, the beginning of the 12th century, Kiev; (5) the Tithe Church of the Assumption of the Virgin, the end of the 10th century, Kiev; (6) the Church of St. Michael of the Vydubytskyi Monastery, 1070s, Kiev; (7–9) the Church of the Assumption of the Virgin in the Monastery of the Caves (Pecherskyi Monastery) in Kiev, 1070s.

Table 1

	Sample	A_s^a (m²/g)	d^b (g/cm³)	r_m^c (μm)	P^d (%)	C_v^e (mm³/g)
HS[f] 1b	Lime mortar matrix	3.82	1.87	0.19	32.71	174.60
HS[f] 3	Ceramic	4.04	1.67	0.39	41.67	249.60
HS[f] 5b	Crushed/brick lime mortar	7.58	1.59	0.23	42.12	261.00
K[g] 3	Crushed/brick lime mortar	5.28	1.51	0.65	46.49	307.94
K[g] 2	Crushed/brick lime mortar	5.79	1.49	0.69	42.38	234.44

[a] Specific surface area.
[b] Bulk density.
[c] Average pore radius.
[d] Total porosity.
[e] Total cumulative volume.
[f] Samples from Hagia Sophia.
[g] Samples from St. Michael Church, Kiev.

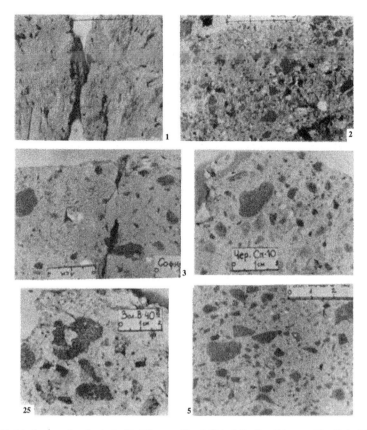

Fig. 8. Mortars: (1) the baked loess mass from the rotunda of the 10th century, Kiev; (2,2b) the Golden Gate, 11th century, Kiev; (3) the Cathedral of St. Sophia in Kiev, 1017–1030; (4) the Cathedral of the Saviour in Chernikiv, 1030s; (5) the Church of St. Michael of the Vydubytskyi Monastery in Kiev, 1070s.

tool for the understanding of earthquake resistant construction techniques and materials. The Byzantine monuments of the 11–13th century in Kiev that were studied provide insights into their effective dynamic properties by which they responded to the severe earthquake which occurred in the area in 1230 AD. The recessed brickworks according to the "concealed course" construction technique of Byzantine monuments in Kiev, as well as the investigated material properties of bricks and lime mortars with ceramic fill of the 11th century St. Michael Church show major similarities with those of the Byzantine monuments in Istanbul (Theodosian Walls and Hagia Sophia — 6–11th century

Fig. 9. Optical microscopy (25 ×). Light coloured ceramic fragments and powder, as well as small rounded quartz grains are embedded in a finely crystallized matrix, well adhered. The calcitic matrix is well and homogeneously processed without any fossils and presents the coherence of hydraulic mortars. The cracks observed to trespass the ceramic fragments and the matrix are filled with secondary crystallized calcitic material.

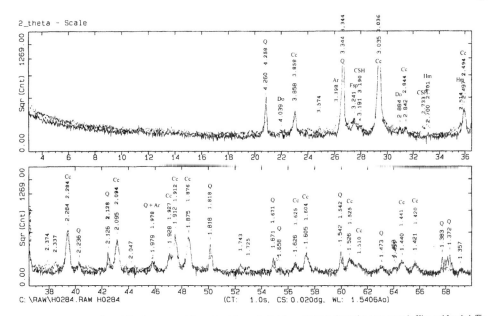

Fig. 10. Comparison between X-ray diffraction spectra of the matrix of the crushed brick samples from Byzantine monuments in Kiev and Istanbul. The spectra of the sample from Kyivan Churches (K) is marked in black and the one of Theodosian walls in Istanbul (W) in brown. Q: quartz; Cc: calcite; Do: dolomite (sample K); Ar: aragonite, $CaCO_3$, 3.40, 1.98, 3.27 (5-453) (sample W); Fsp: feldspars (Sanidin), $NaKAl(Si_3O_8)$, 3.24, 3.21 (10-357) (sample K); CSH: calcium silicate hydrate, $Ca_3(SiO_3OH)_2 \cdot 2H_2O$ (sample K) 3.19, 2.84, 2.23 (9-454), (sample W) 3.19, 2.84, 2.74, 2.23 (9-454); Hm: hematite: Fe_2O_3 (sample K) (13-534); (sample W) 2.69, 1.69, 2.51 (13-534).

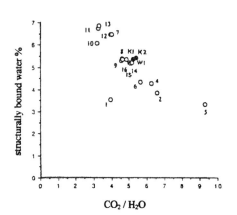

Fig. 11. Structurally bound water percentage vs. CO_2/H_2O ratio (weight loss percentage >600°C/weight loss percentage between 200 and 600°C) referred to the finer fraction (<63 μm): samples 1–16, Hagia Sophia; samples W1/W2, Theodosian Walls; samples K1/K2, St. Michael Church, Kiev.

construction phases), giving evidence to earthquake resistant construction techniques and materials. The results indicated mortars with considerable mechanical strength along with longevity. The model used proved to be resistant to the environmental pollution and to the presence of salt, while the gel phase as binder allows for greater energy absorption, shows self-healing effects and the compatibility of the mortar to the original ones allows for continuous stresses and strains. Some differences noticed regard only the ceramic technology affected by the local traditions, while major similarities of construction techniques and materials prove the research hypothesis [9] of architecture and construction techniques transfer from Byzantium to Ukraine. Hence, didactics on proper restoration techniques and materials are deduced aiming at the present safety of historic structures in the face of future earthquakes.

Acknowledgements

Acknowledgments are attributed to Dr K. Bisbikou, PhD candidate A. Bakolas and PhD candidate P. Michailidis for the analytical measurements.

40

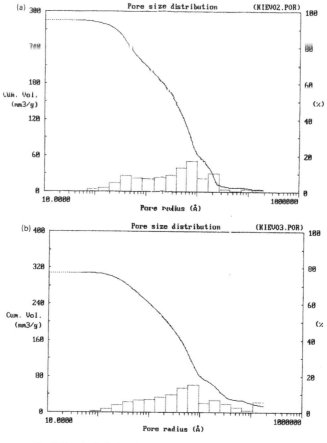

Fig. 12. Pore size distribution of samples K2/K3 from St. Michael Church, Kiev.

References

[1] Cakmak AS, Erdik M, Moropoulou A. A joint program for the protection of the Justinian Hagia Sophia. In: Proceedings of the Fourth International Symposium on the Conservation of Monuments in the Mediterranean Basin, Rhodes, vol. 4, 1997.

[2] Bakolas A, Biscontin G, Zendri E. Characterization of mortars from traditional buildings in seismic areas. PACT Revue du groupe europeen d'etudes pour les techniques physiques, chimiques, biologiques et mathematiques appliquees a l'archcologie 1998;56:17–29.

[3] Durukal E, Cakmak AS, Yuzugullu O, Erdik M. Assessment of the earthquake performance of Hagia Sophia. PACT Revue du groupe europeen d'etudes pour les techniques physiques, chimiques, biologiques et mathematiques appliquees a l'archeologie 1998;56:49–57.

[4] Cakmak AS, Moropoulou A, Mullen CA. Interdisciplinary study of dynamic behaviour and earthquake response of Hagia Sophia. Soil Dyn Earthquake Engng 1995;14(9):125–33.

[5] Moropoulou A, Cakmak AS, Biscontin G. Crushed brick lime mortars of Justinian's Hagia Sophia. In: Vandiver PB, Druzik JR, Merkel JF,

Stewart J, editors. Materials issues in art and archaeology V, vol. 462. Materials Research Society, 1997.

[6] Karger MK. Drevnii Kiev. Ocherki po istorii material noi kul tury, vol. 2, 1961.

[7] Brunov NI. Retsensila na knigu: Karger MK. Arkheologicheskiie issledovaniia drevnego Kieva. Otschety I materialy (1938–1947 gg.). Byzantine Annual, vol. 7, 1953. p. 296–306.

[8] Movchan II. Drevniie Vydubychi, Kiev, 1982.

[9] Lohvyn N. Mikhailovskaia tserkov na Vydubychah v Kieve. Soviet Archaeology 1986;4:266–72.

[10] Moropoulou A, Biscontin G, Bakolas A, Bisbikou K. Technology and behavior of rubble masonry mortars. Construction and Building Materials 1997;11(2):119–29.

[11] Touliatos P. The historic construction as a complete structural system, its decay in time and compatible restoration interventions. In: Proceedings of the INCOMARECH Conference, Athens, 1998.

[12] Mark R, Cakmak AS. Mechanical test of material from the Hagia Sophia Dome. Dumbarton Oaks Papers, 1994.

[13] Moropoulou A, Biscontin G, Theoulakis P, Bisbikou K, Theodoraki A, Chondros N, Zendri E, Bakolas A. Study of mortars in the

Medieval City of Rhodes. In: Thiel J, editor. Conservation of stone and other materials. RILEM-UNESCO, E & FN Spot, Paris, 1993.

[14] Bakolas A, Biscontin G, Moropoulou A, Zendri E. Characterization of structural yzantine mortars by thermogravimetric analysis. Thermochim Acta 1998;321((1–2)):151–60.

[15] Vitruvius P. The ten books on architecture. Morgan MH, trans. New York: Dover, 1960.

[16] Goodbrake CJ, Young JF, Brger RL. Am Ceram Soc 1979;62:488–91.

[17] Moropoulou A, Bakolas A, Bisbikou K. Thermal analysis as a method of characterizing ancient ceramic technologies. Thermochim Acta 1995;2570:743–53.

[18] Shafer J, Hilsdorf HK. Ancient and new lime mortars — the correlation between their composition, structure and properties. In: Thiel J, editor. Conservation of stone and other materials. RILEM UNESCO, E & FN Spot, Paris, 1993.

[19] Theoulakis P, Moropoulou A. Microstructural and mechanical parameters determining the susceptibility of porous building stones to salt decay. Construction and Building Materials 1997;11(1):65–71.

[20] Karaveziroglou M, Papayianni I. Compressive strength of masonry with thick mortar joins. In: Thiel J, editor. Conservation of stone and other materials. RILEM-UNESCO, E & FN Spot, Paris, 1993.

Papers from

Computers and Geotechnics

Influence of the kinematic testing conditions on the mechanical response of a sand

Donatella Sterpi*

Department of Structural Engineering, Politecnico (Technical University) of Milan, P. Leonardo da Vinci 32, 20133, Milan, Italy

Received 2 February 1999; received in revised form 20 October 1999; accepted 25 October 1999

Abstract

The influence of the laboratory testing conditions on the mechanical characterization of a compacted sand is investigated on the basis of standard triaxial and of plane strain compression tests. The two sets of experimental results lead to the independent calibration of two constitutive models. In fact, while the triaxial tests suggest a perfectly plastic behaviour for the sand, the plane strain results show marked strain softening effects, in particular for the denser samples or for high values of the confining pressure. The perfectly plastic and the strain softening models are then applied to the stability analysis of a cantilever sheet pile. The numerical results enable the drawing of some conclusions on the effect that the material model, chosen and calibrated on the basis of a particular laboratory test, could have on the predicted performance of geotechnical structures. © 2000 Elsevier Science Ltd. All rights reserved.

Keywords: Plane strain tests; Strain softening; Retaining structure; Finite Element analysis; Compacted sand

1. Introduction

The majority of stress analyses in geotechnical engineering require an accurate mechanical characterization of the involved geological media, in order to evaluate their deformability and shear strength parameters and to single out possible particular aspects of their behaviour. To this purpose, in addition to the field investigation, laboratory tests under axisymmetric, i.e. triaxial, conditions are customarily performed, regardless of the actual kinematic regime of the problem at hand. It is

* Tel.: +39-02-23994311; fax: +39-02-23994220.
E-mail address: sterpi@stru.polimi.it

Reprinted from *Computers and Geotechnics* 26 (1), 23-41 (2000)

recognized, however, that in many cases, concerning e.g. retaining structures, tunnels, beam foundations, etc. a better approximation of the actual stress/strain regime in the field is often obtained by assuming plane strain conditions in the section normal to the main dimension of the structure.

The mentioned discrepancy between field and testing conditions, and the unavoidable limits of the constitutive models adopted in practice, may lead to a poor accuracy of the results of calculations, which becomes relevant when the material behaviour is appreciably influenced by the kinematic boundary conditions applied during the laboratory tests [1,2].

The effects of this drawback might be particularly relevant when dealing with 'stiff' soils, showing the so called "strain softening" behaviour [3,4], represented by a loss of their overall shear strength after a peak load level has been attained.

This problem is discussed here with reference to the mechanical characterization of a sand, based on laboratory tests carried out both in conventional triaxial and in plane strain conditions [5].

In the following, the laboratory investigation is first described, providing some details on the adopted plane strain apparatus and on the relevant testing procedure. The test results suggest two different interpretations of the non-linear stress–strain law of the sand. In fact, the conventional triaxial tests did not show an appreciable reduction of the shear resistance with increasing shear deformation, whereas the plane strain data were markedly characterized by strain softening effects.

On the basis of the experimental results two relatively simple material models were calibrated. The first one is based on the triaxial test data only and, as a consequence, considers the soil behaviour as elastic perfectly plastic. The second model, based on the plane strain results, extends the previous constitutive law to include softening effects.

To get some insight into the influence that the choice of the constitutive model has on the results of stress analyses, an example concerning the process of excavation and loading of a cantilever sheet wall is finally considered. The comparison between the numerical results based on the two mentioned models allows the drawing of some conclusions about the influence of the laboratory testing conditions, and of the consequent characteristics of the adopted constitutive law, on the predicted performance of relevant geotechnical engineering structures.

2. Experimental investigation

The soil chosen for this study is a siliceous medium uniform sand ($D_{50} = 0.34$, $C_u = 2.5$) from an alluvial deposit in the Kobe prefecture (Hyogo), Japan. The grain size distribution of this soil is shown in Fig. 1 and the variation of its void ratio and dry unit weight with the relative density is reported in Table 1.

The laboratory investigation included two sets of isotropically consolidated, drained compression tests performed, respectively, by a conventional triaxial apparatus and by a plane strain equipment. In both cases, loose ($D_r = 30\%$) and dense ($D_r = 80\%$) specimens were tested, under various effective confining pressures,

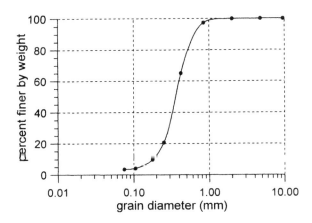

Fig. 1. Grain size distribution of "Kobe" sand.

Table 1
Variation of void ratio e and of dry unit weights γ_d with the relative density D_r for Kobe sand (grain unit weight γ_g)

γ_g (kN/m^3)	e	γ_d (kN/m^3)	D_r (%)
26.3	0.911	13.7	0
	0.814	14.4	30
	0.653	15.8	80
	0.588	16.5	100

namely: 100, 200 and 400 kPa. In addition, direct shear tests were performed on both sets of samples, with values of the effective normal stresses equal to the above cell pressure values.

The 14 cm high samples were prepared by the moist tamping technique [6], using suitably shaped molders having 7 cm diameter, for the triaxial axisymmetric tests, and 4×8 cm rectangular base, for the plane strain tests.

After placing the molder on the lower base of the corresponding apparatus, seven layers of humid sand, 2 cm thick, were subsequently compacted. The weight of the sand of each layer was calibrated in order to reach the chosen relative density at the end of the moist tamping procedure.

The two halves of the mold were then removed and, after positioning the rubber membrane, the cells were assembled. The samples were saturated, first, by flow of carbon dioxide and, subsequently, by de-aerated water, in order to attain a degree of saturation greater than 98%. All tests have been performed under displacement control, keeping the back pressure equal to 200 kPa.

The plane strain apparatus, derived from the original design by Vardoulakis and Drescher [7], imposes plane strain conditions to the prismatic sample by means of two parallel rigid vertical walls, 8 cm apart. The normal contact stress between sample and walls, referred to in the following as out-of-plane stress, is monitored by three load cells placed within one of the walls (Fig. 2).

48

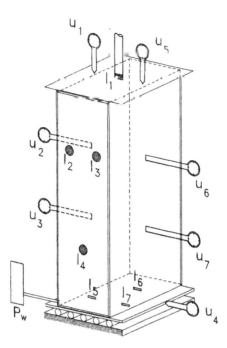

Fig. 2. Location of the measuring devices in the plane strain apparatus: load cells (1); displacement transducers (*u*); pore pressure transducer (P_w).

The lower loading cap is allowed to move, by a sliding base, along the in-plane horizontal direction, corresponding to the smaller dimension of the specimen. The stress and strain state of the sample are controlled during the tests by the devices indicated in Fig. 2. They consist of seven displacement transducers, seven load cells and one pore pressure transducer.

The samples tested under axisymmetric conditions and in plane strain regime showed rather different responses in terms of both stress–strain diagrams and failure mode. A "ductile" behaviour was observed under axisymmetric triaxial conditions (dashed lines in Fig. 3). Failure occurred at fairly constant deviatoric stress and was associated to a "barreling" deformation of the samples. On the contrary, under plane strain conditions the stress–strain curves presented a peak (solid lines in Fig. 3) which was particularly marked for the dense samples and/or for high values of the confining pressures [8,9].

Under plane strain conditions, in the vicinity of the peak level of vertical stress, a strain localization is onset within a narrow zone. With increasing axial deformation this zone develops into a shear band, leading to a brittle failure characterized by a loss of the overall load-carrying capacity of the sample.

The horizontal movement of the lower slide and the lateral displacements recorded by the horizontal transducers are reported in Fig. 4 for one of the plane strain tests. It can be observed that, after an initial settling, the slide undergoes a negligible movement until the axial deformation corresponding to the peak vertical stress is attained. Beyond this point, the displacement of the slide increases roughly linearly with increasing vertical deformation (Fig. 4a).

The transducers 3 and 7, located on the lower part of the sample (Fig. 4b), recorded quite similar displacements. The slight difference between the corresponding

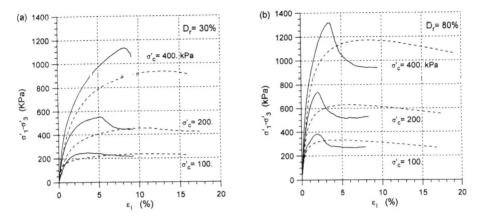

Fig. 3. Variation of the stress difference $(\sigma'_1 - \sigma'_3)$ with increasing axial deformation for triaxial (dashed lines) and plane strain (solid lines) tests on loose (a) and dense (b) samples.

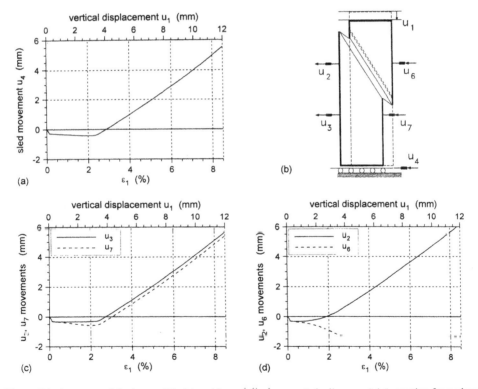

Fig. 4. Displacement of the lower slide (a) and lateral displacements (c,d) vs axial deformation for a plane strain test ($D_r = 80\%$, $\sigma'_c = 200$ kPa). Collapse mechanism and location of the displacement transducers (b).

diagrams (Fig. 4c) is due to the horizontal deformation developed during the early stages of the test.

The displacement transducers 2 and 6, located on the upper part of the sample, showed quite a different behaviour (Fig. 4d). In fact, after recording the deformation before the peak, only transducer 2 detected a large lateral displacement similar to the one shown in Fig. 4c. On the contrary, no further displacements were measured by transducer 6.

This suggests that, beyond the peak condition, a mechanism develops characterized by a rigid movement between two portions of the sample that slide along a shear band (Fig. 4b). This was confirmed by experimental evidence. The almost constant ratio between the horizontal and vertical displacement rates of the two portions is related to the inclination of the shear band and to the effect of dilatancy within it [10].

From the results of these tests it seems reasonable to conclude that only under plane strain conditions a marked difference can be observed between peak and ultimate states. In fact, the particular kinematic conditions imposed by the plane strain apparatus permit the "spontaneous" development of shear bands, which is partially inhibited under axisymmetric (triaxial) conditions.

It should be observed that, beyond the peak, the strain and stress fields are no longer homogeneous within the sample. Therefore the diagrams in Fig. 3, derived from the imposed vertical displacement and from the recorded vertical load, should be considered merely as a measure of the overall response of the specimen [11]. A stress–strain curve derived from a load-displacement experimental diagram might suffer of a magnification of the shear strength loss. In fact, in case the strain localization develops in bands which reduce the size of the sample reacting area, the actual current stress is greater than the stress merely derived as the ratio between load and cross sectional area. However, for the described tests, the derived stress–strain curves were not affected by the mentioned problem, since the observed shear bands did not lead to a reduction of the cross sectional area.

The shear strength and deformability characteristics of the sand, summarized in Tables 2 and 3, are derived from the triaxial (TX) and plane strain (PS) tests, with reference to Mohr–Coulomb failure criterion and to Hooke linear elastic relationships. In addition, Table 2 includes the results of the direct shear tests (DS). Two

Table 2

Shear strength parameters obtained from triaxial (TX), plane strain (PS) and direct shear (DS) tests

Test	D_r [%]	ϕ [°]	ϕ_p [°]	ϕ_r [°]	c
TX	30	32.5	–	–	
	80	36.8	–	–	
PS	30	–	35.7	31.9	0
	80	–	39.0	33.1	
DS	30	30.2	–	–	
	80	–	36.3	32.6	

Table 3
Deformability parameters from triaxial (TX) and plane strain (PS) tests

Test	D_r [%]	Secant Young modulus E_{S50} [MPa]			Poisson ratio v		
		σ'_c [kPa]			σ'_c [kPa]		
		100	200	400	100	200	400
TX	30	23	19.4	28.8	0.11	0.20	0.16
	80	42.7	60	73.4	0.34	0.3	0.3
PS	30	21.6	39	40	0.29	0.32	0.26
	80	63	81	100	0.21	0.25	0.23

values of the friction angle are reported in Table 2 for the tests in which a marked difference was observed between peak and residual conditions, denoted respectively by subscripts p and r.

As to the deformability parameters, the following relationships have been assumed as an acceptable approximation of the experimental diagrams ε_{vol} vs ε_1 and $(\sigma'_1 - \sigma'_3)$ vs ε_1, obtained under axisymmetric, Eq. 1(a), and plane strain, Eq. (1)b, conditions

$$\frac{\delta\varepsilon_{vol}}{\delta\varepsilon_1} = 1 - 2v, \qquad \frac{\delta(\sigma'_1 - \sigma'_3)}{\delta\varepsilon_1} = E \tag{1a}$$

$$\frac{\delta\varepsilon_{vol}}{\delta\varepsilon_1} = \frac{1 - 2v}{1 - v}, \qquad \frac{\delta(\sigma'_1 - \sigma'_3)}{\delta\varepsilon_1} = \frac{E}{1 - v^2} \tag{1b}$$

The values of the secant modulus E_{S50} reported in Table 3 correspond to 50% of the maximum stress difference $(\sigma'_1 - \sigma'_3)$. Note that the assumption of moduli which are secant instead of tangent to the initial loading path might lead to some apparent inconsistency. In this case for instance, E_{S50} for loose sand samples at lowest confining pressure ($\sigma_c = 100$ kPa) is higher than the corresponding value at confining pressure of 200 kPa. For loose samples, tangent or unloading/reloading moduli would be more appropriate. The analyses discussed in the following refer however to dense sand samples, which showed a marked softening behaviour. For dense sand, secant parameters seemed more appropriate and were, therefore, adopted in the calculations.

The dilatancy of the sand was investigated, in particular, by direct shear tests. Loose samples did not show any appreciable dilatancy, whereas the dilatancy of dense samples, in peak conditions, ranged between 5 and 11°, depending on the effective normal stress applied during the tests.

As expected, the peak friction angle obtained from plane strain tests is greater than the friction angle derived from conventional triaxial and direct shear tests [12–14]. In addition, plane strain conditions lead to an overall stiffness of the sample generally higher than the one evaluated under axisymmetric conditions.

3. Calibration of the elasto-plastic constitutive models

It was observed that the triaxial and the plane strain tests on dense samples ($D_r = 80\%$) led to markedly different results. This suggested an independent interpretation of the two sets of experimental data. In particular, an elastic perfectly plastic model was considered for the conventional triaxial tests, which showed an overall "ductile" behaviour of the samples. On the contrary, since "brittle" failure and shear band formation were observed in the plane strain tests, these data were interpreted through an elasto-plastic, strain softening constitutive law.

In both cases Drucker–Prager yield criterion was adopted, which is expressed by the following relationship when compression is assumed as positive,

$$\sqrt{J_2} = \alpha \cdot I_1 + K \tag{2}$$

In Eq.(2), I_1 and J_2 are the first stress invariant and the second invariant of the deviatoric stresses, and the material parameters α and K represent the frictional and cohesive components of the shear resistance. The plastic flow rule is non associated, since the observed dilatancy angle is appreciably smaller than the friction angle of the sand.

For the standard triaxial case the shear strength parameters α and K were related to cohesion c and friction angle ϕ by considering the conical Drucker–Prager surface circumscribing, in the principal stress space, Mohr–Coulomb pyramid. In fact, failure is met during a triaxial test when the corresponding stress path ("a" in Fig. 5) reaches one of the "external" vertices of the pyramid. In this case, the following analytical expressions between the two sets of shear strength parameters are easily worked out

$$\alpha = \frac{2\sin\phi}{\sqrt{3}(3 - \sin\phi)}, \qquad k = \frac{6c\sin\phi}{\sqrt{3}(3 - \sin\phi)}. \tag{3}$$

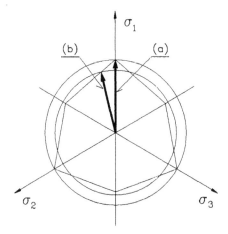

Fig. 5. Stress paths in the octahedral plane for a triaxial (a) and a plane strain (b) test.

Fig. 6 presents a comparison between the load-displacement diagrams recorded during the triaxial tests, under different values of the confining pressure, and their elastic ideally-plastic approximation based on Eq. (2).

As to the calibration of the strain softening law based on the plane strain results, two alternative hypotheses could be formulated for the initiation of softening [15]. The first one considers the phenomenon as a sort of structural instability, a consequence of a bifurcation from the expected (trivial) load-displacement path, caused by the loss of uniqueness of the solution of the governing equations [16–18]. In the case of elastic perfectly-plastic behaviour, the condition of bifurcation is expressed in terms of current stress components.

In the second case, softening is considered as an intrinsic property of the material, which undergoes a loss of its shear strength when a particular condition on the irreversible strain components is fulfilled.

In literature, these two hypotheses have been frequently assumed in order to derive suitable finite element procedures, able to capture the onset of localization and the consequent coalescence and spread of shear bands. In geological media, structural and constitutive aspects of strain localization might be simultaneously present and, in general, cannot be separately investigated.

However, in the following, strain localization will be regarded to as a consequence of the material constitutive properties only. This assumption derives from a previous series of finite element simulations of plane strain tests, that showed a limited discrepancy between the results obtained with the two mentioned procedures, upon an accurate calibration of the constitutive parameters [19].

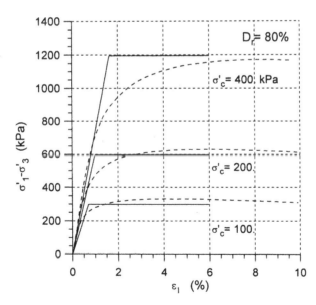

Fig. 6. Comparison between numerical (solid lines) and experimental (dashed lines) results for triaxial tests on dense samples ($D_r = 80\%$).

Softening is assumed to initiate at a material point when the square root of the second invariant of the corresponding plastic deviatoric strains γ^* reaches a given limit. In particular, with reference to the criterion expressed by Eq.(2), a linear decrease of parameters α and K with increasing γ^* is assumed, from their peak (α_p, K_p) to their residual (α_r, K_r) values (Fig. 7). Note that in the present case, concerning a granular soil, parameters K'_p and K'_r vanish.

The peak and residual parameters α_p and α_r have been evaluated by interpolating the experimental data in the space of the stress invariants since, unlike in the axisymmetric case, in plane strain conditions it is not possible to work out simple relationships between Drucker–Prager and Mohr–Coulomb constants. In fact, the stress path of a plane strain test ("b" in Fig. 5) does not reach Mohr–Coulomb locus at a characteristic point, such as one of its vertices, because its direction depends on the increment of the out-of-plane stress.

It can also be observed that the representation of the stress paths in the space of the stress invariants could point out possible inaccuracies of the testing procedures. As an example, Fig. 8 shows the stress path (heavy solid line) of a plane strain test which is characterized by an initial slope m closer to the slope m_{TX} of an ideal triaxial test than to the slope m_{PS} of an ideal plane strain test.

The analytically evaluated expressions of m_{TX} and m_{PS},

$$m_{TX} = \sqrt{\frac{1}{3}}, \qquad m_{PS} = \sqrt{\frac{1 - v + v^2}{3(1 + v)}} \tag{4}$$

show that, for a non-vanishing Poisson ratio v, m_{TX} is always greater than m_{PS}.

A possible cause of the response shown in Fig. 8 lies in the lack of contact between the sample and the lateral rigid walls. This occurs at the early stages of the test, when the cell pressure slightly reduces the sample dimensions. The lack of contact leads to a condition, during the early stages of the test, close to the axisymmetric one. This condition lasts until the axial compression, and the consequent lateral expansion of the sample, re-establish the contact with the rigid vertical walls.

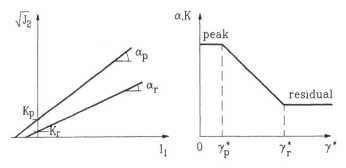

Fig. 7. Peak and residual Drucker–Prager yield surfaces (left) and assumed variation of the shear strength parameters with the square root of the second invariant of the deviatoric plastic strains γ^*(right).

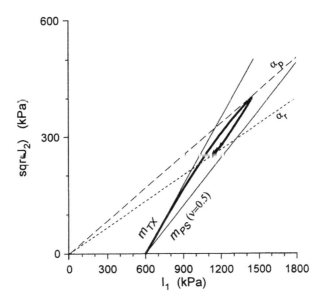

Fig. 8. Stress path of a plane strain test in the stress invariant space (I_1 first stress invariant; J_2 second invariant of deviatoric stress).

If such an effect is present, the interpolation of the experimental data tends to underestimate the value of α, since under axisymmetric conditions the sample generally shows an overall shear resistance smaller than the one in plane strain regime.

The remaining parameters defining the strain softening behaviour, γ^*_p and γ^*_r (Fig.7), are calibrated by a finite element back analysis of the experimental results. It should be noted that these parameters, or more precisely their difference $\Delta\gamma^*$, depend on the material properties as well as on the problem discretization. In fact, in the finite element solution the loss of shear strength, the rate of which is imposed by the difference $\Delta\gamma^*$, is evenly distributed over a zone that depends on the average size of the element [20]. As a consequence, after these parameters are calibrated upon an analysis carried out on a chosen mesh, the analysis on a refined mesh requires a corrected value of $\Delta\gamma^*$. This is determined in such a way that the product between the average element size and the rate of reduction $\Delta\gamma^*$ is the same for the analyses with coarser and refined mesh. Within the frame of a boundary value problem, this provision reduces the dependence of the solution on the discretization, that affects the strain softening analysis.

These calculations, and those described in the next section, were carried out by the finite element code SOSIA2, for non-linear Soil-Structure Interaction Analysis of 2D problems, in which the mentioned strain softening model has been implemented. Details on the solution technique can be found in [21] and will not be recalled here for the sake of briefness.

Fig. 9 shows a comparison between experimental and numerical results in terms of load-displacement curves. Upon calibration, the numerical model is able to correctly

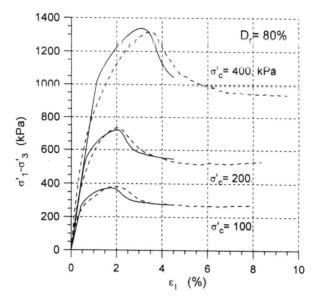

Fig. 9. Comparison between numerical (solid lines) and experimental (dashed lines) results for the plane strain tests on dense samples ($D_r = 80\%$).

reproduce both peak and residual loads. A fair agreement is also observed between the overall patterns of the two sets of diagrams.

The finite element analyses showed also that, in spite of the symmetry in geometry, loading and boundary conditions, a spontaneous process of strain localization develops within a narrow shear band (Fig. 10). Note that this deformation mode occurs even in the presence of initially uniform stress and strain fields within the sample. In fact it showed up under both assumptions of "perfectly rough" and "perfectly smooth" loading bases.

This behaviour can be explained considering that, even when assuming perfectly smooth loading bases, the numerical round off error produces a slightly non symmetrical stress distribution, which is subsequently emphasized by the consequent non uniform reduction of the shear strength parameters. The overall effect is basically similar to that observed when a weak or "trigger" element is randomly introduced in the mesh [22–24]. In addition, this effect can be also induced by the non associated nature of the flow rule, which is a necessary condition for bifurcation in the elastic perfectly plastic case [25,26].

The numerical process turns out to be stable, even in the presence of sharp variations of the mechanical parameters governing the loss of shear strength. The uniqueness of the solution is verified by the fact that repeated analyses lead to the same result, for instance to the development of a non-symmetrical shear band in a symmetrical sample. Since this lack of symmetry in the response depends on the numerical round off errors, the position of the shear band is influenced by particular aspects of the discretization, for example by the node numbering sequence.

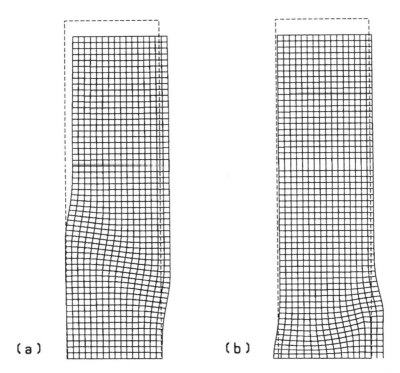

Fig. 10. Deformed grids from the finite element simulation of a plane strain test, assuming perfectly rough (a) and perfectly smooth (b) loading bases (displacements are magnified by a factor of 1).

4. Stability analysis of a cantilever sheet wall

The finite element analysis of the excavation process of a cantilever sheet wall is here discussed to investigate the influence that the constitutive model has on the design of actual geotechnical engineering structures. The problem was studied in the plane strain regime, considered as an adequate approximation of the field conditions, adopting the finite element grid shown in Fig. 11.

The dredge line is located 5 m below the original (horizontal) ground surface. At the beginning of calculations the soil mass is subjected to an uniform stress state characterized by a coefficient of earth pressure at rest equal to 0.5. The excavation process is subdivided into 10 subsequent steps, each corresponding to the removal of a 0.5 m thick layer of soil from the excavation zone. To simulate the working condition of the structure, at the end of this process an uniform surcharge load is applied on the original ground surface, in the vicinity of the wall.

A linear elastic behaviour of the sheet wall is assumed in the calculations, with flexural stiffness EJ equal to 8×10^4 kN*m^2, corresponding to a standard manufactured steel profile. As for the mechanical parameters of the soil mass, the sets calibrated on axisymmetric and on plane strain tests are assumed respectively for the elastic perfectly plastic (EPP) and the softening (EPS) analyses.

58

Two analyses have been carried out considering a penetration depth D of the wall below dredge line respectively equal to 3.5 and to 2.5 m.

The results of the first analysis ($D = 3.5$ m), obtained at the end of excavation (dashed lines) and after the application of a surface load of 70 kPa (solid lines), are shown in Fig. 12 in terms of the horizontal displacements of the wall and of the bending moment distribution.

It can be observed that no critical conditions are reached, although the softening effects produce an increase of the stress level within the structure. In fact, even the EPS analysis leads to acceptable horizontal displacements and to a maximum bending moment appreciably smaller than the critical one, approximately equal to 400 kN*m.

Quite different results are obtained by the two constitutive models when D is reduced to 2.5 m. In particular (Fig. 13), the EPS calculation indicates a possible collapse due to an almost rigid rotation of the wall, while the bending moment does

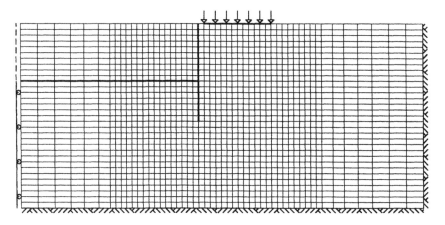

Fig. 11. Finite element mesh adopted for the cantilever sheet wall problem.

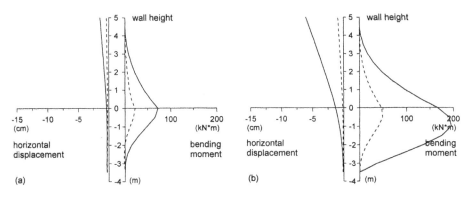

Fig. 12. Horizontal displacements of the wall and bending moment distribution at the end of the excavation (dashed line) and loading (solid line) processes ($D = 3.5$ m), from EPP (a) and EPS (b) analyses.

not present appreciable differences with respect to the one previously evaluated and shown in Fig. 12b.

In order to investigate the influence of the discretization on these results, an analysis on a refined grid was carried out. For this analysis, a corrected value for the rate of shear strength loss was assumed, according to the provision discussed in the previous section.

Fig. 14 shows the increase of the horizontal displacement of the top of the wall during the process of excavation and application of the surface load. While in the EPP analysis this displacement varies almost linearly, in both EPS cases the displacement

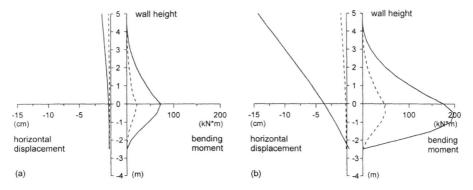

Fig. 13. Horizontal displacements of the wall and bending moment distribution at the end of the excavation (dashed line) and loading (solid line) processes ($D = 2.5$m), from EPP (a) and EPS (b) analyses.

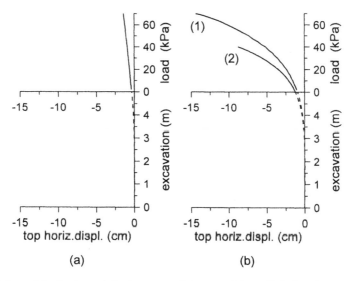

Fig. 14. Variation of the horizontal displacement of the top of the wall during the excavation/loading process ($D = 2.5$ m), from EPP analysis (a) and EPS analyses (b) based on coarse (1) and refined (2) grids.

60

rates increase rapidly, leading to overall unstable conditions. The comparison
between the two softening analyses indicates that some mesh dependence effect is
still present, although limited. It should be also observed [27] that this difference is
partly due to the dependence of the solution on the number of degrees of freedom,
occurring even when no softening effects are present.

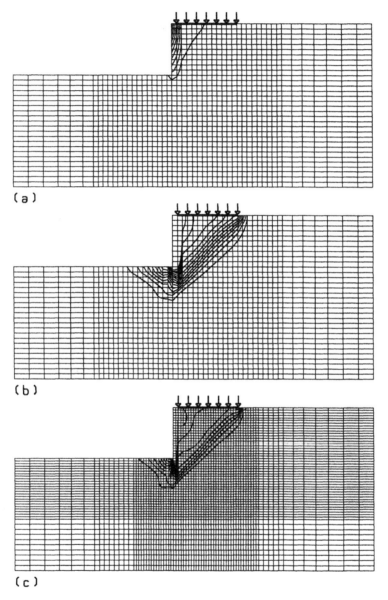

Fig. 15. Contour lines of the square root of 2nd invariant of the deviatoric plastic strains (min.cont. line
value = 0. 1%, $\Delta = 0.2\%$) at the end of the loading process ($D = 2.5$m), from EPP analysis (a) and EPS
analyses based on coarse (b) and refined (c) grids.

Finally, Fig. 15 reports the distributions of deviatoric plastic strains within the soil mass at the end of the analysis. Also in this case the EPS calculations (Fig. 15b,c) clearly show the formation of a collapse mechanism in the sand facing the wall, characterized by the sliding towards the excavation of a wedge of soil within which the non reversible shear strains tend to concentrate. The extent of the failure zone is not markedly affected by the discretization.

5. Concluding remarks

The influence of the kinematic boundary conditions on the results of laboratory compression tests has been discussed considering, in particular, a standard triaxial apparatus and a plane strain device. Two series of tests, performed on samples of compacted sand, show that the behaviour of dense samples strongly depends on the kinematic conditions characterizing the adopted equipment.

This dependence not only affects the values of the mechanical parameters derived from the experimental data, but plays also a major role in defining the overall response of the samples. In fact, a "ductile" behaviour is observed under standard triaxial conditions, while a marked strain softening behaviour characterizes the results of plane strain tests. This difference leads, in turn, to the formulation of different elastic plastic constitutive laws, if the two sets of experimental data are separately considered. In particular, an elastic perfectly plastic and a strain softening constitutive model have been independently calibrated on the basis, respectively, of the triaxial and of the plane strain results.

The assumption of a strain softening model, instead of a simpler elastic perfectly plastic model, has a remarkable influence on the prediction of the structure behaviour and on its design. To get some insight into these practical implications, the stability of a retaining wall has been studied through a finite element technique. The numerical results show that the two analyses lead to substantially different behaviours for the wall. In particular, the one based on the results of plane strain tests predicts the attainment of a possible unstable condition at a load level which would not present any appreciable danger according to the calculations based on the triaxial results.

This indicates that the choice on the use of particular laboratory equipment could eventually lead to relevant differences in the structure design, or even that possible critical conditions could be overlooked if only 'standard' experimental results are available. A better knowledge of the actual behaviour of the geotechnical medium, reached through the additional use of more sophisticated equipment, could permit a reduction of the factor of safety which would involve an economical saving, perhaps larger than the increase of cost of the laboratory investigation.

Acknowledgements

The financial support of the Ministry of the University and Research of the Italian Government is gratefully acknowledged.

References

[1] Cornforth DH. Some experiments on the influence of strain conditions on the strength of sand. Géotechnique 1964;14:143–67.

[2] Vaid YP, Campanella RG. Triaxial and plane strain behaviour of natural clay. J Geotech Eng Div Proc ASCE 1974;100(GT3):207–24.

[3] Prevost JH, Hoeg K. Soil mechanics and plasticity analysis of strain softening. Géotechnique 1975;23:279–97.

[4] Chu L, Lo SCR, Lee K. Strain softening and shear band formation of sand in multi-axial testing. Géotechnique 1996;46:63–82.

[5] Vardoulakis IG, Graf B. Calibration of constitutive models for granular materials using data from biaxial experiment. Geotechnique 1985;35:299–317.

[6] Ladd RS. Preparing test specimen using undercompaction. Geotech Testing J 1978;1:16–23.

[7] Vardoulakis IG, Drescher A. Development of biaxial apparatus for testing frictional and cohesive granular media. Final Report to the National Science Foundation, NSF Grant n. CEE 84-06500, University of Minnesota, USA. 1988.

[8] Tatsuoka R, Sakamoto M, Kawamura T, Fukushima S. Strength and deformation characteristics of sand in plane strain compression at extremely low pressures. Soils and Foundations 1986;26:65–84.

[9] Finno RL, Harris WW, Mooney MA, Viggiani G. Shear bands in plane strain compression of loose sand. Géotechnique 1997;47:149–65.

[10] Han C, Drescher A. Shear band in biaxial tests on dry coarse sand. Soils and Foundations 1993;33:118–32.

[11] Read HE, Hegemier GA. Strain softening of rock, soil and concrete — a review article. Mech of Materials 1984;3:271–94.

[12] Lee KL. Comparison of plane strain and triaxial tests on sand. J Soil Mech and Found Div Proc ASCE 1970;96(SM3):901–23.

[13] Marachi ND, Duncan JM, Chan CK, Seed HB. Plane strain testing of sand. Laboratory shear strength of soil, ASTM STP 1981;740:294–302

[14] Desrues L, Lanier L, Stutz P. Localization of the deformation in tests on sand sample. Eng Fracture Mechanics 1985;21:909–21.

[15] Sterpi D. An analysis of geotechnical problems involving strain softening effects. Int J Numer Anal Meth Geomech (in press).

[16] Hill R. A general theory of uniqueness and stability in elastic-plastic solids. J Mech Phys Solids 1958;6:236–49.

[17] Rice JR. The localization of plastic deformation. In: Koiter, editor. Proceedings of the 14th International Conference on Theoretical and Applied Mechanics. North Holland: Delft, 1976.

[18] Vardoulakis IG, Sulem J. Bifurcation analysis in geomechanics. Chapman & Hall,1995.

[19] Cividini A, Gioda G, Sterpi D. Numerical modeling of plane-strain compression tests. In: Picu-Krempl, editor. Proceedings of the 4th International Conference on Constitutive Laws for Engineering Materials, Troy NY, USA, 1999. p. 86–89.

[20] Pietruszczak St, Mroz Z. Finite element analysis of deformation of strain-softening materials. Int J Numer Meth Eng 1981;17:327–34.

[21] Cividini A, Gioda G. Finite element analysis of direct shear tests on stiff clays. Int J Num and Anal Meth in Geomech 1992;16:869–86.

[22] Shuttle DA, Smith IM. Numerical simulation of shear band formation in soils. Int J Num and Anal Meth in Geomech 1988;12:611–26.

[23] Chan AHC, Jendele L, Muir Wood D, Drescher A. Numerical analyses of strain localization using randomly distributed material properties. In: Pande-Pietruszczak, editor. Proceedings of the 4th International Symposium numer models in geomech, 1992. pp. 349–355.

[24] Asaoka A, Noda T. Imperfection-sensitive bifurcation of cam-clay under plane strain compression with undrained boundaries. Soils and Foundations 1995;35:83–100.

[25] Rudnicki JW, Rice JR. Conditions for the localization of deformation in pressure-sensitive dilatant materials. J Mech Phys Solids 1975;23:371–94.

[26] Vermeer PA. A simple shear band analysis using compliances. In: Vermeer-Luger, editor. Proceedings of the IUTAM Conference on Deformation and failure of granular materials, Delft, 1982.

[27] Sterpi D, Cividini A, Donelli M. Numerical analysis of a shallow excavation in strain softening rock. In: Fujii, Balkema, editor. Proceedings of the 8th International Congress on Rock Mechanics, ISRM, Tokyo Vol. 2, p. 545–9.

Formulation of anisotropic failure criteria incorporating a microstructure tensor

S. Pietruszczak[a],*, Z. Mroz[b]

[a]*Department of Civil Engineering and Engineering Mechanics, McMaster University,
1280 Main St West, Hamilton, ON, Canada L8S 4L7*
[b]*Institute of Fundamental Technological Research, Polish Academy of Sciences, Warsaw, Poland*

Received 26 July 1999; received in revised form 19 October 1999; accepted 5 November 1999

Abstract

Anisotropy is inherently related to microstructural arrangement within a representative volume of material. The microstructure can be represented by a second order tensor whose eigenvectors specify the orientation of the axes of material symmetry. In this paper, failure criteria for geomaterials are formulated in terms of the stress state and a microstructure tensor. The classical criteria for isotropic materials are generalized for the case of orthotropy as well as transverse isotropy. The proposed approach is illustrated by a simple example demonstrating the sensitivity of the uniaxial strength of the material to the orientation of the sample relative to the loading direction. © 2000 Elsevier Science Ltd. All rights reserved.

1. Introduction

Many geomaterials, such as rocks and soils, display anisotropy which is related to specific features of material fabric, such as bedding, layering, crack pattern, contact arrangement, etc. The specification of the conditions at failure for such materials constitutes an important problem and numerous criteria have been proposed in the past. An extensive review, within the context of rock mechanics, can be found for example in [1,2]. One of the approaches to formulate failure criteria is to define the conditions at failure by invoking linear as well as quadratic terms of stress components referred to the coordinate system associated with the axes of symmetry of the material. An

---

* Corresponding author. Tel.: +1-905-525-9140; fax: +1-905-524-2121.
E-mail address: pietrusz@mcmail.cis.mcmaster.ca (S. Pietruszczak).

Reprinted from *Computers and Geotechnics* **26 (2)**, 105-112 (2000)

example of such an approach is an extension of the well-known Hill's criterion [3], as proposed in [4,5]. A more rigorous approach, which makes use of general representation theorems and employs mixed invariants of stress and microstructure tensors, was initiated in [6] and subsequently extended in other papers, e.g. [7]. Recently, the problem was also formulated by invoking the notion of a fabric tensor specifying the directional distribution of lineal/areal porosity [8]. Furthermore, the question of deformation induced anisotropy was addressed in [9].

Yet another approach to formulate the failure criteria involves the notion of the existence of a critical plane, or the weakest orientation, along which the failure function reaches a maximum. Some examples of this approach can be found in [10–12]. In general, the disadvantage of most of the existing phenomenological formulations lies in the fact that they require a large number of material functions/parameters to be identified. Furthermore, the correlation of these parameters with the material microstructure is rather ambiguous.

In this paper a simple approach is advocated in which the failure criterion incorporates a set of material parameters which are explicit functions of mixed invariants of stress and microstructure tensors. Such an approach leads to a simple experimental verification providing, at the same time, a direct correlation with the material microstructure. In the next section the general approach is outlined, followed by a simple example which illustrates the main features of the formulation.

2. Formulation of anisotropic failure criteria

Assume that the failure criterion depends on stress and microstructure tensors, σ_{ij} and a_{ij} respectively, so that

$$F = F(\sigma_{ij}, a_{ij}) = F(\text{tr}\,\sigma, \text{tr}\,\sigma^2, \text{tr}\,\sigma^3, \text{tr}\,a, \text{tr}\,a^2, \text{tr}\,a^3, \text{tr}(\sigma a), \text{tr}(\sigma^2 a), \text{tr}(\sigma a^2),$$
$$\text{tr}(\sigma^2 a^2)) = 0 \tag{1}$$

The microstructure tensor a_{ij} is a measure of material fabric, i.e. it describes, for instance, the spatial distribution of voids or cracks, or the arrangement of intergranular contacts. The principal triad of a_{ij} is specified by the unit vectors v_i, s_i, t_i, and the spectral representation of a_{ij} is

$$a_{ij} = a_1 v_i v_j + a_2 s_i s_j + a_3 t_i t_j = a_1 m_{ij}^{(1)} + a_2 m_{ij}^{(2)} + a_3 m_{ij}^{(3)} \tag{2}$$

where a_1, a_2, a_3 are the principal values of the microstructure tensor and m_{ij}'s are the structure orientation tensors. Let us note that the vector t_i and the tensor $m_{ij}^{(3)}$ can be expressed in terms of v_i and s_i, so that the representation (2) can be written as

$$a_{ij} = a_3 \delta_{ji} + (a_1 - a_3) m_{ij}^{(1)} + (a_2 - a_3) m_{ij}^{(2)} \tag{3}$$

where δ_{ji} is the Kronecker's tensor. Define the microstructure tensor a_{ij} to be coaxial with the axes of orthotropy of the material. We shall now specify the unit vector l_i

characterizing the loading orientation with respect to these axes. Consider for this purpose the principal triad and specify the traction components acting on the set of planes normal to it. The magnitudes of these traction components are

$$L_1 = \left(\sigma_{11}^2 + \sigma_{12}^2 + \sigma_{13}^2\right)^{1/2}; \quad L_2 = \left(\sigma_{12}^2 + \sigma_{22}^2 + \sigma_{23}^2\right)^{1/2};$$
$$L_3 = \left(\sigma_{13}^2 + \sigma_{23}^2 + \sigma_{33}^2\right)^{1/2} \tag{4}$$

so that the generalized loading direction l_i can be defined as

$$L_i = L_1 v_i + L_2 s_i + L_3 t_i; \quad l_i = L_i / (L_k L_k)^{1/2} \tag{5}$$

Let us note that

$$L_k L_k = L_1^2 + L_2^2 + L_3^2 = \sigma_{ij}\sigma_{ij} = tr\sigma^2 \tag{6}$$

and

$$L_1^2 = v_i \sigma_{ij} v_k \sigma_{kj} = tr\left(m^{(1)}\sigma^2\right)$$
$$L_2^2 = s_i \sigma_{ij} s_k \sigma_{kj} = tr\left(m^{(2)}\sigma^2\right) \tag{7}$$
$$L_3^2 = t_i \sigma_{ij} t_k \sigma_{kj} = tr\left(m^{(3)}\sigma^2\right)$$

that is L_1, L_2 and L_3 are expressed as mixed invariants of the stress and structure orientation tensors.

Consider now the projection of the microstructure tensor a_{ij} on the direction l_i, i.e.

$$\eta = a_{ij}l_i l_j = a_1 \frac{tr\left(m^{(1)}\sigma^2\right)}{tr\sigma^2} + a_2 \frac{tr\left(m^{(2)}\sigma^2\right)}{tr\sigma^2} + a_3 \frac{tr\left(m^{(3)}\sigma^2\right)}{tr\sigma^2} \tag{8}$$

The scalar parameter η, as defined above, specifies the effect of load orientation with respect to the material axes. It is a homogeneous function of degree zero, so that stress magnitude does not affect the value of η.

The failure criterion (1) can now be assumed in the form

$$F\left(\sigma_{ij}, a_{ij}\right) = F\left(\sigma_{ij}, \eta\right) = F\left(tr\sigma, tr\sigma^2, tr\sigma^3, tr\left(m^{(1)}\sigma^2\right), tr^{(m)}\sigma^2\right) = 0 \tag{9}$$

so that the effects of anisotropy are described solely by the scalar parameter η, which reflects the orientation-dependent nature of the material strength. This assumption reduces the complex representation (1) to the representation in terms of stress invariants and the two independent mixed invariants [Eq. (8)] of stress and structure orientation tensors.

Consider, for instance, the Drucker–Prager condition, which for a cohesionless soil takes the form

$$F = \bar{\sigma} - \eta\sigma_m = 0 \tag{10}$$

where $\bar{\sigma} = J_2^{1/2} = (tr\mathbf{s}^2)^{1/2}$, $\sigma_m = tr\boldsymbol{\sigma}/3$ and $\mathbf{s}$ represents the stress deviator. For an isotropic material $\eta = $ const., but for an anisotropic one η is specified by the representation (8). Similarly, the Mohr–Coulomb condition

$$F = \bar{\sigma} - \eta g(\theta)\sigma_m = 0 \tag{11}$$

is reformulated by introducing the orientation-dependent parameter η, [Eq. (8)]. In Eq. (11), g is a function of Lode's angle θ, the latter defined in terms of the third and the second invariant of the stress deviator.

The case of transverse isotropy can easily be obtained from the general formulation by assuming $a_2 = a_3 = a$. In this case,

$$a_{ij} = a_1 m_{ij}^{(1)} + a\left(m_{ij}^{(3)} + m_{ij}^{(3)}\right) = a\delta_{ij} + (a_1 - a)m_{ij}^{(1)} \tag{12}$$

and from Eq.(8) it follows that

$$\eta = a + (a_1 - a)\frac{\text{tr}\left(\mathbf{m}^{(1)}\boldsymbol{\sigma}^2\right)}{\text{tr}\boldsymbol{\sigma}^2} \tag{13}$$

so that only two parameters a and a_1 specify the directional dependence of the strength characteristics.

Finally, it should be noted that one can formally define a deviatoric measure of the material microstructure Ω_{ij}, as

$$\Omega_{ij} = \left(a_{ij} - \frac{1}{3}\delta_{ij}a_{kk}\right)\bigg/\left(\frac{1}{3}a_{kk}\right) \tag{14}$$

Then, substituting Eq. (14) in Eq. (8) results in

$$\eta = \eta_0\left(1 + \Omega_{ij}l_i l_j\right); \quad \eta_0 = a_{kk}/3 \tag{15}$$

Representation (15) is analogous to that employed in [13] and describes a scalar-valued function defined over a unit sphere. Here, η_0 is the orientation average of η and Ω_{ij} describes the bias in its spatial distribution. For an orthotropic material there are two distinct eigenvalues of Ω_{ij}, whereas for an isotropic material Ω_{ij} vanishes. Thus, according to Eq. (15), the anisotropy is described as a perturbation of the parameter η around its average value η_0 with the bias defined in terms of Ω_{ij}.

3. An example: directional variation of uniaxial compressive strength

In order to illustrate the framework outlined in the previous section, let us focus on the issue of sensitivity of the material response to the orientation of the sample relative to the loading direction. For this purpose, consider a soil specimen subjected to drained uniaxial compression under an initial confining pressure. Assume that the material is transversely isotropic and the sample is tested under plane strain constraint.

Let the conditions at failure be described by the Mohr–Coulomb criterion, which for a cohesionless soil may be written in terms of principal stress as

$$F = 1/2(\sigma_I - \sigma_{III}) - 1/2(\sigma_I + \sigma_{III})\eta = 0; \quad (\sigma_I \geqslant \sigma_{II} \geqslant \sigma_{III}) \tag{16}$$

where $\eta = \sin\phi$ is assumed to be an orientation-dependent function defined according to Eq. (15). Referring to Fig. 1, identify the direction of the major principal stress σ_1 as that along the y-axis. Moreover, assume that one of the principal material directions, say t_i, is along the z-axis, which at the same time defines the orientation of the intermediate principal stress. Assume that the sample is initially under a hydrostatic stress σ_o and it is brought to failure by increasing the vertical stress by $\Delta\sigma_y$. In this case, according to Eq. (16)

$$\sigma_I = \sigma_y = \sigma_0 + \Delta\sigma_y; \ \sigma_{III} = \sigma_x = \sigma_0 \ \Rightarrow \ \zeta = \frac{\Delta\sigma_y}{2\sigma_0} = \frac{\eta}{1-\eta} \tag{17}$$

where ζ is a dimensionless parameter defining the uniaxial compressive strength at a given initial confining pressure.

The variation of the anisotropy parameter η is described by Eq. (15), i.e

$$\eta = \eta_0\left(1 + \Omega_1 l_1^2 + \Omega_2 l_2^2 + \Omega_3 l_3^2\right) \tag{18}$$

where l_1, l_2 and l_3 are defined relative to the principal material triad [see Eqs. (4,5)]. Given that $\Omega_1 = \Omega_3$ and also $\Omega_1 + \Omega_2 + \Omega_3 = 0$; $l_1^2 + l_2^2 + l_3^2 = 1$, Eq. (18) can be expressed as

$$\eta = \eta_0\left(1 + \Omega_1 - 3\Omega_1 l_2^2\right) \tag{19}$$

whereas from Eq. (4)

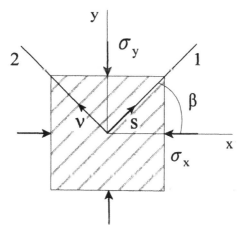

Fig. 1. Geometry of the problem.

$$l_2^2 = \left(\sigma_x^2 \sin^2 \beta + \sigma_y^2 \cos^2 \beta\right) / \left(\sigma_x^2 + \sigma_y^2 + \sigma_z^2\right) \tag{20}$$

where β specifies the orientation of the principal material axes, as shown in Fig. 1.

Assume now that the material is elastic–perfectly plastic, in which case σ_z is the intermediate principal stress. Take, for simplicity

$$\sigma_z = \alpha\left(\sigma_x + \sigma_y\right) = \alpha\left(2\sigma_0 + \Delta\sigma_y\right) \tag{21}$$

where α is a constant. Substituting now Eqs. (21) and (20) in Eq. (19), and utilizing Eq. (17), one obtains after some algebraic manipulations

$$\zeta^3\left(b_4 - b_4 b_1 - b_2 c^2\right) + 4\zeta^2\left(b_3 - b_4 b_1 - b_3 b_1 - 2b_2 c^2\right) \\ + \zeta\left(2b_3 - 6b_3 b_1 - 4b_2 c^2 - b_2\right) - \left(2b_3 b_1 + b_2\right) = 0 \tag{22a}$$

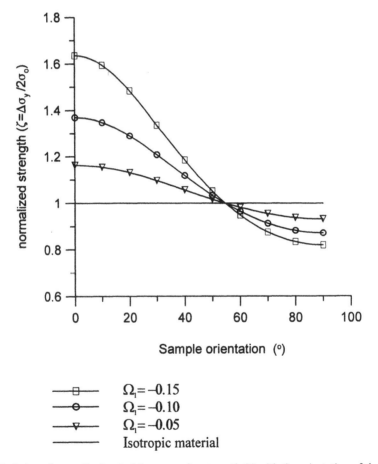

Fig. 2. Variation of normalized uniaxial compressive strength (ζ) with the orientation of the sample.

where

$$b_1 = \eta_0(1 + \Omega_1); b_2 = -3\eta_0\Omega_1; b_3 = 1 + 2\alpha^2; b_4 = 1 + \alpha^2; c = \cos\beta \qquad (22b)$$

Thus, for a given η_o and Ω_1 the normalized uniaxial compressive strength ζ is an explicit function of the orientation of the sample β.

Fig. 2 shows the variation of $\zeta = \Delta\sigma_y/2\sigma_o$ as predicted by Eq. (22). The results correspond to $\eta_0 = 0.5, \alpha = 0.25$ and $\Omega_1 = \Omega_3 = -0.05, -0.10$ and -0.15, respectively. For all simulations, the compressive strength is maximum for the vertical sample and progressively decreases for inclined samples. It should be noted that for an isotropic material ($\Omega_1 = \Omega_3 = 0$) there is $\eta = \eta_o = $ const. and the results based on Mohr–Coulomb criterion (16) are independent of the intermediate principal stress. For an anisotropic material, however, the compressive strength is sensitive to the value of the intermediate principal stress, as the latter affects the evolution of the anisotropy parameter η. Thus, for an elasotoplastic material, the variation of ζ will be affected by the form of the constitutive relation adopted for the description of the deformation process.

4. Final remarks

The paper presents a simple framework for the description of inherent anisotropy in geomaterials. The formulation incorporates a scalar parameter η, which is identified with the projection of the microstructure tensor on a suitably defined loading direction. It has been demonstrated that this approach reduces the complex general representation of the failure function to that employing basic stress invariants and the second mixed invariant of stress and structure orientation tensors. Within such a framework, the inherent anisotropy is described as a perturbation of the parameter η around its average value. An example has been provided illustrating the orientation-dependent nature of strength characteristics in a cohesionless Mohr–Coulomb material.

In the present paper the anisotropy parameter has been defined according to Eq. (15). In general, more complex functional relations may be employed, including incorporation of other mixed invariants. A more comprehensive study aimed at quantitative description of directional strength characteristics in various anisotropic geomaterials, including materials with oriented cracks, will be provided in a separate paper.

References

[1] Duveau G, Shao JF, Henry JP. Assessment of some failure criteria for strongly anisotropic materials. Mech Cohesive Frict Materials 1998;3:1–26.
[2] Kwasniewski MA. Mechanical behaviour of anisotropic rocks. In: Hudson J, editor. Comprehensive rock engineering. Oxford: Pergamon Press, 1993.
[3] Hill R. The mathematical theory of plasticity. Oxford University Press, 1950.

[4] Tsai SW, Wu E. A general theory of strength of anisotropic materials. Journ Comp Mat 1971;5:58–80.

[5] Pariseau W.G. Plasticity theory for anisotropic rocks and soils. In: Proceedings 10th Symposium on Rock Mechanics, AIME, 1972.

[6] Boehler JP, Sawczuk A. Equilibre limite des sols anisotropes. Journ de Macanique 1970;3:5–33.

[7] Nova R. The failure of transversely anisotropic rocks in triaxial compression. Int Journ Rock Mech Mining Sci & Geomech Abstr 1980;17:325–32.

[8] Pietruszczak S. On inealstic behaviour of anisotropic frictional materials. Mech Cohesive Frict Materials 1999;4:281–93.

[9] Mroz Z, Jemiolo A. Constitutive modelling of geomaterials with account for deformation anisotropy. In: Onate E. et al., editors. The finite element method in the 90's. Springer Verlag, 1991. p. 274–84.

[10] Walsh JB, Brace JF. A fracture criterion for brittle anisotropic rock. Journ Geoph Res 1964;69:3449–56.

[11] Hoek E. Strength of jointed rock masses. Geotechnique 1983;33:187–205.

[12] Hoek E, Brown ET. Empirical strength criterion for rock masses. Journ Geotech Eng Div, ASCE 1980;106:1013–35.

[13] Kanatani K. Distribution of directional data and fabric tensor. Int J Eng Sci 1984;22:149–61.

Numerical study of rock and concrete behaviour by discrete element modelling

F. Camborde [a,*], C. Mariotti [a], F.V. Donzé [b]

[a]*Laboratoire de Détection et de Géophysique Commissariat à l'Energie Atomique 91680,
Bruyères-le-Chatel, France*
[b]*GEOTOP-Université du Québec a Montréal C.P. 8888, succursale "Centre Ville" Montréal,
Québec, Canada, H3C 3P8*

Received 20 July 1999; received in revised form 30 March 2000; accepted 10 May 2000

Abstract

A discrete element method (DEM) is used to model both the dynamic and non-linear behaviour of rocks, with a particular focus on a Villejust quartzite and a MB50 type microconcrete. This numerical method has been chosen because, with a description of the medium as an assembly of discrete elements, damage and fragmentation phenomena can be easily simulated. Compact and isotropic synthetic media are generated automatically and are used to investigate the mechanical behaviour of these low-porosity materials. In the case of microconcrete, the model considers the two-phase, aggregate and cement medium, at a mesoscopic scale. The adjustment of local parameters is obtained by static uniaxial compression simulations. Quantitative results of dynamic tests of plate impacts and hemispherical explosions obtained with the DEM shows how highly appropriate this method is. © 2000 Published by Elsevier Science Ltd. All rights reserved.

1. Introduction

Concrete is made of an aggregate of grains and cement with a certain porosity and at this mesoscopic scale, it exhibits an important amount of heterogeneities [1]. The use of continuous approaches, such as the finite element method, to study the fracturing process in this kind of material, gives a good estimate of the weakening zone, but fails to characterise the localisation of the fractures. Indeed, these methods

---

* Corresponding author.

Reprinted from *Computers and Geotechnics* **27 (4)**, 225-247 (2000)

poorly describe the behaviour of discontinuous zones, even if the degradation process is introduced by internal variables, which account for dilatancy, contractance, plasticity, damage and compaction phenomena [2,3].

A discontinuous approach, such as the DEM, which is based on a particle's description of the medium, is thus more suitable to study fracturing in rocks and concrete.

Up to now, DEM has mainly been used to study granular materials flows or jointed rock mass problems [4–7], where the media response involves large displacements. However, the use of this method to study the degradation process of cohesive materials has seen a great increase recently [8–11].

Since 1994, the "Laboratoire de Detection et de Geophysique" of the French Atomic Agency, has been involved in the study of the whole degradation phenomena during dynamic solicitations of these materials. While supporting a university research project in the development of a numerical code SDEC (for "Spherical Discrete Element Code") based on the DEM, quantitative results concerning dynamical problems, such as impacts, have already been obtained [12,13].

The goal of the present work is to improve this numerical method in order to have a better description and understanding of the behaviour of cohesive materials at very high strain rates.

Before examining quantitatively the fragmentation and fracture processes during dynamic solicitations, it is first necessary to show that the model based on the DEM can reproduce the behaviour of cohesive materials during uniaxial compression tests. For this purpose, comparisons of experimental and numerical results will be made. Then, it will be shown that the model is able to predict the behaviour of cohesive materials during dynamic solicitations.

In this paper, the numerical model will be described first. Then the numerical results will be compared to the experimental measurements, for the case of a uniaxial compression test on MB50 concrete. Finally, another comparison between numerical results and experimental measurements will be made for the case of a shock wave propagation during dynamic tests on Quartzite.

2. Numerical model

2.1. Description of a zero porosity numerical medium

The difficulty in simulating geomaterial behaviour is mainly related to the heterogeneities present in the medium. In the DEM, the "meshing" process used to represent the material consists in defining a geometric assembly of particles or discrete element (DE) and the associated interaction network used to determine the displacement of these elements. In the present study, as one of the material under study is very compact (porosity is less than 4% for Villejust Quartzite), the structure described by the DE must represent this level of compactness. Considering an assembly of circular DE (spherical in 3D) is interesting because the contact determination operation is easy to formulate. In a 2D case, it is possible to generate a

compact medium with a uniform monodisperse assembly with a coordination number (average number of contact) of 6. If the elements interact via elastic forces, the isotropy of such a medium can be checked numerically and analytically [14]. However, if fractures occur during compression tests (numerically, fractures occur when a predefined rupture threshold is reached then the local interaction force is set to zero), preferential fracture directions appear every 60° [15]. It is possible to overcome this limitation by determining the fracture thresholds so that the same amount of energy is released in all directions [16]. Nevertheless, a non-uniform distribution of the DE and the contacts is preferable and it can then be seen that preferential directions do not exist during the fracturation process [1,17].

Models using circular DE have a porosity greater than that of a rock, and as long as the compaction phenomenon remains negligible (below 150 MPa for concrete), they present a good agreement with the material behaviour. However, in the case of high pressure and strain rates, the compaction phenomenon is not negligible anymore and because it has a great influence on stress wave attenuation, the use of circular DE seems inappropriate.

In the present study, to overcome this limitation, an isotropic medium with no remaining porosity is generated by means of polygonal DE. The medium is first described with an ordered mesh which is then modified by a random value chosen between λ_{min} and λ_{max} (λ_{max} is equal to the initial spacing between each node). Voronoï cells can then be built simply and very rapidly using a 2D-discrete function representation [18]. Interactions between neighbouring elements are determined by a Delaunay Triangulation. The resulting assembly of polygonal DE is shown in Fig. 1. A medium with zero porosity is thus obtained, in which the fabric tensor is perfectly isotropic [19].

2.2. Interactions laws

Two types of interaction forces between DE are defined: shear and normal forces. Initially, all the DE are linked pairwise and interact via a cohesive interaction force. This means that the interaction can transmit not only compression and shear forces,

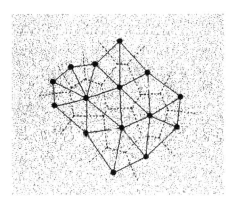

Fig. 1. Discrete element shape, Delaunay triangulation (continuous line).

but also tensile forces [10]. If new contacts between DE occur during the deformation process, there should be compressive and shear forces allowed, but the contact itself is cohesionless and therefore cannot take tensile forces.

Both cohesive and frictional forces have an elastic phase with normal and shear components characterised by the stiffnesses k_n and k_s respectively (Fig. 2). The normal component force F_n, is directed from one centroid of a DE to the other. The interaction force modulus is determined through the relative displacement between the two elements. The stiffnesses k_n and k_s, are kept constant in this elastic domain and are the same for all links. For each link, the forces are calculated incrementally [Eq. (1)],

$$\Delta \vec{F}_i^{(t)} = k_n \Delta U_{ni} + k_s \Delta U_{si}$$
$$\vec{F}_i^{(t)} = \vec{F}_i^{(t-1)} + \Delta \vec{F}_i^{(t)} \tag{1}$$

where, for the lth link between two DE, F_i is the interaction force, U_{ni} and U_{si} are the relative normal and tangential displacements, respectively.

2.3. Modelling the degradation processes

2.3.1. Local behaviour law

Based on the physical observation of the fracturation process in concrete, the basic assumptions of the model are:

- Overall damage process in concrete is due to accumulated surface energy dissipation in tension and shearing [20], leading to failure planes discontinuities.
- Concrete under uniaxial tension is modelled as a brittle material.
- Failure planes are modelled by special frictional contact laws.

To consider directly these assumptions, the fracturation process is introduced at the cohesive interaction force level. A local shear strength threshold of Mohr–Coulomb type is introduced in the model. This threshold is defined by a cohesion C

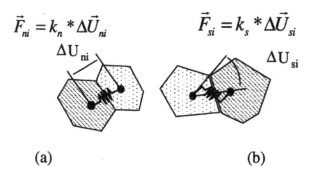

$$\vec{F}_{ni} = k_n * \Delta \vec{U}_{ni} \qquad \vec{F}_{si} = k_s * \Delta \vec{U}_{si}$$

(a) (b)

Fig. 2. Types of laws: Elastic normal (a) and shear (b) interaction forces.

and the friction angle ϕ between σ_n, σ_s as shown in Fig. 3a. The local value σ is calculated from the force F_i and a local influence surface S [21] so that $\sigma = F_i/S$. Thus, a space distribution in the local fracture energy is implicitly introduced in this model because of local geometric heterogeneity. This distribution is very close to a normal distribution law and corresponds to the distribution advised for concrete [22]. Hence, in the case of shear fracture, a frictional zone is created by changing the initially cohesive interaction force into a frictional one. A brittle tensile strength threshold is added. In the case of tensile fracture, the interaction is simply suppressed,

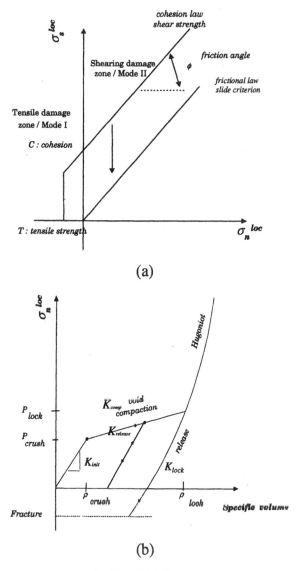

Fig. 3. (a) and (b). Damage laws.

i.e. the interaction force is set to zero. If compression is applied after fracturation has occurred, only a new frictional contact is considered and is limited by a cohesionless Mohr–Coulomb slide criterion.

It is assumed that the DE is the smallest unbreakable entity of the model, so that all fractures processes are located at the surface of the elements with non-existing preferential direction.

2.3.2. Spherical behaviour

When the loading of a concrete structure generates high local stresses, the volume decreases and the density increases, because of the collapse of the porous structure of the cement matrix. This situation is rarely encountered under quasi-static conditions. It is a typical situation of loading by impact and shock. The model as it has been defined so far, has no geometrical porosity. However, in concrete, voids are present at two different levels: capillary porosity, often located at the grain-cement interface, and microporosity included in the cement hydrate. These two types of porosity are at a lower scale than the scale of DE and links. It is thus impossible to model geometrical porosity directly. The proposed solution is to take explicitly the effects of porosity collapse into account with a local variable at the level of each link. These are translated explicitly into parameters K_ρ of the law $P = \int K_\rho \partial \rho$ (P the pressure, K_ρ the bulk modulus and ρ the specific volume) in terms of a variation of the local stiffness k_n. Thus, a synthetic approach [3,23,24] will be adopted, where the spherical behaviour is described by the following three concepts (see Fig. 3b):

- elastic part and volume plasticity threshold,
- porosity collapse,
- plastic hardening, covering all the non-linear phenomena intrinsic to the aggregate [25].

The first region is linear elastic defined by the initial elastic value $k_n = K_{init}$. The volume plasticity threshold is controlled by the value ρ_{crush} which corresponds physically to the beginning of pore collapse.

The second region $\rho_{crush} < \rho < \rho_{lock}$ is a transition region. The air present in the voids is expelled from the concrete, producing a plastic volumetric compaction which is introduced using a linear plasticity law with a tangent stiffness such that, $k_n = K_{comp}$. Unloading occurs along a modified path with, $k_n = K_{release}$ that is interpolated from adjacent regions K_{init} and K_{lock}.

For $\rho < \rho_{lock}$, the materials is completely compacted and the bulk modulus of the material increases substantially with pressure. The bulk modulus is defined by $K_\rho = \rho(\partial P/\partial P)$ which is proportional to Young's modulus and the stiffness k_n. In dense materials, thermodynamics state of the medium is characterised by pressure, internal energy and specific volume terms. Shock wave kinematic parameters are D, the celerity and u, the particle velocity jump. The shock wave conservation equations (Hugoniot relations) are used to determine shock velocity and compressive-strain properties. Two of them are expressed by Eqs. (2) and (3):

$$P = \rho_{\text{lock}} Du \tag{2}$$

$$\rho = \rho_{\text{lock}} D/(D - u) \tag{3}$$

In addition, experimental observations show that there exists a linear relation [26] between D and u:

$$D = a + bu \tag{4}$$

where a and b are the Hugoniot's parameters. An analytical development of $K_\rho = \rho(\partial P/\partial \rho)_\rho$ is then possible. Given the analytical knowledge of K_ρ the analytical ratio $K_\rho(\rho)/K_\rho(\rho_{\text{lock}})$ is calculated and the stiffness evolution k_{n}, can be evaluated (see the Appendix).

Finally,

$$K_\rho(\rho)/K_\rho(\rho_{\text{lock}}) = (\rho/\rho_{\text{lock}})(1 + 2bL)(1 + (b - 1)L)^2 \tag{5}$$

where,

$$L = (\rho - \rho_{\text{lock}})/(\rho - b(\rho - \rho_{\text{lock}})) \tag{6}$$

L is a dimensionless value. At the local level, ρ can be identified as the reciprocal of the distance l/d_{ij} (d_{ij}, distance between the DE, i and j). The second parts of Eqs. (5) and (6) are dimensionless, so that the macroscopic term ρ can be replaced by the local term l/d_{ij},. Finally, we can define the following relation:

$$k_{\text{n}}(1/d_{ij}) = (d_{\text{lock}}/d_{ij})(1 + 2bL)(1 + (b - 1)L)^2$$
$$\text{for } \rho > \rho_{\text{lock}} \text{ (i.e. } d < d_{\text{lock}}) \tag{7}$$

2.4. Resolution scheme

Transitional law of motion is applied to each DE. Explicit integration of the Newton's equation is done by a time-centered scheme (Eq. 8).

$$\ddot{\vec{x}}^{(t)} = \frac{\sum\limits_{i=\text{link}} \vec{F}_i^{(t)}}{m_{\text{DE}}}$$
$$\dot{\vec{x}}^{(t+\Delta t/2)} = \dot{\vec{x}}^{(t-\Delta t/2)} + \ddot{\vec{x}}^{(t)} * \Delta t \tag{8}$$
$$\vec{x}^{(t+\Delta t)} = \vec{x}^{(t+\Delta t)} + \dot{\vec{x}}^{(t+\Delta t/2)} * \Delta t$$

with $\ddot{\vec{x}} \left\{ \begin{array}{c} a_{\text{x}} \\ a_{\text{y}} \end{array} \right\}$ is the acceleration, $\dot{\vec{x}} \left\{ \begin{array}{c} v_{\text{x}} \\ v_{\text{y}} \end{array} \right\}$ is the velocity and $\vec{x} \left\{ \begin{array}{c} d_{\text{x}} \\ d_{\text{y}} \end{array} \right\}$ is the displacement of a DE. The rotation of the DE is neglected here in order to simplify the code and also because the major part of the energy is dissipated in the others mechanisms for the plate–plate impacts and divergent spherical explosions that will be simulated. The use of an implicit calculation scheme implies small time steps for the integration. If it is too large, some numerical instabilities could be created and thus render the simulation impossible. In a linearly elastic case such as $M\ddot{x} = F^{ext}$, the critical integration time step [27] is $\Delta t_c = 2/\omega_{max}$ where ω_{max} is the natural resonance frequency of the system. The evaluation of ω_{max} is based on the minimum DE weight M_{min} and the maximum stiffness k_n of the medium, i.e. k_{max}. Eqn 9 gives the maximum value for the time step.

$$\Delta t_c \leqslant \gamma^* 2^* \sqrt{\frac{M_{min}}{k_{max}}} \qquad (9)$$

The constant γ is to account for the interactions with the neighbourhood that each element has [28]. It is set to 0.1 and the time step is evaluated at each iteration.

This scheme is presently used in the framework of fast dynamics. Beside this, a conjugate gradient type scheme has been used to treat the problem of static compressions [29]. There are other methods, such as a super-relaxation method [30] or a conjugate gradient Fourier acceleration [31] which could be used. However the conjugate gradient type scheme provides a five-fold time saving compared to a super-relaxation method. This observation has previously been made on a stochastic cracking model [32].

3. Uniaxial compression tests

With the use of the simulation of uniaxial compression tests and the comparison between the experimental and the numerical results, the calibration of the model on the behaviour of concrete at low pressure conditions is refined. The physical parameters involved are the local elastic stiffnesses (k_n k_s) and the deviatoric parameters (C, ϕ, T).

3.1. Identification of the local parameters

The model defined in this study has few elastic and fracture parameters. Since the material data available are usually macroscopic (E, ν, fracture surface), it is not possible to act directly on them, but only on the micro parameters of local laws. Nevertheless, micro–macro analytical approaches on elastic parameters and on fracture parameters are still lacking. There are a number of references concerning purely granular media [33–36], but they are difficult to apply to the medium under study here, because the phenomena which generate non-linearity are different (fracture in tension and shearing and sliding phenomena in case of cohesive materials and only sliding phenomena in case of granular 9 medium). Therefore, a direct calibration is presently achieved when the medium has gone beyond its elastic limit.

3.1.1. Elastic parameters

The elastic parameters are determined using Voigt's law, which establishes a relation between Poisson's ratio and the ratio k_s/k_n. In the two dimensional cases, a direct relation between (E, v) and (k_n/k_s) has been formulated by Kusano [37] [Eq. (10)].

$$
\begin{aligned}
k_n &= \frac{1}{\sqrt{3}} \frac{E}{(1+v)^*(1-2v)} \\
k_s &= \frac{1}{\sqrt{3}} \frac{E^*(1-4v)}{(1+v)^*(1-2v)}
\end{aligned}
\tag{10}
$$

It can be checked that, for $k_s = k_n$ $v = 0$ is obtained.

3.1.2. Local fracture parameters

Different shapes of fracture surface criteria are used depending on the nature of the material under study. For metals, a vonMises criterion is often used,

$$
F_{plasticity}(q, k) = q - k
\tag{11}
$$

$$
q = \left[\left((\sigma_1 - \sigma_2)^2 + (\sigma_1 - \sigma_3)^2 + (\sigma_2 - \sigma_3)^2 / 2 \right) \right]^{1/2}
\tag{12}
$$

where q is the deviator and k is twice the apparent cohesion C in (τ, σ) space, $k = 2^*C$. However, in view of the asymmetric character of the behaviour of rocks and concrete, it is not applicable. A generalisation of the von Mises criterion was proposed by Drucker and Prager [38] to take this phenomenon into account. This consists of adding a term related to the pressure P which is expressed in Eq. 13:

$$
F_{DP}(P, q, k, m) = q - mP - k
\tag{13}
$$

where m and k are parameters of these perfectly plastic models. These macroscopic parameters can be determined by triaxial tests at two different low confining pressures. The knowledge of the failure point for each of these tests enables the construction of the fracture surface obtained in the space of the first and second invariant of the stress tensor. Transformed in the 2D case, the biaxial compression simulation shows that the model built in this manner correctly reproduces the linear fracture surface of Drucker–Prager (Fig. 4, local friction 0°). The cohesion at the microscopic and macroscopic scales has the same value (Fig. 5). The transition from m to ϕ (i.e. the transition from microscopic to macroscopic scale, see; Fig. 4) is linear and depends on the value of Poisson's ratio.

3.2. Simulation of MB50 uniaxial compression

With appropriate values of the elastic and fracture parameters, the behaviour of the material under static load can be reproduced accurately. The simulation is performed in a 2D axisymetrical configuration with 5000 DE in 2 h of CPU time for a

82

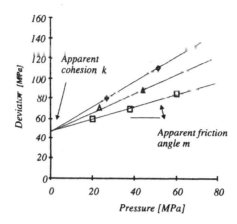

Fig. 4. Local friction angle 0° □, 10° △, 20° ◇.

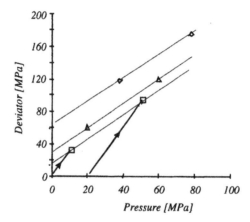

Fig. 5. Local cohesion 8□ 16 △, 32 ◇ MPa.

300 MHz processor. The four loading phases observed experimentally [39] are well reproduced by the numerical model and are represented in Figs. 6 and 7:

- Phase I : The behaviour is elastic linear and reversible. A typical cone of triaxial stress state observed in the case of uniaxial compression on cylindrical specimen, is obtained, even in the 2D axisymetric case (Fig. 8). The volume is decreasing with respect to the experimental results.
- Phase 2 : A specific characteristic of rocks and concretes is the dilatancy phenomenon (volume increase shortly preceding the failure of the material). Dilatancy has been shown experimentally [40, 41] to be caused by the opening of microcracks in the direction of loading. These are mode I type fractures. In the numerical simulations, it has been observed that the fracturing process starts with the creation of a large amount of cracks in mode I and these are oriented in the loading direction (Fig. 9). This is in good agreement with experimental

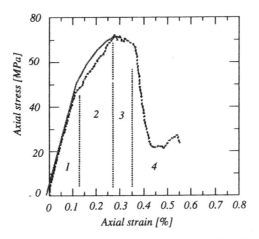

Fig. 6. Axial strain–axial stress, experiment, —, calculation.....

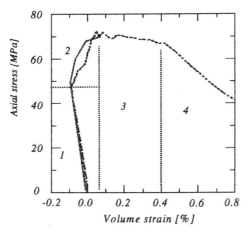

Fig. 7. Volume strain–axial stress, experiment —, calculation.....

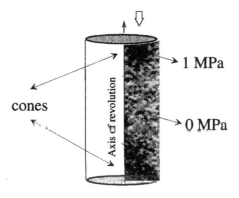

Fig. 8. Radial stress field.

84

results. The dilatancy phenomenon is comparable in all respects to the experimental data (Fig. 7). It starts at 70% of critical stress. The important parameter, in this context, is the tensile strength l.

- Phase 3 : The crack appearance illustrated in Fig. 10 shows substantial localisation along conical slip planes. This localisation, with the presence of binding cones typical in this type of experiment, is caused by a triaxial state of the radial stresses at the head (Fig. 8) resulting from the infinite friction conditions at the plates as during experiments. The localisation occurs in mixed modes I and II. A view of the fragments is shown in Fig. 11 (radial deformation ×60). A large fragment is observed detaching from the test specimen and, closer to the axis of revolution, smaller fragments generated by friction of the crack edges.
- Phase 4 : Friction phenomena are mainly localised at the cracking bands.

Thus, the model cannot only reproduce the quantitative behaviour of the material, but also provides a description of the different degradation processes stage under static load. Moreover, it only needs five parameters to define the complete deviator behaviour.

4. Dynamic tests

In the first series of tests, a sustained uniaxial solicitation is presented while in the second series, the non-sustained solicitation shows how well the dynamic behaviour can be reproduced.

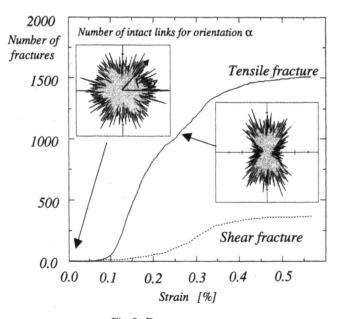

Fig. 9. Damage process.

In experiments, a velocity effect has been observed [42] on compression strength of the material. The compression strength increases when the deformation rate increases. Two effects might be causing this phenomenon: the viscosity of the material or an inertial structural effect. Using the finite element method, a strength increase term is accounted for a viscosity term in the damage laws [43]. However, a more recent study using the DEM, has shown that this velocity effect could be explained by an inertial structural effect, and that there is no need for it to take viscosity into account in the constitutive law for this type of material [13]. Thus, static compression tests have been used to adjust the deviator behaviour involved in both static and dynamic responses of the material.

4.1. Plate–plate impact

4.1.1. Experiments

The plate impact tests were performed on a quartzite. In this analysis, the main interest is in the response of the material under high pressure conditions. This enables the spherical behaviour to be described, and mainly the compaction phenomena. The

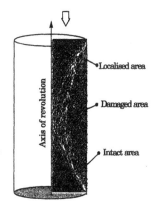

Fig. 10. Distribution of remaining links.

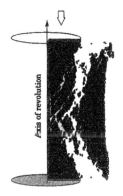

Fig. 11. Distribution of fragmented regions.

plate impact tests (Fig. 12) reproduce the phase of uniaxial loading by a shock wave and are used to determine a point on the state curve $P = f(\rho)$.

The measured parameter is the material velocity at the interface between the quartzite and the window material (PMMA, LiF, Pyrex). The wave path taken during the shock is shown in Fig. 12. During impact at velocity V (impactor velocity), two shock waves propagate, one in the impactor, the other in the quartzite, where the equilibrium pressure is P_e and the material velocity is u_e. At the impactor–quartzite interface, a non-gradient pressure-material velocity zone is established. So the pressure in the impactor is also P_e and the material velocity $V - u_e$.

At the free edge of the impactor, the shock wave front is reflected and generates an expansion wave which returns to the impactor–quartzite interface. The tensile wave is then transmitted to the quartzite. Behind the shock wave propagating in the quartzite, the material is compacted and therefore stiffened.

Because of this, the tensile wave from the impactor can overtake and reduce the amplitude of the shock. Tests were carried out with different types of impactor (copper, aluminium) and different block sizes (Table 1)

4.1.2. Simulations

Macroscopic and microscopic material parameters are shown in Table 2. The material under study had a tensile elastic limit of 25 MPa and a uniaxial compressive elastic limit of 250 MPa. Spherical behaviour law parameters are determined by identification on a pressure–density experimental curve. All the tests which were carried out were performed with the same parameters for the quartzite.

Results of quartzite–copper impacts are shown in Figs. 13 and 14. The equilibrium particle velocity at the moment of the impact calculated by the model is 540 m.s^{-1} (Fig. 13) and the pressure 6.2 GPa. The result of the analytical calculation [44] based on the impact polar theory is 6.4 GPa. The difference can be explained by the porosity,

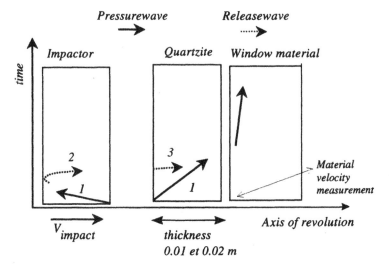

Fig. 12. Plate impact diagram field.

Table 1
Plate–plate impact tests

Test	Impactor		Quartzite		Nature
	Nature	Velocity (ms^{-1})	Size (mm)	Specific volume (Kg m^{-3})	
A	Cu	702	10.39	2500	PMMA
B	Cu	700	20.34	2500	PMMA
C	AU4G	808	5.02	2500	Pyrex
D	AU4G	811	10.02	2500	Pyrex
E	Cu	2130	10	2500	Lif
F	Cu	1657	10.33	2500	Lif

Table 2
Experimental data and local parameters

Macro data		Micro parameters	
E	70 GPa	k_n	5.5×10^9 N m^{-1}
v	0.2	k_s	5×10^8 N m^{-1}
ρ	2600 kg m^{-3}	C	6.5×10^7 Pa
σ_{HEL}	1.6 GPa	ϕ	40°
$\sigma_{frac}^{tension}$	25 MPa	T	35×10^7 Pa
$\sigma_{frac}^{compression}$	250 GPa	P_{crush}	1.5 GPa
Average porosity	6%	ρ_{lock}	6%
a	3700	a	3700
b	1.79	b	1.79

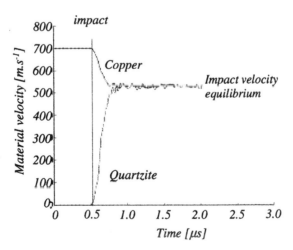

Fig. 13. Impact velocity equilibrium.

88

which is not included in the analytical calculation. The initiation of the shock given by the present model is thus satisfactory.

The results in Figs. 14 and 15 show that the level of the elastic precursor, its velocity, and the velocity of the plastic shock wave are well reproduced by the

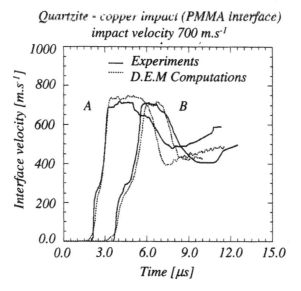

Fig. 14. Results A and B.

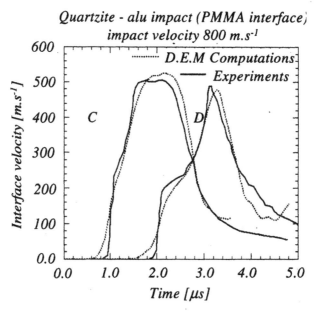

Fig. 15. Results C and D.

model. Similarly, the plateau times determined by the velocity of the tensile wave returning from the impactor are good. This means that the tensile wave passes through a compacted, and therefore stiffened, medium and then catches up with the shock front. The larger the size of the sample, the shorter this plateau (Fig. 14, test B). This feature is also particularly well reproduced by the numerical simulations. A result of a test in which the tensile wave catches up completely with the shock front is shown in Fig. 15, test D.

Before performing the divergent hemispherical explosion tests, a high-velocity impact experiment provides information on the capabilities of the model at very high pressure (20 GPa) in a range where the behaviour of the material approaches that of a fluid. There is no elastic precursor because of the shock intensity (25 GPa) (Fig. 16).

These plate impact simulations show that the model provides a very good description of the behaviour phases of a material such as quartzite under sustained strong shock.

4.2. Divergent hemispherical explosions

The hemispherical explosion tests (Fig. 17) strengthen the validity of the model. They involve a non-sustained shock wave as shown in Fig. 18. Different sample sizes were tested (Table 3) The pressures in the region close to the explosive reach 20 GPa. The kinematic parameter is the velocity at the quartzite–PMMA interface. Numerical simulations (Fig. 19) show that the velocity peak and the arrival time of the shock wave are predicted correctly.

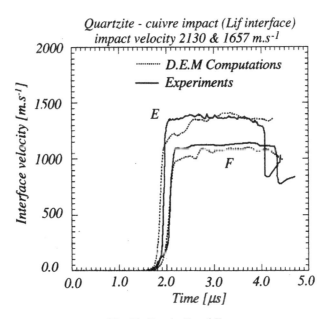

Fig. 16. Results E and F.

Table 3
Hemispherical explosions

Test	Quartzite Thickness (mm)	Specific volume (kg m^{-3})	Window Material
I	35.3	2500	PMMA
J	48.8	2500	PMMA

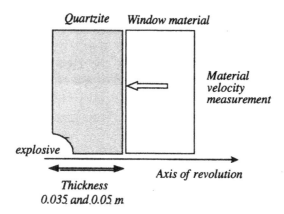

Fig. 17. Divergent hemispherical explosion.

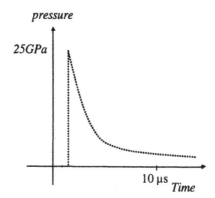

Fig. 18. Incident explosion pressure.

The phases corresponding to a peak, a plateau and then a gradual decrease due to the release of shock wave in the velocity, are comparable with the experiments. Thus, the use of the DEM reveals important details, such as the compaction field, i.e. the amount of irreversible volume reduction. Experiments show that the material is compacted mostly in the region close to the explosive. The results obtained in the numerical simulations and in the experiments are shown in Fig. 20.

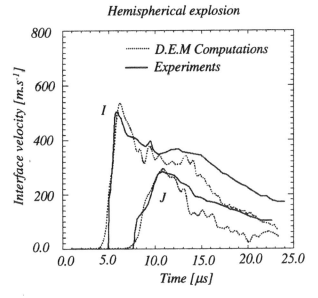

Fig. 19. Interface velocity.

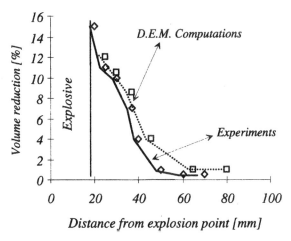

Fig. 20. Compaction zone.

5. Conclusion

A model based on a DEM has been presented to study both the static and the dynamical behaviour of cohesive geomaterials. The detailed study of the fracturing processes is possible because of the local description of the material's behaviour.

In the case of static behaviour under low pressure, it can reproduce, in both a descriptive and a quantitative manner, the creation and propagation of damage, the

92

relation between post-peak stress and strain, as well as the localisation of fractures (including their nucleation and propagation).

Concerning the very high dynamic solicitations, it has been able to reproduce, the evolution of the compaction of the material in a quantitative way In this pioneering field of high strain rate deformation, it has shown its great potential to treat very complex problems, which have been, until now, poorly described by classical numerical methods.

Such a modelling approach can offer a helpful tool of investigation for a large variety of problems, such as explosive stability during a shock, study of armour-plate, or even accidental impact on various concrete structures.

Appendix.

Calculation of the change in bulk modulus. With,

$$K_\rho(\rho) = \rho^* \frac{\partial P}{\partial \rho} = \rho^* \frac{\partial P^*}{\partial u} \frac{\partial u}{\partial \rho} \tag{14}$$

According to the Rankine–Hugoniot conservation expressions:

$$P = \rho_{\text{lock}} D u \tag{15}$$

$$\rho = \rho_{\text{lock}} D/(D - u) \tag{16}$$

with

$$D = a + bu \tag{17}$$

where, u is the particle velocity, D is the wave velocity, a and b are the Hugoniot coefficients.

From (15) and (17) we obtain $P = \rho_{\text{lock}}(a + bu)u$ whence

$$\frac{\partial P}{\partial u} = \rho_{\text{lock}}(a + 2bu) \tag{18}$$

From (16) and (17) we obtain

$$u = (a(\rho - \rho_{\text{lock}}))/(\rho - b(\rho - \rho_{\text{lock}})) \tag{19}$$

therefore

$$\frac{\partial u}{\partial \rho} = \frac{(a + (b - 1)u)^2}{a} \tag{20}$$

From Eqns. (14), (18) and (20), it can then be calculated:

$$K_\rho(\rho) = (\rho\rho_{lock}/\mathrm{a})(a + 2bu)(a + (b-1)u)^2 \tag{21}$$

for $\rho = \rho_{lock}$, Eq. (16) yields $u = 0$, so that

$$K_\rho(\rho_{lock}) = (\rho_{lock}a)^2 \tag{22}$$

Finally,

$$K_\rho(\rho)/K_\rho(\rho_{lock}) = (\rho/\rho_{lock})(1 + 2^*b^*L)(1 + (b-1)L)^2 \tag{23}$$

where

$$L = (\rho - \rho_{lock})/(\rho - \mathrm{b}(\rho - \mathrm{b}\rho_{lock})) \tag{24}$$

substituting ρ by $1/d$ and ρ_{lock} by $1/d_{lock}$, the law governing stiffness change is obtained.

References

[1] Potyondi DO, Cundall PA, Lee CA. Modelling rock using bonded assemblies of circular particles. In: Rock mech. Rotterdam: Balkema 1996. p. 1937–44.
[2] Schatz JF. The physics of SOC and tensor. Lawrence Livermore Laboratory, 1973. (UC- 51352).
[3] Mariotti C, Tranchet JY. Behaviour of a quartzite submitted to a spherical divergent shock-wave: experiments and modeling. JGR, submitted for publication.
[4] Cundall PA, Strack ODL. A discrete numerical model for granular assemblies. Géotechnique 1979;29:47–65.
[5] Casaverde L, Iwashita K, Tarumi Y, Hakuno M. Distinct element analysis for rock avalanche. Structural Eng/Earthquake Eng 1989;6:131–40.
[6] Omachi, T., Arai, Y. Dynamic failure of rockfill models simulated by the distinct element method. Numerical Methods in Geomechanics 1988; 1855–60.
[7] Ghaboussi J, Barbosa R. Three-dimensional discrete element method for granular materials. Int J of Num and Analytical Methods In Geo 1990;14:451–72.
[8] Hakuno M, Yamamoto T. Simulation analysis of dynamic non-linear behavior of underground structures by the extended distinct element method. In: Computer methods and advances in geomechanics, beer, booker and carter, 1991. p. 481–6.
[9] Donzé FV, Mora P, Magnier SA. Numerical simulation of faults and shear zones. Geophys J Int 1994;116:46–52.
[10] Meguro K, Hakuno M. Application of the extended distinct element method for collapse simulation of a double-deck bridge. Struc Eng/Earthquake Eng 1994;10(4):175–85.
[11] Donzé F, Magnier SA. Formulation of a three-dimensional numerical model of brittle behavior. Geophys J Int 1995;122l:790–802.
[12] Magnier SA, Donzé FV. Numerical simulation of impacts using a discrete element method. Mech Cohes-Frict Mater 1998;3:257–76.
[13] Donzé FV, Magnier SA, Daudeville L, Mariotti C, Davenne L. Numerical study of the compressive behavior of concrete at high strain rates. ASCE Journal of Engineering Mechanics 1999;125:1154–63.
[14] Potapov AV, Hopkins MA, Campbell CS. A two-dimensional dynamic simulation of solid fracture. International Journal of Modern Physics C 1995;6(3):371–98.

[15] Meftah W, Biarez J, Evesque P, Lateb G. La localisation dans les milieux granulaires bidimensionnels géométriquement ordonnés, In: 2nd Annual Conference Réseau de Laboratoires G.E.O., Aussois, France, 1995.

[16] Jirásek M, Bazant ZP. Particle model for quasibrittle fracture and application to sea ice. Journal of Engineering Mechanics, September 1995,121(9):1016–25.

[17] Donzé FV, Magnier SA, Bouchez J. Modeling fractures in rock blasting. Int J Rock Mech 1997;34:1153–63.

[18] O'Connor R.M. A distributed discrete element modeling environment — algorithms, implementation and applications. Ph.D thesis, Department of Civil Environmental Engineering, 1996.

[19] Shukla A, Sadd MH. Studies of the effect of microstructure on the dynamic behaviour of granular and particulate media. First Year Report, US Air Force Office of Scientific Research, March 1994, contract no. F49620-93-1-0209.

[20] Guéguen Y, Palciauskas V. Introduction a la physique des roches. Paris: Hermann, 1992.

[21] Preece DL. Discrete element modeling of rock blasting in benches with joints and bedding planes — initial development. In: 11th Proc of Conf on explosive and blasting research, 1995, (CONF-9502 142). p. 324–33.

[22] Rossi P, Wu X. A probabilistic model for materials behaviour analisys and appraisal of the structure of concrete. In: Num models in frac of concrete. Wittman, 1993. p. 2071–21.

[23] Holmquist TJ,.Johnson GR. A computational constitutive model for concrete subjected to large strains, high strain rates, and high pressures. In: 14th International Symposium on Ballistics, September 1993, Québec. p. 591–600.

[24] Shao JF, Henry JP. Development of an elastoplastic model for porous rock. International Journal of Plasticity 1991;7:1–13.

[25] Tranchet JY. Comportement de deux matériaux fragiles polycristallins sous l'effet de la propagation d'une onde spherique divergente. PhD thesis, Université de Bordeaux, France, 1994.

[26] Grady DE. Impact compression properties of concrete. Experimental Impact Physics Division 1433, 1993 (SAND-93-0013C, DE93 007599).

[27] Bathe KJ. Finite element procedures in engineering analysis. Englewood Cliffs (NJ): Prentice-Hall Inc:, 1982.

[28] Hart R. An introduction to distinct element modeling for rock engineering. In: Proc of the 7th Int. Congress on Rock Mechanics, 1991, p. 245–61.

[29] Dudkin LM, Rabinovich I, Vakhatinsky I. Iterative aggregation theory. New York: Marcel Dekker, 1987.

[30] Allen DN, de G. Relaxation method. New York: McGraw-Hill, 1954.

[31] Batrouni GG, Hansen A, Nelkin M. Fourier acceleration of relaxation processes in disordered systems. Phys Rev Lett 1986;57:1336–43.

[32] Meakins P, Li O, Sander LM, Yan H, Guinea F, Pla O, Louis E. Simple stochastic models for material failure. In: Disorder and fracture. New York: Plenum Press, 1990. p. 119–39.

[33] Ching SC, Anil M. Packing structure and mechanical properties of granulate. Journal of Engineering Mechanics 1990;116(5):1077–93.

[34] Emeriault F, Cambou B, Mahboudi A. Homogenization for granular materials: non reversible behaviour. Mechanics of Cohesive-Frictional Materials 1996;1:199–218.

[35] Sidoroff F, Cambou B, Mahboudi A. Contact force distribution in granular media. In: Shen H.H. et al., editors. Advances in micromechanics of granular materials, vol. 16. Elsevier, 1993. p. 31–40.

[36] Cambou B. From global to local variables in granular materials. In: Powders and grains, 2nd Int Conf Balkema: Thornton, 1993, p. 73–86.

[37] Kusano N, Aoyagi T, Aizawa J Ueno, Morikawa H, Kobayashi, N. Impulsive local damage analyses of concrete structure by the distinct element method. Nuclear Engineering and Design 1992;138:105–10.

[38] Drucker DC, Prager W. Soil mechanics and plastic analysis or limit design. Quarterly of Appl Math 1952;10:157–75.

[39] Andreev GE. Brittle failure of rock materials — test results and constitutive models. Balkema, 1998.

[40] Brace WF. Volume change during fracture and frictional sliding, a review. Pageoph 1978;116:603–14.

[41] Critescu N. Rock rheology. Kluwer Academic Publishers, 1989.

[42] Bischoff PH, Perry SH. Compressive behaviour of concrete at high strain rates. Materials and Structures 1991;24:425–50.

[43] Le vu O. Etude et modélisation du comportement du beton sous sollicitation de grande amplitude. PhD thesis, Ecole Polytechnique, France, 1998.

[44] Défournaux, Cours de détonique. Ecole Speciale Militaire, France, 1988.

Papers from

International Journal of Rock Mechanics and Mining Sciences

A three dimensional numerical model for thermohydromechanical deformation with hysteresis in a fractured rock mass[*]

V. Guvanasen[a,*], Tin Chan[b,*]

[a]*HydroGeoLogic Incorporated, 1155 Herndon Parkway, Suite 900, Herndon, VA 20170, USA*
[b]*Atomic Energy of Canada Limited, H16 E22, 700 University Avenue, Toronto, Ontario, M5G 1X6, Canada*

Accepted 7 October 1999

Abstract

A three-dimensional finite-element solution to the problem of coupled thermohydromechanical deformation, groundwater flow, and heat transport in deformable fractured porous media is presented in this paper. The governing equations are based on Biot's consolidation theory for poroelastic materials, extended to the non-isothermal environment. The normal and lateral deformations in joints are simulated by a new joint element. The new joint element is based on the Bandis–Barton models and is capable of simulating normal and lateral deformations with dilatancy, contractancy, and hysteresis due to irrecoverable damages/rubblization. A three-dimensional finite-element code Model Of Transport In Fractured porous media (MOTIF) has been developed based on the theoretical framework presented herein. Verification results with experimental data and analytical solutions are presented in this paper. An application example with flow of fluid through a non-isothermally deforming joint is also presented. Results indicate that non-isothermal deformation could play a major role in the transport of fluid and waterborne substances in fractured rocks. Crown Copyright © 2000 Published by Elsevier Science Ltd. All rights reserved.

1. Introduction

The concept of nuclear fuel waste disposal in geologic formations is generally considered the most viable and is being investigated by many countries with nuclear waste disposal programs. For example, in Canada, the concept of emplacing corrosion-resistant waste canisters in a sealed vault constructed at a depth of 500 to 1000 m in plutonic rock in the Canadian Shield has been assessed by the Atomic Energy of Canada Limited [1]. The potential migration of radionuclides in the groundwater between the vault and the biological environment may be affected by several processes, including: convection, hydrodynamic dispersion, sorption, and radioactive decay. In addition, the transport processes in such plutonic rock are complicated by the fact that fracture networks are present in the rock. The host rock in which a nuclear waste repository may be located will be subjected to in situ, excavation-induced, thermomechanical, and, at some time in the future, glacial stresses. These stresses can induce changes in hydraulic properties in the unfractured rock and along fracture planes, which, in turn, could lead to a change in groundwater flow pattern and might affect mechanical stability (see Fig. 1). In this paper, a three-dimensional solution to coupled thermohydromechanical deformation, groundwater flow, and heat transport is presented. Special joint elements are incorporated to simulate pre-existing fractures in rock. Included in the new joint elements are: normal and lateral (shear) deformations with hysteresis due to rubblization of joint asperities, and dilatancy (or contractancy) associated with relative lateral movements of joint planes. An example test case and discus-

---

[*] This paper is dedicated to the late Professor Neville G. W. Cook, to whom one of the authors (T.C.) owes so much for his mentoring in rock mechanics.

[*] Corresponding authors. Tel.: +1-703-478-5186 (V. Guvanasen); +1-416-592-5296 (T. Chan).

E-mail addresses: dua@hgl.com (V. Guvanasen); chant@aece.ca (Tin Chan).

Nomenclature

A	elemental area (fracture element)	M	bulk hydroelastic modulus for fluid
a	empirical constant for the joint normal stress constitutive relation	M_{rr}	mass matrix
a_s	empirical constant for the joint shear stress constitutive relation	M_L	net mass flux per unit joint surface area from the solid matrix
a_j	initial joint aperture under self-weight	N_J	basis function associated with node J
a_{ijkl}	dispersivity tensor	n_c	number of connected nodes in an element
b	empirical constant for the joint normal stress constitutive relation	n_i	outward unit vector normal to S
		n_p	total number of nodes
b_s	empirical constant for the joint shear stress constitutive relation	p	fluid pressure
		p_{ref}	reference fluid pressure
C_f	fluid compressibility	Q	specific volumetric fluid generation rate
C_{Ik}	k-th empirical coefficient for variable I ($I = V_m, K_{ni}, a_j, V_{irr}, G_i, \tau_1, u_1, u_{peak}, f_{wthr}$)	Q_H	specific enthalpy production rate
		Q_{Hm}	net thermal flux per unit area from the solid matrix
C_{ijkl}	elastic constant tensor	Q_{si}	quantity (displacement, stress, stiffness) weighted by α_{sh}
c_f	specific heat capacity of fluid		
c_s	specific heat capacity of solid	$Q_{si,1}$	quantity (displacement, stress, stiffness) associated with the continuous model
d_i	dilation angle		
$d_{n,mob}^{dilat}$	mobilized angle of dilation	$Q_{si,2}$	quantity (displacement, stress, stiffness) associated with the pre-peak model
E	Young's modulus		
E_m	mechanical joint aperture	q_i	Darcy velocity in the i-th direction
e_h	hydraulic joint aperture	R	Biot's isothermal hydroelastic constant
e_{ij}	strain tensor	S	elemental surface area (solid element)
F_I	load vector	S_{IJ}	stiffness matrix
F_i	body force term in the i-th direction	T	temperature
$f_{u_{peak}}$	fraction of u_{peak} at which the joint stiffness is estimated	$\tilde{T}$	approximate solution for temperature
		T_J	temperature at node J
f_{wthr}	weathering factor	T_{ref}	reference temperature
$f_{\tau_{UL}}$	factor used in modifying the stiffness	t	time
g	gravitational acceleration	U_{ij}	isothermal hydroelastic constant tensor
H	Biot's volumetric hydroelastic strain coefficient	u	lateral joint displacement along the joint plane
H_{ij}	hydrodynamic dispersivity tensor	u_i	deformation in the i-th direction
h	reference hydraulic head	$\bar{u}_i$	approximate solution for deformation in the i-th direction
$\bar{h}$	approximate solution for reference hydraulic head		
		u_{iJ}	displacement in the i-th direction at node J
h_J	reference hydraulic head at node J	u_{peak}	lateral joint displacement along the joint plane at peak shear stress
JCS_n	joint (wall) compressive strength		
JRC_{mob}	mobilized joint roughness coefficient	V_m	joint maximum closure
JRC_n	joint roughness coefficient	V_{irr}	irrecoverable closure
JRC_{peak}	joint roughness coefficient at peak shear stress	v	joint normal displacement
		W_J	weighting function associated with node J
K_n	normal joint stiffness	W_{ij}	thermoelastic constant tensor
K_{ni}	initial joint normal stiffness (at $\Delta v = 0$)	x_i	coordinates in the i-th direction
K_{ni}^{mod}	modified joint normal stiffness		
k_{ij}	permeability tensor	*Greek symbols*	
K_{ij}^F	stiffness tensor of fracture	α	Biot's hydroelastic coupling coefficient
K_s	shear joint stiffness	α_{sh}	weighting factor, $0 \leq \alpha_{sh} \leq 1$, for selecting constitutive relationships for joint shear stress-related variables
K_s^{UL}	joint stiffness during unloading of shear		
L	length along fracture element perimeter		
L_n	sample length	β_f	volumetric thermal expansion coefficient for fluid
l_i	outward unit vector normal to L		

β_s	linear thermal expansion coefficient for solid	θ_0	initial porosity
		λ_{ij}^T	thermal conductivity tensor
Δp	$p - p_{ref}$	ζ	change in volumetric fluid content per unit volume
Δp_{ave}^F	average fluid pressure increase in fracture		
ΔT	$T - T_{ref}$	μ	friction angle
$\Delta u_{j\,ave}^F$	average fracture opening in the j-th direction	μ_f	dynamic viscosity
		ν	Poisson's ratio
Δu_r	increment of lateral displacement along the r-axis on the joint plane	ξ_i	coordinates in the i-th direction along the fracture plane
Δu_s	increments of lateral displacement along the s-axis on the joint plane	ρ_f	fluid density
		ρ_0	reference fluid density
Δv	increment of normal displacement along the axis normal to the joint plane	ρ_s	solid skeleton density
		σ_c	unconfined compressive strength of intact rock
$\Delta \tau_r$	increments of shear stress along the r-axis on the joint plane	σ_n	joint effective normal stress
$\Delta \tau_s$	increment of shear stress along the s-axis on the joint plane	τ	joint shear stress
		τ_{ij}	stress tensor
$\Delta \tau_{ij}^F$	incremental skeletal stress acting on the fracture walls	τ_{ij}^0	initial stress tensor
		τ_{peak}	peak shear stress
$\Delta \sigma_n$	increment of effective normal stress along the axis normal to the joint plane	ϕ_r	residual friction angle

sion are included in the paper. Although the formulation of the model is three-dimensional with three-dimensional continuum elements and two-dimensional joint elements, the verification problems are necessarily simple in geometry so that analytical solutions are available for comparison. Additional benchmarking of the model presented in this paper as well as a field-test simulation have been previously reported [12,26].

2. Mathematical model

Generally, groundwater flow through fractured media is described mathematically in one of three ways:

1. by considering each fracture as a discrete hydraulic conduit;
2. by assuming a hydraulically equivalent, anisotropic porous medium and using an appropriate porous-medium model; and
3. by a combination of (1) and (2).

A model based on discrete fractures only provides a theoretically correct flow pattern if all fractures and connections between fractures are considered. While such a model may be relatively easy to construct from a theoretical point of view, the mass of input data required, the computational effort, and the difficulty in obtaining data for many real-world problems make it generally difficult, if not impossible, to apply to regional groundwater flow regimes.

One way of circumventing this problem is to represent the fracture system by an equivalent porous medium. An equivalent porous medium is a fictitious, fluid-transmitting medium with material properties selected so that, under specific hydrological and geometrical conditions, the hydraulic behavior of the medium is, in important respects, the same as the hydraulic behavior of the actual fractured rock. The validity of the equivalent porous medium concept may be site-specific and flow-path-specific. If it is assumed that, for the purposes of the problem at hand, the domain can be represented by an anisotropic porous medium, then the major problem becomes that of determining the permeability tensor. The mathematics of calculating directional permeabilities from fracture orientation and aperture data was first discussed by Romm and Pozinenko [30]. Extensive work in this area has been done by several others; for example, Louis and Parnot [29] and Long et al. [28]. A similar approach with two overlapping equivalent porous media was proposed by Barenblatt et al. [4]. In their approach, flow of a homogeneous slightly compressible liquid in interconnected porous and fractured media is separated into two overlapping continua, each filling the entire flow domain. Flow takes place throughout each continuum, with exchange of fluid taking place between the two continua according to certain rules. Additional details of this type of model are given by Warren and Root [34], Closmann [13], and many others. This double-continuum concept has been

102

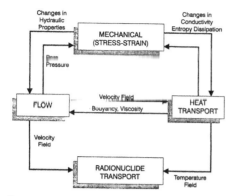

Fig. 1. Interrelationships between thermohydromechanical processes.

extended to include transport of radionuclides [24], and coupled flow and transport of water, vapor, air, and heat [32]. The theoretical framework presented in this paper can be extended to accommodate groundwater flow and heat transport in deformable double-continuum media; however, such an extension is beyond the scope of this paper.

In many instances, a rock mass may be considered to consist of a background system of relatively small-scale fractures that contain some longer, more conductive discrete fractures and a few major fracture zones of relatively high permeability (see Fig. 2). In this case, the longer discrete fractures and the fracture zones should not be included in the determination of the background equivalent porous medium properties; otherwise, their dominance may render the fracture system unrepresentable by an equivalent porous medium. Fortunately, the frequency of occurrence of these major discrete fractures or fracture zones is normally low. Therefore, they can be effectively treated as dis-

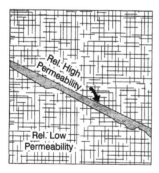

Fig. 2. A typical cross-sectional view of a fractured rock mass traversed by a thin and highly conductive fracture zone.

crete high-permeability planes embedded in the background equivalent porous medium. The approach adopted herein is based on the assumption that, in major discrete fractures or fracture zones, flow is dominant along the axis of the plane of the zones. In addition, it is assumed that the domain is composed of contiguous equivalent porous-medium blocks with embedded connected planes representing the major fracture zones. The main objective of the developmental work reported in this paper is to develop a model that could be used as an investigative tool to study potential thermohydromechanical behaviors of discrete fractures or fracture zones and their interaction with surrounding rock mass. The mechanical models for discrete fractures in this paper are approximated by empirical models that summarize results of laboratory testing and are considered to be realistic for rock joints. Barton et al. [5] and Cook [15] suggested that the in situ behaviors of rock joints may be approximated by laboratory experimental results from the third or later loading–unloading cycles. In situ measurements of transmissivity and in situ stresses in a major fracture zone at AECL's Underground Research Laboratory (URL) in Manitoba, Canada [31] reveal that the zone appears to behave like a single fracture obeying a hyperbolic normal stress–normal closure relationship. The model presented herein will be utilized in the future as an analytical tool for some of the ongoing field and laboratory experiments, and will be modified to reflect additional information as it becomes available. For some applications in which the properties of a major fracture zone may not be known a priori, simpler models (such as those similar to the Coulomb friction model for shear behavior) may have to be assumed.

The matrix-related and fracture-related processes are discussed in the ensuing subsections: the matrix-related processes in Sections 2.1, 2.2, and 2.3; and the treatment of fractures in Section 2.4.

2.1. Thermohydromechanical equilibrium in a solid matrix

Generalized equations describing isothermal hydroelastic equilibrium in poroelastic materials have been given by Biot [9]. These equations can be extended to describe thermohydroelastic phenomena by assuming that: (1) the fluid and solid are in thermodynamic equilibrium, and (2) a change in fluid density has no effect on the solid strain. Based on these assumptions, the following equations can be written tensorially [19]:

$$e_{ij} = (C_{ijkl})^{-1}(\tau_{ij} - \tau_{ij}^0) + \Delta p U_{ij} + \Delta T W_{ij} \tag{1a}$$

$$e_{ij} = \frac{1}{2}\left(\frac{\partial u_i}{\partial x_j} + \frac{\partial u_j}{\partial x_i}\right) \qquad (1b)$$

$$\tau_{ij} = C_{ijkl}e_{kl} - \Delta p\overline{U_{ij}} - \Delta T\overline{W_{ij}} + \tau_{ij}^0 \qquad (1c)$$

$$\overline{U_{ij}} = C_{ijkl}U_{kl} \qquad (1d)$$

$$\overline{W_{ij}} = C_{ijkl}W_{kl} \qquad (1e)$$

with $i, j, k, l = 1, 2, 3$; and where e_{ij} is the strain tensor, u_i the deformation in the i-direction, C_{ijkl} the elastic constant tensor, τ_{ij} the stress tensor, τ_{ij}^0 the initial stress tensor, U_{ij} the isothermal hydroelastic constant tensor, W_{ij} the thermoelastic constant tensor, x_i the coordinates in the i-th direction, p the fluid pressure, p_{ref} the reference fluid pressure, $\Delta p = p - p_{ref}$, T the temperature, T_{ref} the reference temperature, and $\Delta T = T - T_{ref}$. A list of nomenclature is provided at the beginning of the paper.

In the case of isotropic materials, Eq. (1c) may be recast as:

$$\tau_{ij} = \frac{E}{1+v}e_{ij} + \frac{Ev}{(1-2v)(1+v)}\delta_{ij}e_{kk}$$
$$+ \frac{E}{3H(1-2v)}\delta_{ij}\Delta p + \frac{E}{1-2v}\beta_s\delta_{ij}\Delta T \qquad (1f)$$

where E is Young's modulus, v Poisson's ratio, H Biot's volumetric hydroelastic strain coefficient, and β_s the linear thermal expansion coefficient for solid. The coefficient of the pressure differential term corresponds to the hydroelastic coupling coefficient α of Biot [8]. The stress expression in Eq. (1c) is used in the following quasi-static equilibrium equation in which the acceleration term is assumed negligible:

$$\frac{\partial \tau_{ij}}{\partial x_j} + F_i = 0 \qquad (2)$$

with $i, j = 1, 2, 3$; and where F_i is the body force term in the i-th direction.

2.2. Non-isothermal flow equation in poroelastic media

Based on the principle of mass conservation, a modified form of Biot's [9] equation for flow through deformable porous media is written tensorially as

$$\frac{\partial}{\partial t}\rho_f\zeta + \frac{\partial}{\partial x_i}\rho_f q_i - \rho_f Q = 0 \qquad (3a)$$

$$\zeta = \overline{U_{ij}}e_{ij} + \frac{1}{M}\Delta p - (\overline{U_{ij}}W_{ij} + \theta_0\beta_f)\Delta T \qquad (3b)$$

$$\frac{1}{M} = \frac{1}{R} - \overline{U_{ij}}U_{ij} = \frac{\theta_0}{\rho_f}\frac{\partial \rho_f}{\partial p} = \theta_0 C_f \qquad (3c)$$

with $i, j = 1, 2, 3$; and where M is the bulk hydroelastic modulus for fluid, R Biot's isothermal hydroelastic constant, θ_0 initial porosity, C_f the fluid compressibility, q_i Darcy velocity in the i-th direction, t the time, ζ the change in volumetric fluid content per unit volume, ρ_f the fluid density, β_f the fluid volumetric thermal expansion coefficient, and Q the specific volumetric fluid generation rate.

Using Darcy's law [7], Eq. (3a) is rewritten as

$$\frac{\partial \rho_f\zeta}{\partial t} - \frac{\partial}{\partial x_i}\frac{\rho_f k_{ij}\rho_0 g}{\mu_f}\frac{\partial}{\partial x_j}(h + \delta x_3) - \rho_f Q = 0 \qquad (4a)$$

$$\delta = \frac{\rho_f - \rho_0}{\rho_0} \qquad (4b)$$

$$h = \frac{p}{\rho_0 g} + x_3 \qquad (4c)$$

where k_{ij} is the permeability tensor, ρ_0 the reference fluid density, g the gravitational acceleration, μ_f the dynamic viscosity of fluid, and h the reference hydraulic head.

2.3. Heat transport in porous media

Assuming that the interconvertibility of hydro-mechanical and thermal energy is negligibly small, and that the fluid and solid phases are in thermodynamic equilibrium, the equation of heat transport in porous media can be written as [14]

$$\frac{\partial}{\partial t}[\theta\rho_f c_f + (1-\theta)\rho_s c_s]T + \frac{\partial}{\partial x_i}q_i\rho_f c_f T$$
$$- \frac{\partial}{\partial x_i}H_{ij}\frac{\partial T}{\partial x_j} - Q_H = 0 \qquad (5a)$$

$$H_{ij} = c_f\rho_f a_{ijkl}\frac{q_k q_l}{\sqrt{q_m q_m}} + \lambda_{ij}^T \qquad (5b)$$

where $i, j, k, l, m = 1, 2, 3$; and where c_f is the specific heat capacity of fluid, c_s the specific heat capacity of solid, ρ_s the solid skeleton density, H_{ij} the hydrodynamic dispersivity tensor, a_{ijkl} the dispersivity tensor, λ_{ij}^T the thermal conductivity tensor, and Q_H the specific enthalpy production rate. Note that the strain rate term in Eq. (5a) has been dropped. In most materials, especially stiff materials, the strain rate terms in heat transport are insignificant [10].

104

2.4. Treatment of fractures

2.4.1. Equilibrium analysis

The mathematical idealization of fractures has been proposed by many researchers. A fracture is approximated by a pair of surfaces between which normal and shear displacements are permissible. In this paper, the models of Bandis [2] and Barton and Bakhtar [6] are adopted because of the relative ease with which the data can be gathered in the field.

The hydromechanical equilibrium conditions in open joints are governed by joint mechanical stresses and fluid pressure between joint surfaces. The relationships between the joint mechanical stresses and joint deformation are based on the formulation of Leichnitz [27]. In his work, Leichnitz noted that the entire shear process is independent of load paths and that the shear stress and normal displacement are dependent on two common variables: lateral displacement and normal stress. This observation may be written as:

$$\tau = \tau(u, \sigma_n) \tag{6a}$$

$$v = v(u, \sigma_n) \tag{6b}$$

where τ is the joint shear stress, σ_n the joint effective normal stress, u the lateral joint displacement along the joint plane, and v the joint normal displacement.

It is assumed that joint mechanical behavior is isotropic along the joint plane so that the principal shear stress, τ, and the corresponding lateral displacement, u, along the joint plane may be resolved along the directions of the joint plane reference axes, r and s. Combining Eqs. (6a) and (6b), and resolving τ and u in the r- and s-directions, we obtain [21]:

$$\begin{pmatrix} \Delta\tau_r \\ \Delta\tau_s \\ \Delta\sigma_n \end{pmatrix} = \begin{bmatrix} K_s - \mu K_n d_i & 0 & \mu K_n \dfrac{\partial u_r}{\partial u} \\ 0 & K_s - \mu K_n d_i & \mu K_n \dfrac{\partial u_s}{\partial u} \\ -d_i K_n \dfrac{\partial u_r}{\partial u} & -d_i K_n \dfrac{\partial u_s}{\partial u} & K_n \end{bmatrix} $$
$$\times \begin{pmatrix} \Delta u_r \\ \Delta u_s \\ \Delta v \end{pmatrix} \tag{7}$$

where $K_n = (\partial\sigma_n/\partial v)$ is the normal joint stiffness; $K_s = (\partial\tau/\partial u)$ is the shear joint stiffness; $\mu = (\partial\tau/\partial\sigma_n)$ is the friction angle; $d_i = (\partial v/\partial u)$ is the dilation angle; Δu_r, Δu_s the increments of lateral displacements along the r- and s-axes, respectively; Δv the increment of normal displacement; and $\Delta\tau_r$, $\Delta\tau_s$, and $\Delta\sigma_n$ the increments of shear stresses and effective normal stress corresponding to the increments of lateral and normal displacements, respectively. It should be noted here that the effective

normal stress takes into account the presence of fluid, and is equal to the difference between the total normal stress and fluid pressure. This relationship is based on an assumption that, in a given joint, the joint surface is completely exposed to fluid.

The terms appearing in Eq. (7) may be obtained from corresponding constitutive relationships described in the ensuing subsections.

2.4.1.1. Joint closure–normal stress relationship.
The normal displacement–normal stress relationship is based on that of the hyperbolic relationship developed by Bandis et al. [3], shown below:

$$\sigma_n = \left[\frac{\Delta v}{a - b\Delta v} \right] \tag{8a}$$

where a, b are the empirical constants.

The empirical constants a, b may be determined from the following relationship [3,5,6]:

$$a = \frac{1}{K_{ni}}, \quad \Delta v \to 0$$

$$\frac{a}{b} = V_m, \quad \sigma_n \to \infty$$

$$K_{ni} = C_{K_{ni1}} \left(\frac{JCS_n}{a_j} \right) + C_{K_{ni2}} JRC_n + C_{K_{ni3}}$$

$$a_j = \frac{JRC_n}{C_{a_{j1}}} \left(C_{a_{j2}} \left(\frac{\sigma_c}{JCS_n} \right) + C_{a_{j3}} \right)$$

$$V_m = C_{V_{m1}} + C_{V_{m2}} JRC_n + C_{V_{m3}} \left(\frac{JCS_n}{a_j} \right)^{C_{V_{m4}}} \tag{8b}$$

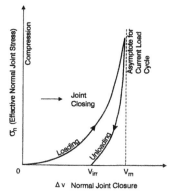

Fig. 3. Constitutive relationship between joint normal stress and normal displacement.

where V_m is the joint maximum closure (mm), K_{ni} the initial joint normal stiffness (at $\Delta v = 0$) (MPa/mm), a_j the initial joint aperture (mm) under self-weight stress, JCS_n the joint (wall) compressive strength (MPa), JRC_n the joint roughness coefficient, σ_c the unconfined compression strength of intact rock (MPa), C_{Ik} the k-th empirical coefficient for variable I ($I = V_m$, K_{ni}, a_j) in the above equations. Units for the above-mentioned variables are specified to be consistent with the units upon which the currently available empirical parameters are based. A diagrammatic interpretation of V_m is presented in Fig. 3. The intersection between the Δv, σ_n-axes, in Fig. 3, corresponds to the condition of no closure and self-weight stress, where a_j (aperture with no closure) and K_{ni} (gradient of σ_n with respect to Δv at the intersection) are determined.

From Eqs. (7), (8a), and (8b), the normal joint stiffness is given by:

$$K_n = K_{ni}\left[1 - \frac{\sigma_n}{V_m K_{ni} + \sigma_n}\right]^{-2} \qquad (9a)$$

Hysteresis during a loading–unloading cycle of a joint may be described by modifying the empirical constants [a and b in Eq. (8a)], thus [16]:

$$\frac{a}{b} = V_m - \Sigma V_{irr}$$

$$a = \frac{1}{K_{ni}^{mod}}$$

$$K_{ni}^{mod} = C_{K_{ni1}}\left[\frac{JCS_n}{a_j - \Sigma V_{irr}}\right] + C_{K_{ni2}}JRC_n$$

$$\frac{V_{irr}}{V_m} = C_{V_{irr1}} + C_{V_{irr2}}\left[\frac{JCS_n}{a_j}\right] \qquad (9b)$$

where V_{irr} is the irrecoverable closure (mm) (shown in Fig. 3), K_{ni}^{mod} the modified joint normal stiffness (MPa/mm), and C_{irrk} the k-th empirical coefficient for the expression for V_{irr}.

If the joint becomes mismatched due to lateral shear displacement, the joint normal stiffness may be further modified by the relationships presented by Gale et al. [16].

2.4.1.2. Shear stress–shear displacement relationships. The relationships may be divided into three phases: the pre-peak phase; the post-peak phase; and the unloading phase. Constitutive relationships for these phases are described below.

1. *Pre-peak hyperbolic relationship.* In the semi-elastic range of shear deformation, the shear stress–shear displacement relationship is given by Kulhaway [25]

and Gale et al. [16]. This relationship is based on the following equations:

$$\tau = \left[\frac{u}{a_s + b_s u}\right] \qquad (10a)$$

where a_s and b_s are the empirical coefficients, and

$$a_s = \frac{1}{\tau_{peak}} - \frac{1}{G_i \tau_{peak}}$$

$$b_s = \frac{1}{G_i}$$

$$\tau_{peak} = \sigma_n \tan\left[JRC_n \log_{10}\left(\frac{JCS_n}{\sigma_n}\right) + \phi_r\right]$$

$$G_i = \max\left(\frac{\tau_{s1}}{u_1}, [C_{G_{i1}}JRC_n + C_{G_{i2}}]\sigma_n^{C_{G_{i3}}}\right)$$

$$\tau_{s1} = \sigma_n \tan[C_{\tau_{s1}}JRC_n - \phi_r] \qquad (10b)$$

$$u_1 = C_{u_1}u_{peak}$$

$$u_{peak} = \frac{L_n}{C_{u_{p1}}}\left[\frac{JRC_n}{L_n}\right]^{C_{u_{p2}}}f_{wthr}$$

$$f_{wthr} = 1 - \frac{JCS_n}{C_{wthr1}}$$

where τ_{peak} is the peak shear stress (MPa) at the corresponding lateral displacement u_{peak}, ϕ_r the residual friction angle (degree), C_{Ik} the k-th empirical

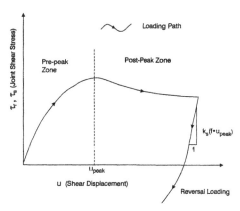

Fig. 4. Constitutive relationship between joint shear stress and shear displacement.

coefficient for variable I ($I = G_i$, τ_{s1}, u_1, u_{peak}, f_{wthr}), L_n the sample length (m), and f_{wthr} the weathering factor. G_i, τ_{s1}, u_1 in the above equations are intermediate expressions. The constitutive relationship between shear stress and lateral displacement is shown in Fig. 4.

Based on Eqs. (7) and (10a), the joint shear stiffness K_s is given by:

$$K_s = \frac{a_s}{[a_s + b_s u_s]^2} \qquad (10c)$$

2. *Continuous pre- and post-peak relationship.* Following the work of Barton and Bakhtar [6] and Gale et al. [16], the relationship may be written:

$$\tau = \sigma_n \tan\left[JRC_{mob} \, \log_{10}\left(\frac{JCS_n}{\sigma_n} \right) + \phi_r \right] \qquad (11)$$

The mobilized *JRC* (JRC_{mob}) may be found directly from an experimentally determined relationship between JRC_{mob}/JRC_{peak} and u/u_{peak} [3]. The joint shear stiffness based on this model must be determined segmentally from the JRC_{mob}/JRC_{peak}–u/u_{peak} curve.

Because of the difference between the two shear constitutive relationships, as well as the possibility of discontinuity at u_{peak} due to different joint shear stiffnesses, a weighting factor, α_{sh} ($0 \leq \alpha_{sh} \leq 1$), is used to allow for the selection of the most appropriate constitutive relationship, thus:

$$Q_{si} = \alpha_{sh} Q_{si,\,1} + (1 - \alpha_{sh}) Q_{si,\,2} \qquad (12)$$

where $Q_{si,1}$ is the quantity (displacement, stress, stiffness) associated with the continuous model, $Q_{si,2}$ the quantity (displacement, stress, stiffness) associated with the pre-peak model.

3. *Unloading phase relationship.* Currently, there are no published constitutive relationships describing the unloading phase. In the current analysis, the following relationship is used for the early stage of unloading:

$$K_s^{UL} = f_{\tau_{UL}} K_s (f_{u_{peak}} u_{peak}) \qquad (13)$$

where K_s^{UL} is the joint shear stiffness during unloading of shear (MPa/mm), $f_{\tau_{UL}}$ the factor used in modifying the shear stiffness, $f_{u_{peak}}$ the fraction of u_{peak} at which the joint stiffness is estimated. Based on laboratory observations, the early unloading stage is approximately linear [16] (see Fig. 4).

4. *Dilatancy/contractancy.* According to Gale et al. [16], the mobilized angle of dilation is given by:

$$d_{n,\,mob}^{dilat} = -\tan(\phi_r) + \frac{\tau}{\sigma_n} \qquad (14)$$

As shown in the above equation, dilation is pre-

ceded by contraction when:

$$\frac{\tau}{\sigma_n} < \tan\,\phi_r, \quad \text{and begins when} \quad \frac{\tau}{\sigma_n} = \tan\,\phi_r \qquad (15)$$

Additional constitutive relationships may be found in Barton and Bakhtar [6] and Gale et al. [16].

2.4.2. Flow analysis

Assuming that hydraulic conductivity in the fracture zone is large compared with the background porous matrix, the flow can be considered dominant along the fracture plane axis. The flow equation for a fracture plane can then be derived by vertically integrating Eq. (3a) along the direction normal to each fracture plane to yield

$$E_m \frac{\partial \rho_f \zeta}{\partial t} - \frac{\partial}{\partial \xi_i} e_h \rho_f \frac{k_{ij} \rho_0 g}{\mu_f} \frac{\partial}{\partial \xi_j}(h + \delta x_3)$$
$$- \rho_f E_m Q - M_L = 0 \qquad (16)$$

with $i, j = 1, 2$; and where E_m is the mechanical joint aperture, e_h is the hydraulic joint aperture, ξ_i the coordinates along the fracture plane, and M_L the net mass flux per unit area from the solid matrix.

The mechanical joint aperture is calculated using the following relationships:

$$E_m = a_j - \Delta v; \quad E_m \geq a_j - V_m \qquad (17)$$

Note that the joint closure, Δv, is positive for closing joints, and negative for opening joints. Assuming that the hydraulic property of each joint plane is isotropic, the joint permeability is calculated using Snow's [33] parallel plate relationship:

$$k_{ii} = \frac{e_h^2}{12} \qquad (18)$$

The empirical relationship between e_h and E_m has been given by Barton and Bakhtar [6] as:

$$e_h = \frac{E_m^2}{JRC^{2.5}}; \quad e_h \leq E_m \qquad (19)$$

Note that the mechanical aperture is used in the terms in Eq. (23) that require the actual dimension of the aperture. If the mechanical aperture is greater than the hydraulic aperture, the mechanical aperture is thought to be divided into two fractions, hydraulically active fraction, and hydraulically inactive fraction. Both fractions are available for storage. The hydraulic aperture in Eq. (25) is the aperture in which the flow is active and is appropriate for the terms involving fluid transmission.

2.4.3. Heat transport analysis

Based on the assumption of flow dominance along the direction of the fracture plane axis, the transport of thermal energy is expressed as

$$\frac{\partial}{\partial t}E_{\mathrm{m}}[\theta\rho_{\mathrm{f}}c_{\mathrm{f}} + (1-\theta)\rho_{\mathrm{s}}c_{\mathrm{s}}] + \frac{\partial}{\partial\xi_i}e_{\mathrm{h}}\rho_{\mathrm{f}}c_{\mathrm{f}}q_iT$$

$$+ \frac{\partial}{\partial\xi_i}E_{\mathrm{m}}H_{ij}\frac{\partial T}{\partial\xi_j} - E_{\mathrm{m}}Q_H + Q_{Hm} = 0 \tag{20}$$

with $i, j = 1, 2$; and where Q_{Hm} is the net thermal flux per unit area from the solid matrix.

3. Finite-element solution

The finite-element solution to the equilibrium, flow, and heat transport equations comprises two types of elements: three-dimensional solid elements representing the background porous matrix; and two-dimensional fracture elements representing rock joints. Eight-noded hexahedral elements are used for solid elements. Each two-dimensional planar element for the equilibrium equation consists of two four-noded quadrilateral sub-parallel planes, whilst the same for the flow and transport analysis consists of a plane along the fracture plane axis. The eight-noded fracture element is modified from the four-noded lineal element of Goodman [17]. A typical ensemble of solid and fracture elements for the equilibrium, flow, and transport equations is shown in Fig. 5.

It is assumed that the solution can be expressed in the form

$$\bar{u}_i(x_i, t) = N_J(x_i)u_{iJ}(t)$$

$$\bar{h}(x_i, t) = N_J(x_i)h_J(t)$$

$$\bar{T}(x_i, t) = N_J(x_i)T_J(t) \tag{21a}$$

with $J = 1, \ldots, n_c$, and where N_J is the basis function

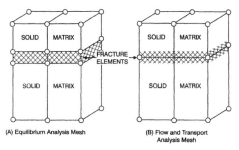

(A) Equilibrium Analysis Mesh   (B) Flow and Transport Analysis Mesh

Fig. 5. Typical finite-element discretization.

associated with node J, h_J the reference hydraulic head at node J, T_J the temperature at node J, u_{iJ} the displacement in the i-th direction at node J, and n_c the number of connected nodes in an element. The overlined variables (u, h, T) represent the approximations of solutions to the respective governing partial differential equations.

Applying the weighted-residuals technique [35], and Green's Theorem to the equilibrium equation, using Eqs. (2) and (21a), one obtains:

$$-\int_A W_I\tau_{ij}n_j\,\mathrm{d}S$$

$$+ \int_V C_{ijmn}\frac{1}{2}\frac{\partial W_I}{\partial x_j}\left[\frac{\partial N_J}{\partial x_m}u_{mJ} + \frac{\partial N_J}{\partial x_n}u_{nJ}\right]\mathrm{d}V$$

$$- \int_V \frac{\partial N_J}{\partial x_j}(\overline{U}_{ij}\Delta p_J + \overline{W}_{ij}\Delta T_J)\mathrm{d}V - \int_A W_I\tau_{ij}^0 n_j\,\mathrm{d}S$$

$$+ \int_V \tau_{ij}^0\frac{\partial N_I}{\partial x_j}\mathrm{d}V - \int_V N_I F_i\,\mathrm{d}V = 0 \tag{21b}$$

where $i, j, m, n = 1, 2, 3$, and $I, J = 1, \ldots, n_c$; and where W_J is the weighting function associated with node J and n_j the outward unit vector normal to the elemental surface $\mathrm{d}S$. A and V represent the elemental area and volume, respectively. The contribution from the mechanical fracture elements is calculated using stresses on the fracture walls and average displacements within each element, thus:

$$\int_A W_I\Delta\tau_{ij}^{\mathrm{F}}n_j\,\mathrm{d}A - \int_A W_I(K_{ij}^{\mathrm{F}}\Delta u_{j\,\mathrm{ave}}^{\mathrm{F}} - \Delta p_{\mathrm{ave}}^{\mathrm{F}})\mathrm{d}A = 0 \tag{21c}$$

where $\Delta\tau_{ij}^{\mathrm{F}}$ is the incremental skeletal stress acting on the fracture walls, K_{ij}^{F} the stiffness tensor of fracture, $\Delta u_{j\,\mathrm{ave}}^{\mathrm{F}}$ the average fracture opening in the j-th direction, and $\Delta p_{\mathrm{ave}}^{\mathrm{F}}$ the average fluid pressure increase in fracture.

Applying the weighted-residuals technique [35], and Green's Theorem to the flow equation, using Eqs. (4a), (16), and (21a), one obtains:

$$\int_V W_I\frac{\partial\rho_{\mathrm{f}}\zeta}{\partial t}\mathrm{d}V + \int_V \rho_{\mathrm{f}}\frac{k_{ij}\rho_0 g}{\mu}\frac{\partial W_I}{\partial x_i}\frac{\partial N_J}{\partial x_j}(h_J$$

$$+ \delta x_{3J})\mathrm{d}V - \int_V W_I\rho_{\mathrm{f}}Q\,\mathrm{d}V$$

$$- \int_A W_I M_L l_i\,\mathrm{d}S$$

$$- \int_S W_I\rho_{\mathrm{f}}\frac{k_{ij}\rho_0 g}{\mu}\frac{\partial(\bar{h} + \delta x_3)}{\partial x_j}l_i\,\mathrm{d}S = 0 \tag{21d}$$

with $i, j = 1, 2, 3$; $I, J = 1, \ldots, n_c$, and

$$\int_A W_I E_m \frac{\partial \rho_f^\zeta}{\partial t} dA + \int_A e_h \rho_f \frac{k_{ij}\rho_0 g}{\mu_f} \frac{\partial W_I}{\partial \xi_i} \frac{\partial N_J}{\partial \xi_j}(h_J$$

$$+ \delta x_{3J}) dA - \int_A W_I E_m \rho_f Q \, dA$$

$$+ \int_A W_I M_I l_i \, dA \tag{21e}$$

$$- \int_L W_I e_h \rho_f \frac{k_{ij}\mu_0 g}{\mu_f} \frac{\partial(\bar{h}+\delta x_3)}{\partial \xi_j} n_i \, dL = 0$$

with $i, j = 1, 2$; and $I, J = 1, \ldots, n_c$ and where l_i is the outward unit vector normal to the elemental length dL. L is the length of the fracture element perimeters.

Applying the weighted-residuals technique [35], and Green's Theorem to the heat transport equation, using Eqs. (5a), (20), and (21a), one obtains:

$$\int_V W_I N_J \left[(\theta \rho_f c_f + (1-\theta)\rho_c c_s) \frac{\partial T_J}{\partial t} \right] dV$$

$$+ \int_V W_I \rho_f q_i c_f \frac{\partial N_J}{\partial x_i} T_J \, dV$$

$$- \int_V W_I Q_H \, dV - \int_A W_I Q_L l_i \, dS \tag{21f}$$

$$+ \int_V H_{ij} \frac{\partial W_I}{\partial x_i} \frac{\partial N_J}{\partial x_j} T_J \, dV - \int_S W_I H_{ij} \frac{\partial \bar{T}}{\partial x_j} l_i \, dS$$

$$= 0$$

with $i, j = 1, 2, 3$; $I, J = 1, \ldots, n_c$; and

$$\int_A W_I N_J E_m \left[(\theta \rho_f c_f + (1-\theta)\rho_s c_s) \frac{\partial T_J}{\partial t} \right] dA$$

$$+ \int_A W_I e_h \rho_f q_i c_f \frac{\partial N_J}{\partial \xi_i} T_J \, dA - \int_A W_I E_m Q_H \, dV$$

$$+ \int_A W_I Q_{Hm} l_i \, dA + \int_A E_m H_{ij} \frac{\partial W_I}{\partial \xi_i} \frac{\partial N_J}{\partial \xi_j} T_J \, dA \tag{21g}$$

$$- \int_L W_I E_m H_{ij} \frac{\partial \bar{T}}{\partial \xi_j} l_i \, dL = 0$$

with $i, j = 1, 2$; $I, J = 1, \ldots, n_c$.

Summing up elemental contributions from Eqs. (21b)–(21g), one obtains

$$S_{mnIJ}^E u_{nJ} + F_{mI}^E = 0 \tag{22a}$$

$$M_{IJ}^F \frac{dh_J}{dt} + S_{IJ}^F h_J + F_I^F = 0 \tag{22b}$$

$$M_{IJ}^H \frac{dT_J}{dt} + S_{IJ}^H h_J + F_I^H = 0 \tag{22c}$$

with $m, n = 1, 2, 3$; $I, J = 1, 2, 3, \ldots, n_p$; and where M_{IJ} is the mass matrix, S_{IJ} the stiffness matrix, F_I the load vector, and n_p the total number of nodes. The superscripts 'E', 'F', and 'H' refer to equilibrium, flow, and heat transport, respectively. In this paper, the weighted-residuals technique using the Galerkin method, with $W_I = N_I$, is employed. Unless otherwise stated, the time-stepping schemes chosen for the flow and heat transport equations are fully implicit and Crank–Nicholson, respectively.

Eqs. (22a)–(22c) form a system of non-linear simultaneous equations, which are solved by the Picard iteration technique [23]. In each Picard iteration loop, the equilibrium, flow, and heat transport equations are sequentially solved. The iteration is terminated when the absolute differences between the displacements, heads, and temperatures from the current and previous iterations are concurrently smaller than a set of predetermined tolerances. For the equilibrium equation, the solution is based on an assumption that in each time level (at a given time t) the whole system is under the condition of pseudo-equilibrium. Within a given Picard iteration, the equilibrium equation is solved iteratively within the Picard iteration.

A three-dimensional finite-element code called MOTIF, based on the above equations and solution techniques, has been developed by the Atomic Energy of Canada Limited. This code can solve four equations simultaneously: the thermohydromechanical equilibrium; the groundwater flow; the heat transport; and the radionuclide transport equations. The solution to the radionuclide transport equation has been presented elsewhere [18]. Details of the mathematical formulation and numerical method used in MOTIF and its application to a hypothetical case of nuclear fuel waste disposal has been presented by the authors [11,18,20].

4. Verification and application examples

Several cases of different loading configurations were used to verify the MOTIF code. The verification cases included: normal displacement with irrecoverable damage (Case 1); shear displacement without and with dilatancy and contractancy (Cases 2 and 3, respectively); and non-isothermal consolidation (Case 4). An application example for the case of fracture closure under the condition of non-isothermal expansion is also given (Case 5). Details of the verification and application cases are presented below. Empirical parameter values used in Verification Cases are summarized in Table 1 for Cases 1, 2, 3, and 5. Elastic, hydraulic, and transport parameter values for all cases are summarized in Table 2.

4.1. Case 1 — multiple-step normal displacement under compression — with irrecoverable damage (hysteresis)

A schematic of this verification case is shown in Fig. 6. In this verification case, the intact rock block is represented by a solid element and the rock joint by a joint element. All the dimensions are provided in the figure. The thickness of the rock block and the width of both rock joint in the direction normal to the page are 1 m. The initial joint aperture at seating stress (1 Pa) is 0.225 mm [estimated from parameters in Eq. (8b)]. As shown in Fig. 6, an external stress is applied to the bottom boundary. By varying the externally applied stress, the stress in the rock joint also varies to equilibrate with the external stress. In the verification process, a series of increasing compressive stresses (loading) and decreasing compressive stresses (unloading) was applied to the bottom boundary of the rock block through a transient mode, with each time level corresponding to a loading stage with a predetermined

stress value. Shown in Fig. 7 is a plot of loading and unloading curves (compressive stress versus joint closure). As can be observed in Fig. 7, the agreement between the MOTIF result and the analytical Bandis–Barton model is excellent. It should be noted here that hysteresis is present in the unloading curve due to the occurrence of irrecoverable damage. During the verification run, it was found that one of the unloading-curve parameters, C_{irr2}, had to be modified to lessen the numerical severity due to the steepness of the slope of the unloading curve.

4.2. Case 2 — multi-step shear displacement — no dilatancy

The configuration of this verification problem is shown in Fig. 8. Two rock joints and a solid rock block were simulated in this verification case. The intact rock block is represented by a solid element and the rock joints by two joint elements. All the dimensions are provided in the figure. The thickness of the rock block and the width of both rock joints in the direction normal to the page are 1 m. The joint aperture is 0.19 mm, determined using parameters in Table 1 and Eq. (8b). The movement in the direction normal to the page is constrained. Dilatancy and contractancy were also suppressed in this simulation. Initial normal compressive stress in the rock joint, and initial compressive stress in the y-direction in the solid block are both equal to 2 MPa and remain constant.

Table 1
Values of empirical coefficients[a]

Parameter	Cases 1 and 5	Cases 2 and 3
$C_{K_{n1}}$	0.02	0.02
$C_{K_{n2}}$	1.75	1.75
$C_{K_{n3}}$	−7.15	−7.15
$C_{a_{n1}}$	5.0	5.0
$C_{a_{n2}}$	0.2	0.2
$C_{a_{n3}}$	−0.1	−0.1
$C_{V_{m1}}$	−0.296	−0.296
$C_{V_{m2}}$	−0.0056	−0.0056
$C_{V_{m3}}$	2.24	2.24
$C_{V_{m4}}$	−0.245	−0.245
$C_{V_{m1}}$	0.848	0.848
$C_{V_{m2}}$	0.002	0.0002
C_{e_1}	2.0	2.0
C_{e_2}	2.5	2.5
$C_{K_{sn1}}$	2,500	2,500
$C_{K_{sn2}}$	500	500
$C_{K_{sn3}}$	0.7	0.7
$C_{G_{n}}$	3.86	3.86
$C_{G_{2}}$	−17.2	−17.2
$C_{G_{n}}$	0.7	0.7
$C_{\tau_{r1}}$	0.0	0.0
C_{μ_1}	0.005	0.005
$C_{u_{peak1}}$	500	3,530
$C_{u_{peak2}}$	0.33	0.33
$C_{JRC_{int}}$	0.9	0.9
C_{wthr}	10^{38}	10^{38}
C_{dilat}	1.0	1.0
ψ_r (°)	30	30
JRC_n	7.5	5.85
JCS_n (MPa)	80.0	161.0
σ_c (MPa)	100.0	210.0
L_n (m)	500.0	0.2
$f_{\tau_{UL}}$	1.0	1.0
α_{sh}	1.0	1.0

[a] Sources: 1. Barton and Bakhtar [6]; 2. Gale et al. [16].

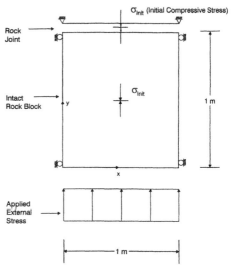

Fig. 6. Case 1: Idealized representation of a rock joint undergoing normal displacement under compression.

Table 2
Elastic, flow, transport parameter values, and discretization details

	Cases 1, 2, and 3	Case 4	Case 5
Flow and transport parameters			
E (Pa)	5.5×10^{10}	5.5×10^{10}	5.5×10^{10}
ν	0.0	0.2	0.2
h (Pa)		3.056×10^{10}	3.056×10^{10}
α [see Eq. (1f)]		1	1
C_f (Pa^{-1})		4.57×10^{-10}	4.57×10^{-10}
θ_0		0.1	0.05
β_s (°C^{-1})		1.0×10^{-5}	1.0×10^{-5}
β_f (°C^{-1})		3.17×10^{-4}	3.17×10^{-4}
c_f (J/(kg °C))		4,200	4,200
c_s (J/(kg °C))		900	900
$\lambda_{11}^T, \lambda_{22}^T, \lambda_{33}^T$ (W/(m °C)		4.2	4.2
a_L (m)[a]		6.0	6.0
a_T (m)[a]		0.6	0.6
ρ_f (kg/m^3)		1,000	1,000
ρ_s (kg/m^3)	2,600	2,600	2,600
Q_H (W/m^3)		10^6	
k (m^2)		10^{-15}	10^{-15}
μ (Pa s)		1.38×10^{-3}	See Appendix A
P_0 (Pa)		10^7	
g (m/s^2)	9.81	9.81	9.81
Discretizations			
$\Delta x_1, \Delta x_2, \Delta x_3$ (m)	1.0	0.1	1.0
Δt (s)		0.5	10,000 min, 100,000 max

[a] Note a_L and a_T are longitudinal and lateral dispersivities, respectively. See Bear [7] for relationships between a_{ijkl}, a_L, and a_T.

The joint initial shear stress in both joints was absent. To simulate the loading stage of a shear test, the left-hand face of the rock block was gradually pushed to the right until the total movement reached 0.17 mm from the original position. The peak shear stress of 1.75 MPa was attained at 0.17 mm. In the verification process, a time series of incremental (both positive and negative) lateral movements was prescribed to the left-hand-side boundary through a transient mode, with each time level corresponding to a loading stage with a predetermined lateral movement.

The total weighting ($\alpha_{sh} = 1$) was also applied to the original continuous model [6]. A JRC_{mob}/JRC_{peak} versus u/u_{peak} curve corresponding to experimental results reported by Gale et al. [16] was used as a basis for the

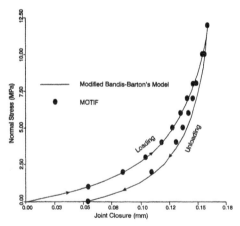

Fig. 7. Case 1: Multiple-step normal joint displacement under compression.

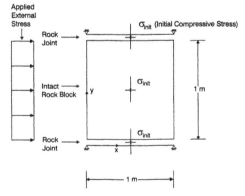

Fig. 8. Case 2: Idealized representation of two rock joints undergoing shear displacement under constant compressive stress.

simulation. Shown in Fig. 9 is a comparison between the MOTIF result and the result from a shear test performed on Sample AECL3 by the Memorial University of Newfoundland [16]. As shown in Fig. 9, the agreement between the MOTIF result and the experimental result is excellent. Note also that each of the solid circles corresponds to the termination of an approximately linear segment or the beginning of the next approximately linear segment in the JRC_{mob}/JRC_{peak} versus u/u_{peak} curve derived from the experiment.

4.3. Case 3 — multiple-step shear displacement with dilatancy

The configuration of this verification problem is shown in Fig. 10. A rock joint and a solid rock block were simulated in this verification case. The intact rock block is represented by a solid element and the rock joint by a joint element. All the dimensions are provided in the figure. The thickness of the rock block and the width of rock joint in the direction normal to the page are 1 m. The initial joint aperture at seating stress is 0.19 mm [determined by parameters in Table 1 and Eq. (8b)]. The movement in the direction normal to the page is constrained. Initial normal compressive stress in the rock joint, and initial compressive stress in the y-direction in the solid block are both equal to 2 MPa and remain constant. The joint initial shear stress in both joints was absent. To simulate the loading stage of a shear test, the left-hand face of the rock block was gradually pushed to the right until the total movement reached 0.170 mm from the original position where the peak shear stress was reached. In the

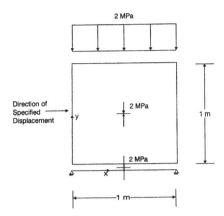

Fig. 10. Case 3: Idealized representation of a rock joint undergoing shear displacement, contractancy, and dilatancy under constant compressive stress.

verification process, a time series of incremental external stresses was prescribed to the left-hand-side boundary through a transient mode, with each time level corresponding to a predetermined value of external stress. Similar to Case 2, the total weighting was also applied to the original Barton model [6]. The JRC_{mob}/JRC_{peak} versus u/u_{peak} curve employed in the verification of Case 2 was also used as a basis for this simulation. Shown in Fig. 11 is a comparison between the MOTIF result and the result from the analytical Bandis–Barton model. As shown in Fig. 11, the agreement between the MOTIF result and the analytical model is excellent.

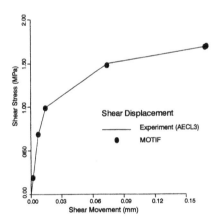

Fig. 9. Case 2: Multiple-step shear joint displacement under compression.

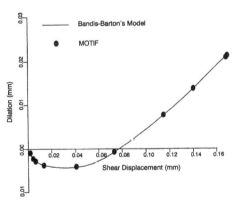

Fig. 11. Case 3: Multiple-step normal joint displacement with contractancy, and dilatancy.

112

4.4. Case 4 — one-dimensional non-isothermal consolidation simulation

This case is an extension of the isothermal case presented by Biot [8]. The soil column is heated uniformly by an internal, uniformly distributed heat source with an intensity of $1,000,000$ W/m^3. All the boundaries surrounding the column are assumed to be adiabatic. Because of the uniformity of heat distribution and the low permeability of the soil column, the lateral heat fluxes (conductive and convective) are assumed to be insignificant.

This case was simulated by assuming that the column was oriented horizontally so that the column was confined in the $x_2(y)$- and $x_3(z)$-directions. The upper surface of the column corresponds to $x_1 = 0$, and the bottom of the column is located at $x_1 = h$. The definition sketch of the problem is shown in Fig. 12.

The governing equilibrium, groundwater flow, and heat transport equations are shown below. In the interest of simplicity, the subscript of the Cartesian coordinates is dropped in the following equation:

$$\frac{E(1-v)}{(1+v)(1-2v)}\frac{\partial^2 u}{\partial x^2} - \alpha\frac{\partial p}{\partial x} - \frac{E\beta_s}{1-2v}\frac{\partial T}{\partial x} = 0 \qquad (23a)$$

$$\frac{k}{\mu_f}\frac{\partial^2 p}{\partial x^2} = \alpha\frac{\partial e_{xx}}{\partial t} + \frac{1}{M}\frac{\partial p}{\partial t} - (3\alpha\beta_s + \theta_0\beta_f)\frac{\partial T}{\partial t} \qquad (23b)$$

$$[(1-\theta)c_s\rho_s + \theta c_f\rho_f]\frac{\partial T}{\partial t} = Q_H \qquad (23c)$$

The above equations are subject to the following boundary and initial conditions:

$$p_0 = -\tau_{xx} = \text{constant for all } x \qquad (24a)$$

$$p = 0, \quad \text{at } x = h \qquad (24b)$$

$$\frac{\partial p}{\partial x} = 0, \quad \text{at } x = 0 \qquad (24c)$$

$$\frac{\partial T}{\partial x} = 0, \quad \text{at } x = 0, \quad \text{and} \quad x = h \qquad (24d)$$

$$u(x, t = 0) = 0 \qquad (24e)$$

$$p(x, t = 0) = \sigma_0 = \frac{a - a_i}{\alpha a}p_0 \qquad (24f)$$

[see Biot [8]; and Eq. (30)]; and

$$T(x, t = 0) = 0 \qquad (24g)$$

From Eqs. (23a) and (24a), one obtains:

$$p_0 = -\tau_{xx} = -\frac{1}{a}\frac{\partial u}{\partial z} + \alpha p + \frac{E\beta_s}{1-2v}T = \text{constant} \qquad (25)$$

From Eqs. (23c) and (24d), one obtains:

$$T = \frac{Q_H t}{(1 - \theta_0)\rho_s c_s + \theta_0\rho_f c_f} \qquad (26)$$

Differentiating Eq. (25) with respect to time, one obtains:

$$0 = -\frac{1}{a}\frac{\partial}{\partial t}\frac{\partial u}{\partial z} + \alpha\frac{\partial p}{\partial t} + \frac{E\beta_s}{1-2v}\frac{\partial T}{\partial t} \qquad (27)$$

Combining Eqs. (23b), (26), and (27), one obtains:

$$\frac{\partial^2 p}{\partial x^2} = \frac{1}{c}\frac{\partial p}{\partial t} - q_H \qquad (28)$$

in which c and q_H are defined in Eq. (30).

Eq. (28) is subject to the following boundary conditions:

$$\frac{\partial p}{\partial x} = 0, \quad \text{at } x = 0 \qquad (29a)$$

[Eq. (24c)]:

$$p = 0, \quad \text{at } x = h \qquad (29b)$$

[Eq. (24b)]:

$$p(x, t = 0) = \sigma_0 \qquad (29c)$$

[defined in Eq. (24f)].

Solving Eqs. (28), (29a)–(29c) using the Laplace transformation, one obtains [22]:

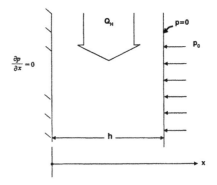

Fig. 12. Case 4: Problem definition for one-dimensional non-isothermal consolidation.

$$p = -4\frac{\sigma_0}{\pi}\sum_{n=1}^{\infty}\frac{(-1)^n}{(2n-1)}\exp\left(\frac{-(2n-1)^2\pi^2 ct}{4h^2}\right)$$

$$\times \cos\left(\frac{(2n-1)\pi(h-x)}{2h}\right) - q_H\frac{h^2}{2}\left[\frac{(h-x)^2}{h^2}\right.$$

$$- 1 - \frac{32}{\pi^3}\sum_{n=1}^{\infty}\frac{(-1)^n}{(2n-1)^3}\exp\left(\frac{-(2n-1)^2\pi^2 ct}{4h^2}\right) \tag{30}$$

$$\left.\times \cos\left(\frac{(2n-1)\pi(h-x)}{2h}\right)\right]$$

where

$$\sigma_0 = \frac{a - a_i}{\alpha a}p_0$$

$$a = \frac{(1-2v)(1+v)}{E(1-v)}$$

$$a_i = \frac{a}{1 + \frac{\alpha^2 a}{\theta_0 C_f}}$$

$$\frac{1}{c} = \alpha^2 a\frac{\mu_f}{k} + \theta_0 C_f\frac{\mu_f}{k}$$

$$q_H = \left[\frac{Q_H}{\theta_0\rho_f c_f + (1-\theta_0)\rho_s c_s}\right]\frac{\mu_f}{k}\left[\alpha\beta_s\frac{2(1-2v)}{1-v} + \theta_0\beta_f\right]$$

The total deformation at $x_1 = 0$, is given by:

$$u_{x=0} = -\int_0^h \frac{\partial u}{\partial x}dx \tag{31}$$

Using Eqs. (26), (27), and (31), one obtains

$$u_{x=0} = ap_0 h - \frac{8}{\pi^2}(a$$

$$- a_i)hp_0\sum_{n=1}^{\infty}\frac{1}{(2n-1)^2}\exp\left(\frac{-(2n-1)^2\pi^2 ct}{4h^2}\right)$$

$$+ a\alpha q_H\frac{h^2}{2}\left[-\frac{2}{3}h\right.$$

$$+ \frac{64}{\pi^4}\sum_{n=1}^{\infty}\frac{1}{(2n-1)^4}\exp\left(\frac{-(2n-1)^2\pi^2 ct}{4h^2}\right)\right]$$

$$- \left[\frac{aE\beta_s h}{(1-2v)}\right]\left[\frac{Q_H t}{(\theta_0\rho_f c_f + (1-\theta_0)\rho_s c_s)}\right] \tag{32}$$

Parameters used in the simulation and discretization details are shown in Table 2. Ten uniform eight-noded

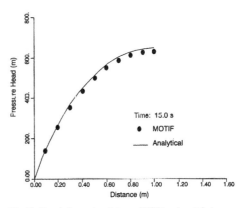

Fig. 13. Case 4: Comparison between MOTIF and analytical pressure profile at $t = 15$ s.

brick elements (along the x_1-direction) were used in the simulation, with each element being of dimension $0.1 \times 0.1 \times 0.1$ m. A spatial distribution of water pressure heads (in m of water) at 15 s is presented in Fig. 13, in which it can also be seen that the MOTIF-calculated pressure profile is in excellent agreement with the analytical solution. At the elapsed time of 15 s, the elevated temperature was 6.59°C.

Fig. 14 shows a plot of soil column displacement immediately adjacent to the porous slab (at $x_1 = 0$) versus time. An inspection of the figure reveals that the MOTIF solution agrees remarkably well with the analytical solution.

4.5. Case 5 — single fracture with thermohydromechanical deformation

This case is an example of possible application of the thermohydromechanical model presented in this

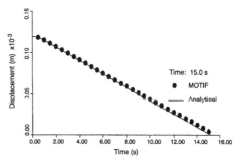

Fig. 14. Case 4: Comparison between MOTIF and analytical solution column deformation at $x_1 = 0$.

114

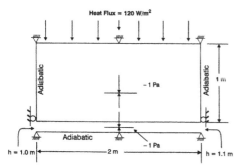

Fig. 15. Case 5: Problem definition for two-dimensional non-isothermal flow and transport with thermohydromechanical deformation in a hypothetical fractured rock system.

paper. A definition sketch of the hypothetical scenario is presented in Fig. 15. A rock block, underlain by a subjacent horizontal joint (or fracture), was subjected to an areally uniform thermal loading on the top surface. A small hydraulic gradient (0.05) was imposed across the joint. The joint properties are identical to those for Case 1 (see Table 1). Initial compressive stresses in both the rock matrix and joint were assumed to be very small (1 Pa). Initial fluid and solid temperatures were 6°C. Other properties are given in Table 2. Fluid dynamic viscosity in the simulation was based on the equation given in Appendix A.

Presented in Fig. 16 are four simulated time-varying variables near the joint outlet (the left boundary in

Fig. 15): joint compressive stress, joint closure, temperature of fluid, and fluid velocity. In the figure, the joint compressive stress increases monotonically with time due to the thermal expansion of solid matrix in response to the constant thermal flux at the top surface of the rock block. In response to the thermal expansion in the adjacent rock matrix, the closure of the joint also increases monotonically until the maximum joint closure is reached at approximately 800,000 s. Prior to 100,000 s, fluid velocity near the joint outlet decreases gradually. Between 100,000 and 400,000 s, fluid velocity in the joint near the outlet decreases dramatically (approximately two orders-of-magnitude) prior to the onset of maximum joint closure. This significant drop in fluid velocity is due to the non-linear relationship between joint hydraulic aperture and velocity in the joint. Temperature in the joint remains relatively constant up to 100,000 s (when the joint fluid velocity begins to decrease significantly), whereupon it begins to rise monotonically. After 1,000,000 s, joint fluid velocity begins to rise gradually due to the decrease in dynamic viscosity associated with elevated fluid temperature in the joint. The results from this example demonstrates that the thermohydromechanical phenomena in fractured rock systems can be very complex.

5. Summary and conclusions

A three-dimensional finite-element solution to the problem of coupled thermohydromechanical defor-

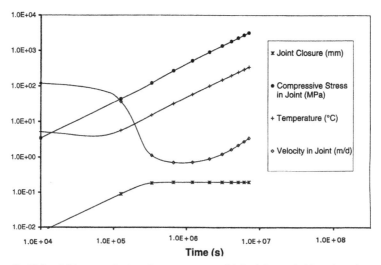

Fig. 16. Case 5: Joint compressive stress, closure, temperature, and fluid velocity near the joint outlet vs time.

mation, groundwater flow, and heat transport in deformable fractured porous media is presented in this paper. The governing equations are based on Biot's consolidation theory for poroelastic materials, extended to the non-isothermal environment. The normal and lateral deformations in joints are simulated by a new joint element. The new joint element is based on the Barton–Bandis model and is capable of simulating normal and lateral deformations with dilatancy, contractancy, and hysteresis due to irrecoverable damages/rubblization. A three-dimensional finite-element code Model Of Transport In Fractured porous media (MOTIF) has been developed based on the theoretical framework presented herein. Verification results with experimental data and analytical solutions are presented in this paper.

An application example with flow of fluid through a non-isothermally deforming joint is also presented. Results indicate that non-isothermal deformation could play a major role in the transport of fluid and water-borne substances in fractured rocks.

Acknowledgements

The Canadian Nuclear Fuel Waste Management Program is funded jointly by AECL and Ontario Hydro under the auspices of the CANDU Owners Group. In the last three years, this program has been funded exclusively by Ontario Hydro's (now Ontario Power Generation) Used Fuel Program.

Appendix A. Equation of state

Fluid dynamic viscosity used in the simulations was based on the following equation:

$$\mu(\text{Pa s}) = 1.984 \times 10^{-6} \ \exp\left(\frac{1825.85}{273 + T}\right) \tag{A1}$$

where T is the current temperature in $^\circ$C.

References

[1] AECL. Environmental impact statement on the concept of disposal in Canada's nuclear fuel waste. AECL-10711, COG-93-1, September, 1994.

[2] Bandis SC. Experimental studies of scale effects of shear strength and deformation of rock joints. Ph.D. thesis, University of Leeds, UK, 1980.

[3] Bandis SC, Lumsden AC, Barton NR. Fundamental of rock joint deformation. Int J Rock Mech Min Sci and Geomech Abstr 1983;20(6):249–68.

[4] Barenblatt GI, Zheltov IP, Kochina IN. Basic concepts in the theory of seepage of homogeneous liquids in fissured rocks PMM. Sov Appl Math Mech 1960;24(5):852–64.

[5] Barton NR, Bandis SC, Bakhtar K. Strength, deformation, and conductivity of rock joints. Int J Rock Mech Min Sci and Geomech Abstr 1985;22(3):249–68.

[6] Barton NR, Bakhtar K. Description and modeling of rock joints for the thermohydromechanical design of nuclear waste values. TR-418, Atomic Energy of Canada Limited, 1985.

[7] Bear J. Dynamics of fluids in porous media. New York, NY: Elsevier, 1972 764 pp.

[8] Biot MA. General theory of three-dimensional consolidation. Journal of Applied Physics 1941;12:155–64.

[9] Biot MA. Theory of elasticity and consolidation for a porous anisotropic solid. Journal of Applied Physics 1955;26:182–5.

[10] Boley BA, Weiner JH. Theory of thermal stresses. McGraw-Hill, 1960.

[11] Chan T. An overview of groundwater flow and radionuclide transport modelling in the Canadian nuclear fuel waste program. In: Proc. Conf. Geological, Sensitivity, and Uncertainty Methods for Groundwater Flow and Radionuclide Transport Modelling, San Francisco, 15–17 September 1987, Battelle Press, 1989. p. 39–61.

[12] Chan T, Khair K, Jing L, Ahola M, Noorishad J, Vuillod E. International comparison of coupled thermo-hydro-mechanical models of a multiple fracture bench mark problem: DECOVALEX Phase I, Bench Mark Test 2. Int J Rock Mech Min Sci and Geomech Abstr 1995;32:435–52.

[13] Closmann PJ. An aquifer model for fissured reservoirs. Soc Pet Eng J 1975;15(4):385–98.

[14] Combarnous MA, Bories SA. Hydrothermal convection in saturated porous media. In: Chow VT, editor. Advances in hydroscience, vol. 10. New York, NY: Academic Press, 1971. p. 231–307.

[15] Cook NGW. Natural joints in rock: mechanical, hydraulic, and seismic behaviour and properties under normal stresses. Int J Rock Mech Min Sci and Geomech Abstr 1992.

[16] Gale JE, McLeod R, Gutierrez M, Dacker L, Makurat A. Integration and analysis of coupled stress-flow: laboratory tests data on natural fractures — MUN and NGI tests. Report submitted to Atomic Energy of Canada Limited by Fracflow Consultants, Inc. and Norwegian Geotechnical Institute, 1993.

[17] Goodman RE. Methods of geological engineering in discontinuous rocks. West Publishing Company, 1977.

[18] Guvanasen V. Development of a finite-element code and its application to geoscience research. In: Proceedings of the 17th Information Meeting of the Nuclear Fuel Waste Management Program, Ottawa, Ontario, February 1984, Atomic Energy of Canada Limited Technical Record, TR-299, Pinawa, Manitoba, 1984.

[19] Guvanasen V, Chan T, Huyakorn PS. Finite-element simulation of fluid and energy transport in fractured porous media. In: Proceedings of the 6th International Conference on Finite Elements in Water Resources, Lisbon, Portugal, 1–5 June, 1986, 1986. p. 261–70.

[20] Guvanasen V, Chan T. A three-dimensional finite-element solution for heat and fluid transport in deformable rock masses with discrete fractures. In: Proceedings of the 7th International Conference on Advances in Computational Geomechanics, Cairns, Australia, 6–10 May, 1991.

[21] Guvanasen V, Chan T. A new three-dimensional analysis of hysteresis thermohydromechanical deformation of fractured rock mass with dilatancy in fractures. In: Proceedings of the 2nd International Conference on Mechanics of Jointed and Faulted Rocks, Vienna, Austria, 10–14 April, 1995. p. 437–42.

[22] Guvanasen V. Unpublished, 1996.

[23] Huyakorn PS, Pinder GF. Computational methods in subsurface flow. Academic Press, 1983.

[24] Huyakorn PS, Lester BH, Mercer JW. An efficient finite-element technique for modeling transport in fractured porous media, 1,

single species transport. Water Resources Research 1983;19(3):841–54.

[25] Kulhaway FH. Geomechanical model for rock foundation settlement. J Geotechnical Division, ASCE 1978;106(GT2):211–28.

[26] Wan H, Shen J. Similation and results of the borehole injection test, test case 6, phase 3 of the DECOVALEX Project AECL-TR-681, COG-95-112-1, March, 1997.

[27] Leichnitz W. Mechanical properties of rock joints. Int J Rock Mech Min Sci and Geomech Abstr 1985;22(5):313–21.

[28] Long JCS, Remer JS, Wilson CR, Witherspoon PA. Porous media equivalents for networks of discontinuous fractures. Water Resources Research 1982;18(3):645–58.

[29] Louis C, Parnot M. Three-dimensional investigation of flow conditions at Grand Maison dam site. In: Proceedings of a Symposium, Stuttgart, Germany, 1972 Percolation through Fissured Rock, Paper T4-F.

[30] Romm ES, Pozinenko BV. Investigation of seepage in fracture rocks, Vnigri Trudy, (in Russian) 1963:214.

[31] Martin CD, Davison CC, Kozak ET. Characterizing normal stiffness and hydraulic conductivity of a major shear zone in granite. In: Barton NR, Stephansson OL, editors. Rock joints. Rotterdam: A. B. Balkema, 1990. p. 549–56.

[32] Nitao JJ. V-TOUGH — An enhanced version of the TOUGH code for the thermal and hydrologic simulation of large-scale problems in nuclear waste isolation. UCID-21954, Lawrence Livermore National Laboratory, Livermore, CA, 1989.

[33] Snow DT. A parallel plate model of fractured permeable media. Ph.D. thesis, University of California, Berkeley, 1965.

[34] Warren JE, Root PJ. The behavior of naturally fractured reservoirs. Soc Pet Eng J 1963;3:245–55.

[35] Zienkiewicz OC. The finite-element method. 3rd ed. London: McGraw-Hill, 1977 787 pp.

Influence of fracture geometry on shear behavior

S. Gentier [a],*, J. Riss [b], G. Archambault [c], R. Flamand [c], D. Hopkins [d]

[a] *BRGM, 3 avenue Claude Guillemin, BP 6009, 45060 Orléans cédex 2, France*
[b] *CDGA Université Bordeaux I, Avenue des facultés, 33405 Talence cédex, France*
[c] *CERM, Université du Québec à Chicoutimi, 555 boulevard de l'Université, Chicoutimi, Québec, Canada, G7H2B1*
[d] *Laurence Berkeley National Laboratory, One Cyclotron road, MS 46A-1123, Berkeley, CA 94720, USA*

Accepted 7 October 1999

Abstract

Laboratory research during the past 10 years has explained many critical links between the geometrical characteristics of fractures and their hydraulic and mechanical behavior. One of the remaining research challenges is to directly link fracture geometry with shear behavior, including behavior in response to changes in normal stress and shear direction. This paper describes results from a series of shear tests performed on identical copies (replicas) of a natural granite fracture. Based on these tests, we developed a method using image processing techniques to identify and quantify damage that occurs during shearing. We find that there is a strong relationship between the fracture's geometry and its mechanical behavior under shear stress and the resulting damage. Using a three-dimensional geostatistical model of the fracture surfaces, we analyze the dependence of the size and location of damage zones on the local geometry and propose an algorithm for predicting areas that are most likely to be damaged during shearing in a given direction. © 2000 Elsevier Science Ltd. All rights reserved.

Keywords: Fracture geometry; Fracture topography; Geostatistics; Image processing; Jointed rock masses; Kriging; Directional gradient; Rock joint; Shear behavior; Shear test; Surface roughness

1. Introduction

Prediction of the hydromechanical behavior of fractured rock masses at the field scale is important for a variety of activities, including oil and geothermal production, underground excavation, and mine design. Understanding changes in the behavior of fractured rock resulting from changes in stress improves our ability to assess the stability of excavations, and to predict changes in fluid flow, and improves our understanding of induced seismicity.

Laboratory research during the past 10 years has made many critical links between the geometrical characteristics of fractures and their hydromechanical behavior under normal stress [1–3]. Understanding the

relationship between fluid flow and geometry for fractures in shear is more complicated than for fractures under only normal stress because of the need to account for the creation of damage zones. Linking fracture geometry to flow and changes in flow that occur with changes in shear stress and direction is, however, beyond the scope of this paper. Before we can understand the hydraulic behavior of fractures in shear, we must first understand the mechanics of shear, in particular, mechanical behavior in response to changes in normal stress and shear direction. An essential step in gaining this understanding is making the link between fracture geometry and shear behavior. In this paper, we take a first step toward this work by describing results from a series of shear tests performed on replicas of a natural fracture. Because the replicas were identical, we were able to study the mechanical response of samples with the same fracture geometry to a variety of shear conditions (three values

* Corresponding author. Tel.: +33-2-38-64-38-88; fax: +33-2-38-64-33-61.
E-mail address: s.gentier@brgm.fr (S. Gentier).

Reprinted from *International Journal of Rock Mechanics and Mining Sciences* **37 (1-2)**, 161-174 (2000)

of normal stress and four shear directions). We show that all mechanical parameters are dependent on shear direction.

Analysis of our mechanical results, in conjunction with images and profiles of sheared fracture surfaces, has helped clarify the relationship between the fracture's geometry and shear behavior. Before the shear tests are performed, profiles of the fracture surfaces are measured, and maps of the topography of each surface are created from the profiles using geostatistical methods. After shearing, images of the fracture surfaces are analyzed to identify damaged areas. Maps of the damage zones are superimposed onto the geostatistically constructed maps of the fracture surfaces to characterize the damage with respect to shear direction, normal stress, and the local geometry of the surfaces. Comparison of the profiles of fracture surfaces obtained before and after shearing, together with the analysis of the images of sheared surfaces, helps us quantify the damage that occurs during shearing and develop a method to predict its magnitude and location based on the initial geometry of the fracture surfaces. To link fracture geometry directly to shear behavior, we propose an algorithm that predicts the locations on the surfaces of a fracture that are likely to be damaged during shearing.

2. Experimental procedures

The work was carried out on identical replicas of a granite sample so that shear tests could be performed on fractures with the same surface topography. The sample was cored across a natural joint of Guéret granite from France [4]. The fracture is located halfway up the cylindrical core sample and perpendicular to its axis; the diameter of the core sample is 90 mm.

2.1. Sample preparation and characteristics

Molds of both walls of the fracture were made by casting with a silicon elastomer. These molds were then used to create replicas from non-shrinking cement mortar. The fracture's upper walls were left gray, and the lower walls were stained pink in order to highlight material exchange during shearing. Mean mechanical properties of the mortar are shown in Table 1 [5].

Table 1
Mechanical characteristics of the mortar used to make fracture replicas: compressive strength (C_0), tensile strength (T_0), friction angle ($\Phi\mu$), and Young's modulus (E)

C_0	T_0	$\Phi\mu$	E
75 MPa	6.6 MPa	35°	30 800 MPa

Table 2
Combinations $(\theta, -\sigma_N, \Delta u)$ used for shear tests

	7 MPa					14 MPa					21 MPa				
−30°	–	0.3	0.7	1.4	–	–	0.3	0.8	–	2.5	–	0.3	0.8	–	–
0°	–	0.3	0.5	1.0	2.0	–	0.3	0.5	1.0	2.0	–	0.3	0.5	1.0	.2
60°	–	–	0.7	–	2.1	–	–	–	1.0	2.5	–	–	–	1.1	.6
90°	–	–	–	1.0	1.4	–	–	0.8	1.2	2.2	–	–	0.8	–	.8
−30°	5.1					4.9					5.0				
0°	5.0					5.0					5.0				
60°	5.0					5.0					5.0				
90°	5.0					5.1					5.0				

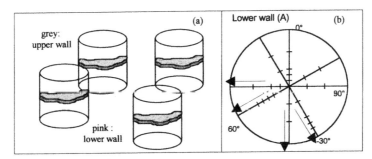

Fig. 1. Schematic representation of the series of replicas (a) and definition of shear directions shown on the lower wall of the fracture (b).

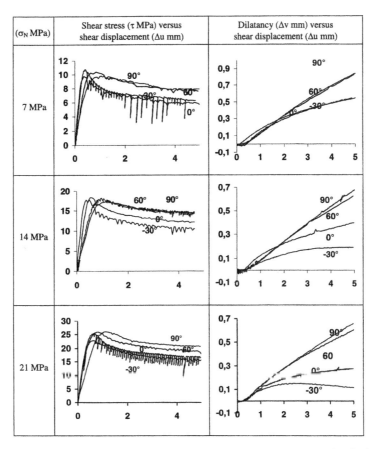

Fig. 2. Direct-shear stress test results: shear stress vs shear displacement and shear stress vs dilatancy in relation to shear direction, θ (see Fig. 1).

Table 3
Dependence of mechanical characteristics on shear direction

σ_N (MPa)	τ_p (MPa)			K_s (MPa/mm)			ΔU_p (mm)			i_p (°)			τ_r (MPa)		
	7	14	21	7	14	21	7	14	21	7	14	21	7	14	21
−30°	9.78	17.82	22.86	37.61	60.10	64.77	0.366	0.443	0.633	12.33	11.80	6.92	6.0	10.5	21.0
0°	10.25	18.37	24.25	33.76	51.40	68.49	0.423	0.502	0.599	14.22	12.32	10.07	6.5	11.5	15.7
+60°	11.41	19.98	25.94	32.84	50.69	59.92	0.536	0.644	0.767	12.50	13.75	10.69	7.4	13.5	17.0
90°	10.05	17.99	25.99	18.95	29.70	33.44	0.905	1.000	1.140	12.49	9.57	9.99	7.0	14.6	20.7

2.2. Shear test device

The experimental device used is a classical shear machine that is servocontrolled in force or displacement. The shear machine consists of two parallelepiped half-boxes: one fixed and one that moves. The lower, fixed half-box, was bound to the frame and rested against the back of a horizontal jack. The upper, moving half-box was pulled by the rod of the horizontal jack. This device had three degrees of freedom:

1. a translation along x (shearing direction);
2. a translation along z (direction perpendicular to the shearing plane); and
3. a rotation around y (direction perpendicular to the shear direction in the shear plane).

The normal stress, σ_N, was applied by a second, vertical jack attached to a moving frame, so the jack could follow the upper moving half-box. An inclinometer tracked movement of the frame continuously during the shear tests to provide a record of any tilting. Two digital servo-regulators controlled the oil pressure in the two hydraulic jacks. Normal displacement, Δv, shear displacement, Δu, inclination of the upper plateau of the moving half-box, and pressure in the horizontal and vertical jacks were recorded during shear tests.

2.3. Shear test program

Shear tests were performed under applied normal stresses, σ_N, of 7, 14, and 21 MPa. For each normal stress, shearing was permitted, in separate tests with different samples, to continue to several different levels of displacement as indicated in Table 2 [5]. The shear velocity was 0.5 mm/min. Each test was repeated for four different directions (0°, −30°, 60°, 90°) denoted by θ (Fig. 1). For each change of variable, a single fracture replica was used. Thus, each mortar copy was used only once, for a single combination of stress, displacement, and shear direction. Table 2 shows the combinations $\{\theta, \sigma_N, \Delta u\}$ that were used for shear tests.

3. Mechanical results

We can divide the shearing process into five phases [6,7]:

1. The elastic phase, with reduction in aperture on the joint plane as a result of σ_N. Contact area and shear stiffness increase in this phase, and there is a reduction in the volume of the void space.
2. The non-linear dilatancy phase (up to peak shear stress), dominated by dilatancy and local redistribu-

tion of stress resulting from deformation and frictional slip at individual contacting asperities (isolated small asperities may break).

3. The peak shear stress phase, with failure of asperities and a maximum dilatancy rate.

4. The post-peak phase, a progressive softening (unstable yielding) phase, in which there is progressive degradation of the fracture surfaces (from microfracturing of asperities, crushing, and failure of asperities, all depending on σ_N). There is an increase in the contact area between the two surfaces of the fracture with a corresponding decrease in normal stress acting on the contact areas. Dilatancy is still increasing but at a lower rate than during phase 3.

5. The residual strength phase (stable sliding) in which shear and normal stresses are relatively stable, but degradation on joint surfaces is still occurring.

Although shear behavior is qualitatively similar for different shear directions, the main characteristics (peak shear stress, τ_p; residual shear stress, τ_r; dilatancy at the peak, i_p; shear stiffness, K_s; horizontal displacement at the peak, ΔU_p; etc.) have all been shown to be dependent on the shear direction (Table 3).

Fig. 2 shows key experimental results. The fracture's mechanical response differs depending on the shear direction. The mechanical behavior falls into two categories. The first corresponds to the $-30°$ and $0°$ directions (parallel to the strike of the fracture), and the second to the $+60°$ and $+90°$ directions (parallel to the dip of the fracture). The fracture's strike and dip are calculated from the mean fracture plane derived from a multivariate linear regression $z = f(x,y)$

described by Riss et al. [8]. Because results of laboratory experiments on samples with identical fracture geometry show that mechanical parameters depend on shear direction, we attempt in the next sections to determine the role of geometrical characteristics, in particular, surface roughness, in shear behavior.

4. Damaged areas

Degradation on the surface of the replicas consisted of gouge material: pink material was found on the gray (upper) surfaces of the fractures and gray material on the pink (lower) surfaces (Fig. 3). However, most gouge material was white as a result of mylonitization (crushing) during shearing. The greater the normal stress and displacement, the more white material resulted. Some gouge material from both joint faces was dislodged without being stuck to the opposite face. Gouge material was not uniformly distributed.

Because the gouge material was mostly white, a black and white CCD (charged–coupled-device) camera was used to make images of the upper and lower surfaces of the fracture after shear testing. All the images were taken under identical conditions with each fracture replica placed in the same position relative to the camera and light sources. The images were digitized using 512 horizontal pixels, 512 vertical pixels, and 256 gray levels (Figs. 4, 5, 6a). Image analysis was then used to quantify damage on the joint surfaces, using the segmentation method described in Riss et al. [9,10]. Results are presented in binary images (black

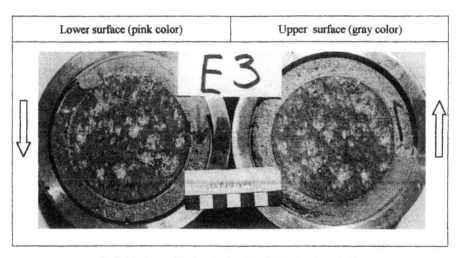

| Lower surface (pink color) | Upper surface (gray color) |

Fig. 3. Color images of the sheared surfaces ($\sigma_N = 21$ MPa, $\Delta u = 5$ mm, $\theta = 0°$).

122

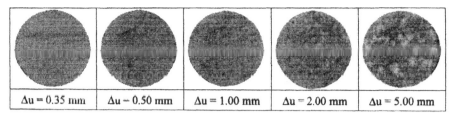

| $\Delta u = 0.35$ mm | $\Delta u = 0.50$ mm | $\Delta u = 1.00$ mm | $\Delta u = 2.00$ mm | $\Delta u = 5.00$ mm |

Fig. 4. Grey-level images showing the evolution of damaged areas with increasing horizontal displacement (constant normal stress $\sigma_N = 21$ MPa, $\theta = 0°$).

and white) that show the size, shape, and spatial distribution of the areas damaged during shearing (Figs. 6b and 7). We conclude that the size, shape, and spatial distribution of damaged areas depend on the shear direction and the degree of stress and horizontal displacement; as would be expected, damage increases as stress and displacement increase.

It does not appear that any major damage occurs prior to the peak shear stress. Damage occurred primarily during the softening and residual phases; however, although visible damage was absent during the first two stages, the locations of contacting asperities during these phases correlated closely to the damaged areas that appeared later. For a specified surface, these damage zones expand in a direction opposite to the direction of displacement of the surface. Damage starts around the areas of the fracture with the steepest slopes. The shape of the damage zones depends on the local geometry of the fracture surface, including the size and shape of asperities. The extension of the damage zones in the shear direction depends on the fracture's horizontal displacement. In any damaged area, asperities can be located on top of each other, making contact at their tips, or in any kind of side-by-side configuration. Because the nature of the asperities

is an important factor for modeling shear behavior, we show in the following section how to get accurate geometrical information about damage zones and how they may be predicted.

5. Three-dimensional modeling of the fracture surface

The spatial features of a fracture can be studied using geostatistical methods [11]. The overall objective of a geostatistical approach is to characterize the spatial variations of the fracture topography and to incorporate them into modeling the fracture, using techniques including kriging. In statistical terms, kriging involves using an unbiased linear estimator with a minimal quadratic error to estimate surface elevation (z).

A set of profiles was recorded in each of the four shear directions, resulting in 7563 (lower surface) and 7556 (upper surface) coordinates $\{x, y, z\}$. After the experimental variograms of elevations z and first derivatives z' were calculated, a theoretical variogram was computed and fit to the experimental data. A good fit to the experimental data was achieved using a theoretical variogram obtained

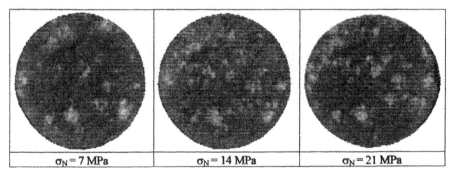

| $\sigma_N = 7$ MPa | $\sigma_N = 14$ MPa | $\sigma_N = 21$ MPa |

Fig. 5. Grey-level images showing the evolution of the damaged areas with increasing normal stress (constant horizontal displacement $\Delta u = 5$ mm, $\theta = 90°$).

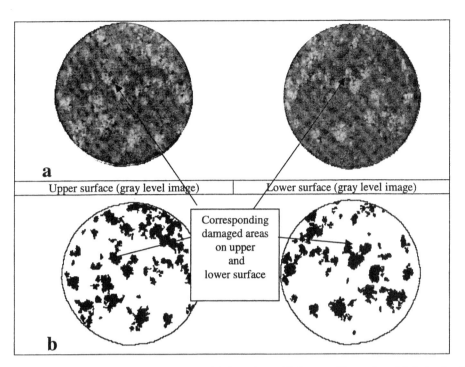

Fig. 6. (a and b) Grey-level (top) and binary images (bottom) of the damaged areas of both upper and lower surfaces of the fracture ($\sigma_N = 21$ MPa, $\Delta u = 5$ mm, $\theta = -30°$).

by superposition of three basic models (Gaussian, spherical, and quadratic rational; see Riss et al. [12] and Gentier et al. [13]). Although kriging based on this theoretical variogram smoothes local variations, it provides a good representation of surface topography (Fig. 8). Furthermore, because $\{x, y\}$ data around the boundaries of the fracture's damaged areas were computed from image analysis of the surfaces, $\{z\}$ values can be calculated from the previous three-dimensional geostatistical reconstruction in such a way that the damaged areas can be imaged in three dimensions (Fig. 9). Additional information (e.g., roughness, local dipping, and local maximum elevation) is available from this three-dimensional representation of the damaged areas. A specific example ($\sigma_N = 21$ MPa, $\theta = +90°$, $\Delta u = 5$ mm, $\Delta v - 0.68$ mm) illustrates general results; Fig. 10a,b shows the relationships among the gradient (i.e., first derivatives z' in the vertical plane of the profile) of the actual profiles of the upper and lower surfaces [z, before (a) and after (b) shearing], and the luminance (light intensity) of the gray-level image of the fracture

along the lower profile. We also added the luminance (light intensity) of the lower profile, shown in part in Fig. 11, from a color photograph of the fracture taken after shearing and then scanned. Damaged asperities with a luminance higher than the threshold that was used to create the binary images (see Fig. 10a,b) correlated well with positive gradients. In fact, there are three types of damage (Fig. 11): (1) actual crushed white material (easy to see on black and white images) slightly shifted in the direction opposite to shearing where gradients are steep; (2) uncrushed material from the opposite surface, detected because it retained its original color, occurring adjacent to areas of maximum gradient; and (3) gouged areas that produce type 2 material. In other words, we encounter successively: gray stuck material, white crushed material, and dark gouged areas along a line parallel but opposite to the shear direction on the lower pink wall (along a line drawn symmetrically on the gray upper wall, we find a pink stuck material, a white crushed material, and dark gray areas).

We conclude that damaged areas predominated

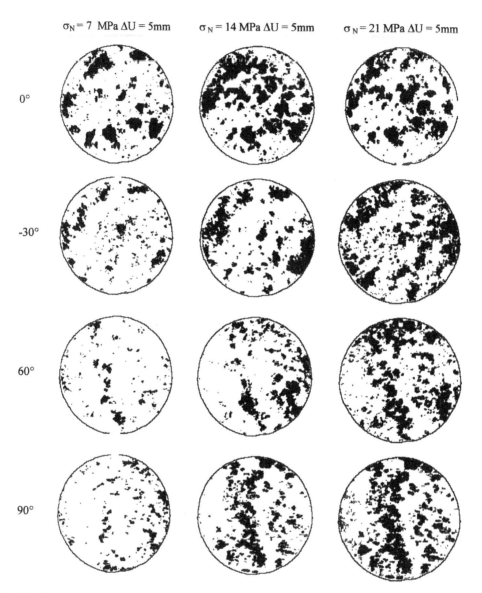

Fig. 7. Size, shape, and spatial distribution of damaged areas occurring during shear tests in relation to shear direction and normal stress ($\Delta u = $ 5 mm).

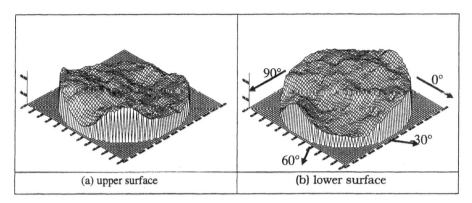

Fig. 8. Three-dimensional reconstruction of the fracture surfaces (Kriging method).

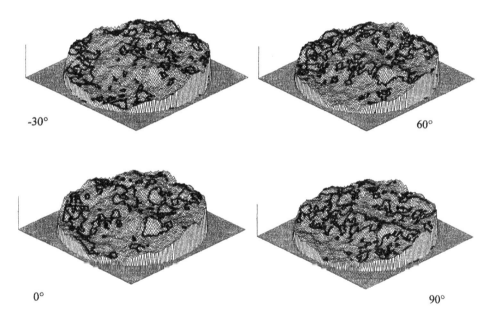

Fig. 9. Damaged zones superimposed on fracture topography ($\sigma_N = 21$ MPa, $\Delta u = 5$ mm). On the left, shearing directions ($-30°$, $0°$) are nearly parallel to the strike direction (N–S) of the fracture; on the right the shearing directions ($60°$, $90°$) are nearly parallel to the dip direction of the fracture (E–W).

126

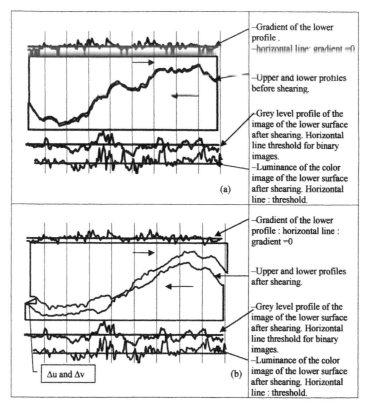

—Gradient of the lower profile .
—horizontal line: gradient =0

—Upper and lower profiles before shearing.

—Grey level profile of the image of the lower surface after shearing. Horizontal line threshold for binary images.
—Luminance of the color image of the lower surface after shearing. Horizontal line : threshold.

(a)

—Gradient of the lower profile : horizontal line : gradient =0

—Upper and lower profiles after shearing.

—Grey level profile of the image of the lower surface after shearing. Horizontal line threshold for binary images.
—Luminance of the color image of the lower surface after shearing. Horizontal line : threshold.

Δu and Δv

(b)

Fig. 10. (a and b) Relationships between damaged areas and topography.

where the local dip direction was close to the shear direction, and also that areas with higher dip values had the highest probability of damage. Assuming this conclusion applies to any stress (σ_N) and shear direction (θ), it is possible to predict, using three-dimensional representation of the fracture geometry, which asperities on a fracture's surface will likely be damaged during shearing.

6. Prediction of the mechanical damage

Mechanical damage to a fracture as a result of shear stress can be predicted using image analysis, which is very effective for assessing the roughness of non-planar surfaces such as fracture walls. We created a gray-level image of the lower wall of the fracture (Fig. 12), assuming that the gray levels $f(z)$ are proportional to the elevations (z) of the fracture surface.

Because the slope of the asperities correlates with the likelihood that they will be damaged during shearing, we analyzed the gray-level image using the directional gradient, which is an image-processing operator that is used to calculate the slope of the surface in a specified direction. The directional gradient $g_\theta(f)$ of a function f in the direction θ is defined by the difference between the 'thickening' and the 'thinning' of the function f (thickening and thinning are classical morphological transformations operating on the gray levels of the images [15]):

$$g_\theta(\mathscr{F}) = (\mathscr{F} \oplus T_\theta) - (\mathscr{F}_0 T_\theta)$$

where T_1 and T_2 form the two-phase structuring element $T_\theta = (T_1, T_2)_\theta$. The procedure for computing the directional gradient is given in [14]. The result of the operation is to change the gray level of the current pixel x_i depending on the difference in gray levels at

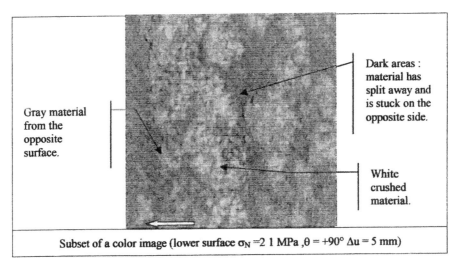

Fig. 11. The three types of damage during shearing.

adjacent pixels x_{i-1} and x_{i+1} in the θ direction; the new gray level at x_i is proportional to the difference between the gray levels at adjacent pixels for differences greater than zero (indicating positive slope), or black for differences less than or equal to zero (indicating negative or zero slope). Fig. 13 (2nd row) shows binary images with the locations of the main directional gradients that were computed for each of the shearing directions. The locations of the maximum gradients correspond well to the locations of damage zones for a normal stress of 21 MPa for all shear directions. It must be noted that the damaged areas occurring under normal stresses of 7 MPa and 14 MPa are subsets of the damage zones that result from a nor-

Fig. 12. Grey-level image of the topography of the lower wall of the fracture. Grey levels $F(z)$ are proportional to the elevations (z) of the fracture surface. Grey levels are computed using the equation $F(z) = aZ + b$. Z values are the elevations calculated at the grid nodes by the geostatistical simulation. $F(z) \in [5250]$, pixel size is 0.5 mm^2, and each gray level represents 33 μm.

mal stress of 21 MPa. When the shear direction is parallel to the dip direction, the damaged areas are strongly correlated to the directional gradient. The same correlation is also apparent when the shear direction is parallel to strike but is less obvious because topographical features in this direction are dispersed and not contiguous. Therefore, we conclude that the directional gradient is a good predictor of the areas of the fracture surface that will be damaged during shearing.

The areas of maximum directional gradient occur either on the sides of asperities or on the sides of crest lines; these locations can be defined readily, but they do not indicate the full extent of future damaged areas. Assuming that the volume of material that will be damaged during shearing is located all around points of maximum gradient but preferentially in the shearing direction, we propose that damaged areas can be inferred by image analysis methods [15] using the following steps (Fig. 14):

1. Computing the gray-level gradient for a given direction, i.e., directional gradient (Fig. 14b and c).
2. 'Thresholding' the image obtained from Step 1 to get a binary image that more clearly shows the areas of highest directional gradient (Fig. 14d).
3. Filtering the binary image using an image-analysis operator (structural element with opening of size 1) to remove small isolated features, followed by reconstruction (Fig. 14e). The filtering procedure removes all features smaller than the defined structural element. The purpose of reconstruction is to

128

recover the deleted areas that are not isolated but are instead part of larger features.

4. 'Dilating' this binary image of maximum gradient to expand the predicted damage zones using a directional structuring element with a size correlated to the horizontal displacement (Fig. 14f). Steps 1–3 identify the locations of steepest gradient; the purpose of dilation is to expand these areas in the shear direction to account for damage that occurs with increasing shear displacement.

5. Determining the symmetrical difference between the images of the gradient and dilation of the gradient (Fig. 14g). The symmetrical difference is defined in terms of basic set theory: for two sets X and Y, the symmetrical difference $X \backslash Y = (X \cup Y) - (X \cap Y)$. For the results presented here, X corresponds to the high-gradient areas determined in Steps 1–3, and Y corresponds to the area after the dilation of X (Step 4). The symmetrical difference $X \backslash Y$ corresponds to the area added during directional dilation. Although

one might expect damage to initiate in the highest gradient areas identified in the first three steps, the best fit with experimental data is obtained when these areas are subtracted from the dilated area by taking the symmetrical difference defined above.

Fig. 14h and i illustrate the results of applying the algorithm to the experimental data for shear direction $\theta = 90°$. The predicted damage areas (Fig. 14h) correlate closely with the actual damage (Fig. 14g) delineated by white gouge material (damage from crushing). We conclude that our algorithm effectively predicts damage to fracture surfaces that occurs during shearing and is the first method to do so. The next step in this research is to define an algorithm for damage prediction that is based on deterministic criteria; for example, setting the threshold used in Step 2 according to the normal stress and the size of dilation (Step 4) according to the horizontal displacement (Δu).

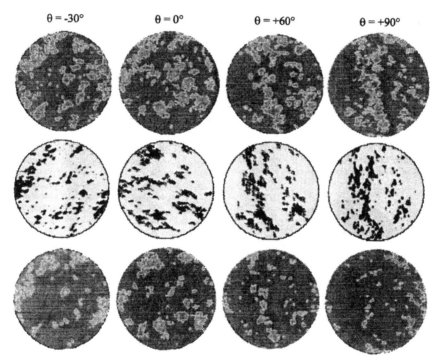

Fig. 13. Directional gradient for the lower surface of the fracture (2nd row). Contoured areas show the damaged areas (white pixels) that occurred during shearing with $\sigma_N = 21$ MPa and $\Delta u = 5$ mm (1st row) and with $\sigma_N = 7$ MPa and $\Delta u = 5$ mm (3rd row).

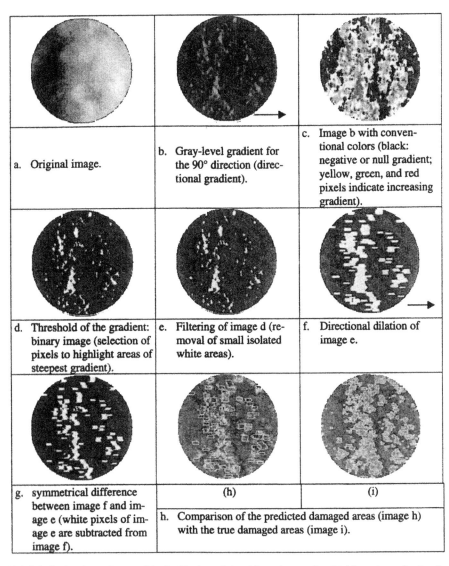

		c. Image b with conventional colors (black: negative or null gradient; yellow, green, and red pixels indicate increasing gradient).
a. Original image.	b. Gray-level gradient for the 90° direction (directional gradient).	
d. Threshold of the gradient: binary image (selection of pixels to highlight areas of steepest gradient).	e. Filtering of image d (removal of small isolated white areas).	f. Directional dilation of image e.
g. symmetrical difference between image f and image e (white pixels of image e are subtracted from image f).	(h)	(i)
	h. Comparison of the predicted damaged areas (image h) with the true damaged areas (image i).	

Fig. 14. (a–i) Application of successive steps of the algorithm for prediction of damaged areas and predicted damaged areas (for shear direction 90°, $\sigma_N = 21$ MPa and $\Delta u = 5$ mm).

7. Conclusions

Laboratory experiments performed on replicas of a natural granite fracture show that the mechanical behavior of fractures under shear stress is strongly related to the geometry of the fracture surfaces. Mechanical parameters measured during experiments varied depending on the direction of shear. Although the mechanical behavior was qualitatively similar for different shear directions, primary parameters all depended on the shear direction; these parameters include peak shear stress, residual shear stress, dilatancy and displacement at peak shear stress, and shear stiffness.

Analysis of images of damage zones on fracture surfaces after shearing suggests that there is little damage during the first two phases of shearing, prior to the peak shear stress. Damage is initiated in areas of the fracture with the steepest slopes. The damage zones expand during the post-peak softening and residual phases of shear. The size, shape, and spatial distribution of the damaged areas depend on the shear direction as well as on the level of stress and shear displacement. The extension of the damage zones in the shear direction depends on the amount of horizontal displacement during shearing.

Fracture surfaces reconstructed from profiles using geostatistical methods were compared to damage zones created during shearing characterized using image processing techniques. This comparison showed that damage is most likely to occur in areas where the local dip direction is close to the shear direction and in areas with the steepest gradients.

The experimental results and results from the comparison between fracture surface reconstructions and damage zone images were used to construct an image-analysis-based algorithm that predicts where damage will occur. Directional gradients are calculated from the images of the surfaces. It is assumed that damage will occur around areas of maximum gradient but preferentially in the shear direction. The areas of maximum gradient occur either on the sides of asperities or on the sides of crest lines. To achieve a good match to the damage zones observed on laboratory samples, the areas of maximum gradient predicted to be damage zones must be expanded in the shear direction by an amount related to the shear displacement. The location, size, and shape of damage zones predicted by the algorithm agree closely with the damage observed on laboratory specimens.

Results of laboratory experiments, used in conjunction with images of fracture surfaces and modeling that allows reconstruction of fracture topography from surface profiles, provide a powerful tool for linking fracture geometry to the mechanical behavior of fractures under shear stress. Studies such as this one that utilize data obtained from a variety of techniques are likely to be fundamental in formulating a constitutive law of shear behavior that incorporates fracture geometry.

Acknowledgements

BRGM (France) and the University Bordeaux I (France) supported the research described in this paper.

References

[1] Stephansson O, Jing L. Testing and modeling of rock joints. In: Rossmanith HP, editor. Proceedings of Mechanics of Jointed and Faulted Rock (MJFR-2). Rotterdam: Balkema, 1995. p. 37–47.

[2] Barton N, de Quadros EF. Joint aperture and roughness in the prediction of flow and groutability of rock masses. Int J Rock Mech & Min Sci 1996;34:3(4) Paper No. 252.

[3] Gentier S, Hopkins D, Riss J, Lamontagne E. Hydromechanical behavior of a fracture: how to understand the flow paths. In: Rossmanith HP, editor. Proceedings of Mechanics of Jointed and Faulted Rock (MJFR-3). Rotterdam: Balkema, 1998. p. 583–8.

[4] Gentier S. Morphologie et comportement hydromécanique d'une fracture naturelle dans un granite sous contrainte normale. PhD thèse d'université, Orléans, France 1986.

[5] Flamand R. PhD thesis, université du Québec à Chicoutimi, Canada (in preparation).

[6] Archambault G, Gentier S, Riss J, Flamand R. A reevaluation of irregular joint shear behavior on the basis of 3D modelling of their morphology (Part II: joint shear behavior mechanical modelling). In: Rossmanith HP, editor. Proceedings of Mechanics of Jointed and Faulted Rock (MJFR-2). Rotterdam: Balkema, 1995. p. 163–8.

[7] Archambault G, Gentier S, Riss J, Flamand R, Sirieix C. Joint shear behaviour revised on the basis of morphology 3D modelling and shear displacement. In: Aubertin M, Hassani F, Mitri H, editors. Proceedings of the 2nd North American Rock Mechanics Symposium. Rotterdam: Balkema, 1996. p. 1223–30.

[8] Riss J, Gentier S, Flamand R, Archambault G. Detailed description of the morphology of a fracture in regard of its behaviour during shearing. In: Rossmanith HP, editor. Proceedings of Mechanics of Jointed and Faulted Rock (MJFR-3). Rotterdam: Balkema, 1998. p. 205–10.

[9] Riss J, Gentier S, Sirieix C, Archambault G, Flamand R. Degradation characterization of sheared joint wall surface morphology. In: Aubertin M, Hassani F, Mitri H, editors. Proceedings of the 2nd North American Rock Mechanics Symposium. Rotterdam: Balkema, 1996. p. 1343–9.

[10] Riss J, Gentier S, Laffréchine K, Flamand R, Archambault G. Binary images of rock joints: characterisation of damaged zones. Microscopy, Microanalysis, Microstructures 1996;7:521–6.

[11] Chilès JP, Gentier S. Geostatistical modelling of a single fracture. In: Soares A, editor. Geostatistics Troia '92. Dordrecht, The Netherlands: Kluwer Academic Publishers, 1993. p. 95–108.

[12] Riss J, Gentier S, Sirieix C, Archambault G, Flamand R. Sheared rock joints: dependence of damage zones on morphological anisotropy. Int J Rock Mech & Min Sci 1997;34:3(4) Paper No. 258.

[13] Gentier S, Verreault N, Riss J. Roughness and flow paths in a fracture. Acta stereologica 1997;16:3:307–14.

[14] Micromorph . Mathematical and Image Processing, release 1.3, CMM/ARMINES/ENSMP/TRANSVALOR 1997.

[15] Serra J. Image analysis and mathematical morphology. London: Academic Press, 1982.

Constraining the stress tensor in the Visund field, Norwegian North Sea: Application to wellbore stability and sand production

D. Wiprut*, M. Zoback

Department of Geophysics, Stanford University, Stanford, CA 94305-2215, USA

Accepted 7 October 1999

Abstract

In this study we examine drilling-induced tensile wellbore failures in five exploration wells in the Visund oil field in the northern North Sea. We use observations of drilling-induced wellbore failures as well as density, pore pressure, and leak-off test measurements to estimate the magnitudes and orientations of all three principal stresses. Each well yields a very consistent azimuth of the maximum horizontal stress ($100° \pm 10°$), both with depth and laterally across the field. Stress orientations are constrained at depths as shallow as 2500 m and as deep as 5300 m in these wells. We show that the magnitudes of the three principal stresses (S_v, S_{hmin}, and S_{Hmax}) are also consistent with depth and reflect a strike-slip to reverse faulting stress regime. The magnitude of the maximum horizontal stress is shown to be significantly higher than the vertical and minimum horizontal stresses (e.g. $S_v = 55$ MPa, $S_{hmin} = 53$ MPa, and $S_{Hmax} = 71.5$ MPa at 2.8 km depth). Data from earthquake focal plane mechanisms (Lindholm et al., 1995, Proceedings of the Workshop on Rock Stresses in the North Sea, Trondheim, Norway [1]) show similar stress orientations and relative magnitudes and thus indicate a stress field that is relatively consistent throughout the thickness of the brittle crust.

We illustrate how knowledge of the full stress tensor allows one to place bounds on in situ rock strength and determine optimally stable trajectories for wellbore stability and sand production during drilling, after the completion of drilling, and as pore pressure is reduced during oil and gas production. © 2000 Elsevier Science Ltd. All rights reserved.

1. Introduction

Determination of the full stress tensor in oil fields is critical for addressing engineering problems of wellbore stability and sand production as well as geologic problems such as understanding dynamic constraints on hydrocarbon migration and fracture permeability. Controlling wellbore instabilities requires understanding of the interaction between the rock strength and in situ stress. Because in situ stress and rock strength cannot be altered or controlled, the only way to inhibit wellbore failure during drilling is to adjust engineering practice by choosing optimal trajectories and mud weights. Similarly, utilization of an appropriate trajec-

tory can limit sand production by reducing the tendency for failure around a wellbore.

This paper presents an analysis of stress and wellbore stability in the Visund field, which is located in the Norwegian North Sea to the northwest of Bergen, near the western edge of the Viking Graben (Fig. 1). The Visund field sits within the approximately 25 km long, 2.5 km wide Visund fault-block, which is the most easterly major fault block on the Tampen Spur [2]. The state of stress in the Norwegian North Sea is generally characterized by an east–west to northwest-southeast compression, but exhibits appreciable scatter in places (e.g., Müller et al. [3]; Lindholm et al. [1]).

In the following sections we describe how observations of drilling-induced compressive failures (e.g.,[4–6]) and wellbore tensile failures (e.g. [7–10]) can be integrated with other routinely available wellbore information to constrain the full stress tensor.

* Corresponding author. Tel.: +1-650-725-6072; fax: +1-650-725-7244.

E-mail address: wiprut@pangea.stanford.edu (D. Wiprut).

Reprinted from *International Journal of Rock Mechanics and Mining Sciences* **37 (1-2)**, 317-336 (2000)

132

Our approach follows an integrated stress measurement strategy (ISMS), outlined by Zoback et al. [11], and Brudy et al. [12], to constrain the magnitudes and orientations of all three principal stresses. Using our estimates of in situ stress we place bounds on the effective rock strength. This information is then used to determine optimally-stable trajectories for drilling and minimizing sand production.

2. Observations of wellbore failure

Examination of Formation MicroScanner/Formation MicroImager (FMS/FMI) logs [13] provided by Norsk Hydro, and run in seven wells, revealed extensive drilling-induced tensile failures in five of the wells. Examples of this data can be seen in Appendix A. We plot the azimuth of the tensile fractures as a function of depth in Fig. 2. In each case, black data points represent tensile fractures that are aligned with the axis of the wellbore, and data points representing tensile fractures that are inclined with respect to the axis of the wellbore are shown in gray. Error bars for near-axial tensile failures show the variation in azimuth of each fracture; while error bars for inclined tensile failures show the portion of the wellbore circumference spanned by each fracture. Near the center of each plot, bit trips and "wash and ream" operations are shown by horizontal and vertical lines respectively. A bit trip is plotted each time the drill string is run into the hole. This operation may cause a significant rise in the mud pressure at the bottom of the hole due to a piston effect. Washing and reaming the hole involves scraping the hole clean, and may remove evidence of drilling-induced tensile fractures. There is no visible correlation between the occurrence (or absence) of tensile fractures and these special drilling operations, suggesting that the tensile fractures formed (or did not form) during normal drilling operations, rather than as a result of extreme conditions in the well such as tripping the bit or reaming the hole.

All wells except 10 S were drilled nearly vertically to total depth. Near-axial tensile fractures in wells 6, 7, 8, and 11, as well as in the vertical portion of 10 S, suggest that the vertical and two horizontal stresses are principal stresses in this field [9]. Inclined tensile fractures observed in four of the wells (most prevalent in the 10 S well) indicate possible exceptions to this and are discussed in Appendix B.

In order to determine the orientation of the stress field in this region, we focus on the orientation of the tensile fractures aligned with the wellbore axis. We use

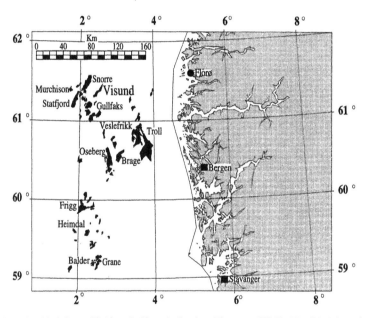

Fig. 1. Map of the northern North Sea modified from the Norwegian Petroleum Directorate, 1997. The Visund field sits on the western edge of the Viking Graben to the northwest of Bergen, Norway.

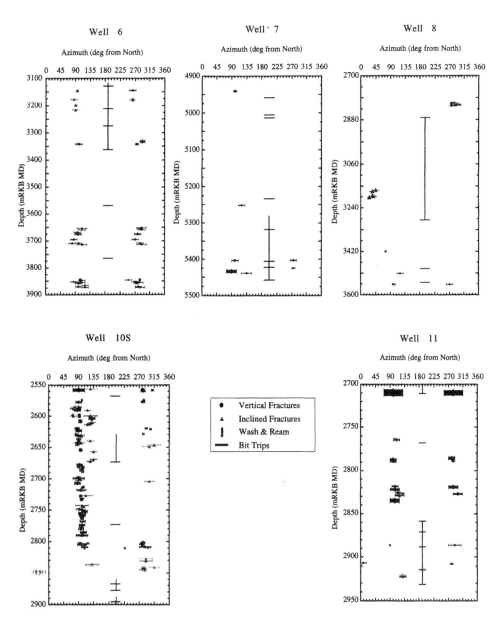

Fig. 2. Azimuth of drilling induced tensile fractures observed in image logs as a function of depth.

134

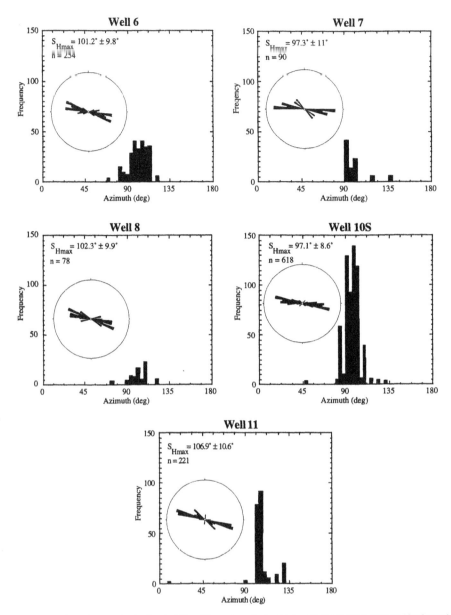

Fig. 3. Histograms and rose diagrams showing the orientation of the maximum horizontal stress. Each rose plot is normalized by the number of data points, and therefore the length of the bars does not reflect the relative frequency of the tensile fractures between wells. The number of 0.2 m observations (*n*) is shown in each case. The statistics follow Mardia [14].

a circular statistical method developed by Mardia [14] to obtain the mean azimuth and standard deviation of the maximum horizontal stress for each well (insets of Fig. 3). The uncertainty in the azimuth represents two standard deviations from the mean. The frequency is calculated by adding the tensile fractures in 0.2 m intervals.

An anomalous stress orientation seen from a well-bore breakout (reported by Fejerskov [15] and later analyzed in this study) and inclined tensile fractures was detected over an approximately 135 m interval of well 8. This stress orientation was not seen in any of the other wells to the north or south, and appears to be the result of a localized stress anomaly due to slip

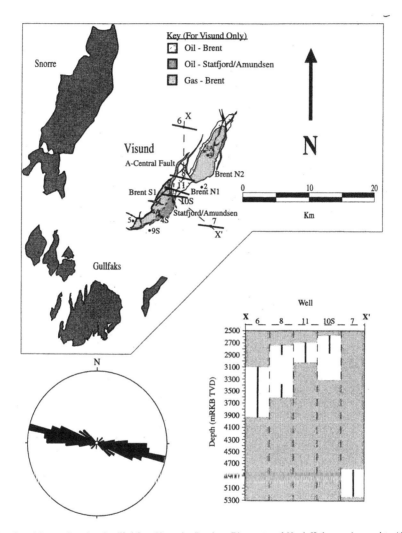

Fig. 4. Map view of S_{Hmax} orientations (modified from Norwegian Petroleum Directorate and Norsk Hydro maps), rose plot with all of the orientation data, and plot showing the depths over which S_{Hmax} is constrained in each well. The gray regions show portions of the wells that were not logged. The orientation of the maximum horizontal stress is consistent both laterally and with depth in this field.

on a preexisting fault penetrated by this well. The subseismic fault responsible for this anomaly can be inferred from the geologic section by noting a repeated sequence of the Brent reservoir sands. The relationship between anomalous stress fields and slip on faults has been noted by other authors (e.g. [16–18]). We therefore assume that this single anomaly is not representative of the tectonic stress field.

Fig. 4 summarizes our findings concerning the orientation of the maximum horizontal stress. The mean orientation of S_{Hmax} for each well is plotted in map view and clearly shows a consistently oriented stress field. The rose diagram shows the orientations of all of the tensile fractures from all of the wells. The plot in the lower right shows the depths over which the orientation of S_{Hmax} is constrained in each well. Gray shaded regions show portions of each well that were not logged. This plot shows that the maximum horizontal stress is constrained to depths as shallow as 2550 m and as deep as 5250 m (RKB TVD). The orientation of the maximum horizontal stress as determined from earthquake focal plane mechanisms [1] is similar to our findings (azimuths between 90° and 120°). Our data, coupled with these findings, indicate that the orientation of the maximum horizontal stress is consistent across the Visund field as well as throughout the thickness of the brittle crust.

The scatter in stress directions observed in some parts of the northern North Sea (determined from wellbore elongations) has led some investigators to conclude that shallow stress directions are decoupled from the deeper regional stress field (e.g. [19]). We have shown that close examination of reliable data reveals a consistently oriented stress field, and we further demonstrate in Appendix B that we can use this stress field to explain the occurrence of inclined drilling-induced tensile fractures.

3. In situ stress and rock strength

We utilize the interactive software package, Stress and Failure of Inclined Boreholes (SFIB), developed by Peska and Zoback [20], to constrain the maximum horizontal stress magnitudes and to put limits on rock strength. Estimation of the maximum horizontal stress requires prior knowledge of the vertical stress, the minimum horizontal stress, the pore pressure, the mud weight, and the change in temperature at the wellbore wall during drilling. Final well reports provided by Norsk Hydro contain this information. We analyze each well individually so that estimates of the maximum horizontal stress are not affected by data from wells in different pore pressure compartments or with slightly differing overburden stresses. The vertical stress, S_v, used in this study was derived from inte-

grated density logs. Because these logs are seldom run up to the sea floor, the density must be estimated in the shallow subsurface. The resulting errors in the overburden gradient are negligible in wells 10 S and 11 since only a small amount (approximately 30 m) of sediment is not accounted for by the density logs. In wells 6, 7, and 8, the density logs were only run in the deeper portions (below approximately 2500 m) of the holes. The densities used in the upper portions of these holes are estimated from shallow measurements in nearby holes. Overburden gradients from wells 6, 7, and 8 nevertheless provide overburden stresses which are similar to those found in wells 10 S and 11. We derive the minimum horizontal stress, S_{hmin}, from Leak-off Tests (LOT) and Formation Integrity Tests (FIT) conducted in each well (see Gaarenstroom et al. [21]). In all of the wells the depth at which we wish to constrain the maximum horizontal stress is below the deepest LOT or FIT. We therefore assume a linear stress gradient between LOTs in each well and extrapolate this trend to the depth of investigation. The pore pressure, P_p, was obtained from Repeat Formation Tests (RFT). In order to use a reliable pore pressure value the stress analysis in each well was conducted as close as possible to an RFT depth. Pore pressure data was not available for well 7. Consequently, we do not constrain the maximum horizontal stress in this well. We derive a mud weight value, P_m, from the maximum equivalent circulating density (ECD), which takes into account frictional effects between the wellbore wall and the mud as well as the mud density. We use a static mud weight in well 6 because an ECD value was unavailable. Use of the highest mud density value is required since it is impossible to determine the precise mud weight at which the tensile cracks initiated. Although the tensile fractures may have formed at mud pressures lower than the ones we use, our utilization of the highest mud pressure allows us to calculate a reasonable lower bound for the maximum horizontal stress. An upper bound for the maximum horizontal stress is derived from our analysis of rock strength and compressive failures in these wells. The amount of cooling at the wellbore wall was derived from temperature gradient plots. Each well showed cooling between 20° and 30°C at the depths of investigation. While the temperature change of the wellbore wall was considered in these calculations, it had little effect on the estimation of S_{Hmax}.

Fig. 5 shows the analysis for the maximum horizontal stress, S_{Hmax}. The plots represent the allowable stress state at a given depth constrained by: (1) Mohr–Coulomb frictional faulting theory for the crust assuming a coefficient of sliding friction of 0.6 [22] (four sided polygon); and (2) compressive and tensile wellbore wall failures (thick short dashes and long dashes respectively) (see Moos and Zoback [8], for an explanation). Stress magnitudes that fall above the short dashed tensile failure contour (assumed to be zero in this study) indicate stress states consistent with the occurrence of drilling-induced tensile fractures, while values that fall below indicate no tensile failures should occur. Similarly, for a given rock strength, stress magnitudes that fall above one of the long dashed compressive failure contours indicate breakouts should occur, while those that fall below indicate no breakouts should be observed. We constrain the maximum horizontal stress only where tensile fractures were observed in each of these wells, mean-

ing the stress state must be such that it falls above the short dashed tensile failure line in our figures. No breakouts were observed in any of the wells at the depths where we constrain the maximum horizontal stress. This observation is utilized below to place a lower bound on the in situ rock strength. While coefficients of sliding friction may be as high as 1.0 in some rocks, faults in sediments tend to have lower coefficients of friction. We consider a coefficient of friction of 0.6 to be an upper bound in this case, as the sediments are poorly cemented and consolidated.

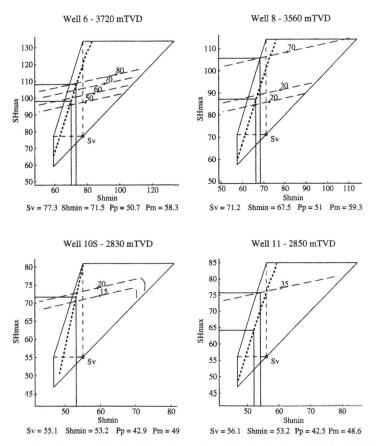

Fig. 5. Stress polygons showing the constrained values of the maximum horizontal stress in each of the wells. Each analysis is developed for a specific depth shown above the figure. The relevant stress inputs (in MPa) are shown below the figures. Tensile failure contours are thick short dashes, and breakout contours are long dashes. Each breakout line represents a different rock strength value, as shown, assuming that a breakout width of at least 40° would be required in order to be detected by the wide pads of the FMI tool.

138

3.1. Constraining S_{Hmax} in wells 10 S and 11

Wiprut et al. [23] presented a stress analysis of well 10 S using near-axial tensile fractures. The points of departure of this study from the previous study are three-fold: (1) we constrain the maximum horizontal stress in multiple wellbores; (2) we perform a more comprehensive analysis of in situ rock strength; (3) we analyze inclined tensile fractures observed in well 10 S.

Well 10 S presents the best opportunity to reliably constrain the magnitude of the maximum horizontal stress. As the well increases in deviation and begins to encounter a different stress field a transition from a stress state in which tensile fractures form, to one in which they do not, occurs. Fig. 6 shows that tensile fractures form continuously in this well to a deviation of approximately 35°, and then abruptly stop. Because tensile failures are present at the depth of investi-

gation, and the tensile cracks stop just below 2830 mTVD (as shown in Fig. 6), we expect the value of S_{Hmax} at 2830 m to be approximately equal to the value indicated by the short dashed contour line (71.5 ± 4.5 MPa) (Fig. 5). The steep slope of the short dashed tensile failure lines in Fig. 5 means a small uncertainty in the minimum horizontal stress creates a large uncertainty in the maximum horizontal stress. We illustrate in Appendix B that the occurrence of inclined tensile fractures is easily explained by this stress tensor, using only minor (±10°) perturbations to the mean orientation of the stress tensor and slight increases in the mud pressure.

Each compressive failure contour in Fig. 5 represents a different uniaxial compressive rock strength (UCS) value, as shown, assuming that a breakout width of at least 40° would be required in order to be detected by the wide pads of the FMI tool. Well 10 S

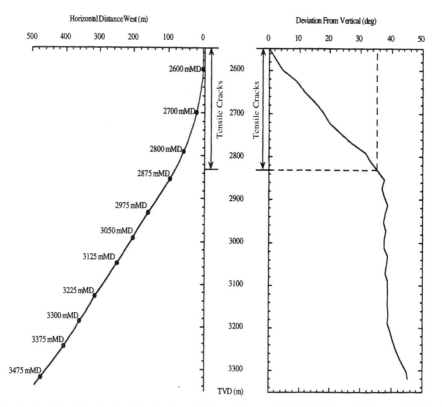

Fig. 6. Section view of the drilling direction (left) and wellbore deviation as a function of depth (right). Measured depths are plotted along the wellbore path. Tensile fractures were detected from 2550 to 2850 m measured depth (2550–2830 m true vertical depth). The tensile fractures stopped at a deviation of 35° at 2830 mTVD.

is the only well in which knowledge of the rock strength is not needed to constrain S_{Hmax} because the stress is constrained by the tensile failure contour. Since no breakouts were detected in this well, the stress values determined above imply the apparent UCS of the rock (assuming a coefficient of internal friction of about 1.0) is greater than 18 MPa. Laboratory strength measurements conducted by Norsk Hydro on cores taken throughout the region give an average value for the uniaxial compressive rock strength of approximately 25 MPa in the caprocks at this depth. The rock strength may also be constrained by the breakout width if breakouts occur, as weak rock will have wider breakouts than strong rock [5].

In well 11 we use a reasonable upper bound on uniaxial compressive rock strength derived from laboratory tests on core, the upper bound of the minimum horizontal stress, and the lack of breakouts at this depth, in order to constrain the upper bound for the maximum horizontal stress. Since the upper bound on the rock strength determined from the Norsk Hydro laboratory tests on core is approximately 35 MPa at this depth, and in this formation, the maximum horizontal stress must fall below the 35 MPa breakout line. We use the lower bound of S_{hmin}, with the tensile failure contour, to constrain the lower bound of S_{Hmax}. The mean value from this range is our assumed value of S_{Hmax}, and the range is the uncertainty (70 ± 6 MPa) (Fig. 5).

3.2. Constraining S_{Hmax} in wells 6 and 8

In well 6, the least principal stress at the depth of interest is 71.5 ± 1.5 MPa. To account for the existence of drilling-induced tensile fractures, this requires lower bound values of S_{Hmax} to be 103.5 ± 5.5 MPa (Fig. 5). As there are no breakouts observed in this well, these values imply UCS values that exceed 50–80 MPa. While laboratory UCS tests done on the Etive sands (which occur only 5 m below the depth at which the stress analysis is done in well 6) indicate maximum strengths of only 30 MPa, the sandstone at the depth analyzed is more well cemented with silica than is the Etive. A theoretical estimate of strength in clean arenites provides a potential strength of 90 MPa for an average porosity of 15% [24]. Although the sandstone in question is not clean, it is reasonable to assume that the UCS is 50–80 MPa. We assume that the lower bound value of S_{Hmax} (103.5 ± 5.5 MPa) is approximately correct because if the value were even slightly higher (as permitted by Coulomb faulting theory, Fig. 5), the corresponding rock strengths would be unreasonably high.

In well 8 the upper bound of the maximum horizontal stress is constrained using the upper bound of S_{hmin} and Coulomb faulting theory. Since no breakouts were observed in this well at 3560 mTVD, the upper bound values of S_{hmin} and S_{Hmax} predict that the rock strength is greater than 70 MPa. In this well we use the lower bound of S_{Hmax} as our assumed value of the maximum horizontal stress, and the upper bound as the extent of our uncertainty since both the rock strength and the friction coefficient may be either higher or lower than shown in the figure. The maximum horizontal stress in well 8 is constrained to be 105 MPa S_{Hmax} 87 MPa (Fig. 5) although the upper bound is only limited by the assumed frictional strength of the crust.

Fig. 7 shows a summary of our stress results for the Visund field. We compiled overburden, LOT, and RFT data from all 13 wells drilled in the Visund field. Each data point in this plot is derived as strictly as possible from the most reliable data from each well. The data for the minimum horizontal stress is derived from our analysis of leak-off test curves. Reported leak-off tests without pressure–time curves are not considered in this compilation. The well number is shown next to each data point. The vertical stress is derived using an overburden gradient averaged across the entire field. Small differences in the overburden gradients were detected, but the resulting stress profiles from these gradients were sufficiently similar such that we could neglect any differences. The pore pressure data is compiled from RFTs conducted in all of the wells in the field. Our constrained values for the maximum horizontal stress are shown in gray.

Note that the maximum horizontal stress is significantly larger than the vertical and minimum horizontal stresses, contrary to what other studies have found [25]. However, our stress magnitudes are consistent with the strike-slip to reverse faulting stress field observed from earthquake focal plane mechanisms in the North Sea [1]. Further evidence of recent reverse faulting deformation was seen in sub-seismic faults inferred from repeated sequences of Brent sands in the lithology logs of several wells. The existing deformation data therefore supports our prediction of high horizontal stresses in this region of the North Sea.

Table 1 provides a summary of the stress data found in each of the seven wells. Wells 4A and 4 S did not have any interpretable tensile fractures as a result of poor image quality and low wellbore coverage. The tensile fractures observed in well 7 appear at depths far below any pore pressure or leak-off test measurements. We therefore do not provide estimates of stress magnitudes in well 7.

4. Application to wellbore stability

Exploratory wells drilled in Visund prior to the spudding of well 10 S frequently encountered drilling

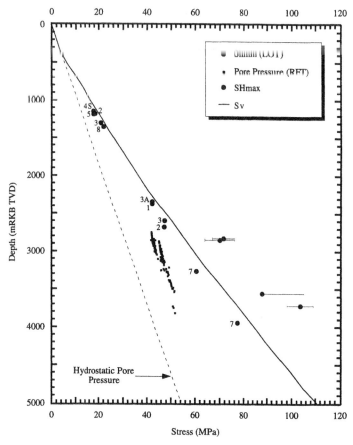

Fig. 7. Stress vs depth in Visund. Each stress point in this plot is derived as strictly as possible from the most reliable data in the field. Each leak-off test data point shows the well in which the test was run.

Table 1

Well	Log	SH Azimuth	Depth (mTVD)	SH Magnitude (MPa)	Sh Magnitude (MPa)	Sv Magnitude (MPa)
4A	FMS	–	–	–	–	–
4 S	FMS	–	–	–	–	–
6	FMS	$101.2° \pm 9.8°$	3720	103.5 ± 5.5	71.5 ± 1.5	77.3
7	FMS	$97.3° \pm 11°$	–	–	–	–
8	FMI	$102.3° \pm 9.9°$	3560	$87 + 18$	67.5 ± 1	71.2
10 S	FMI	$97.1° \pm 8.6°$	2830	71.5 ± 4.5	53.2 ± 1	55.1
11	FMI	$106.9° \pm 10.6°$	2850	70 ± 6	53.2 ± 1	56.1

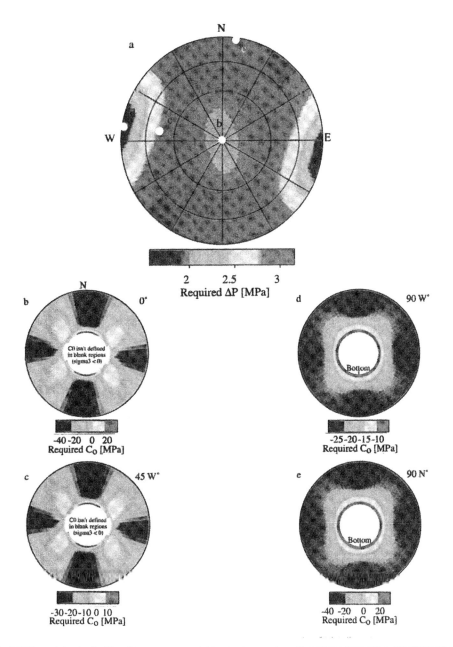

Fig. 8. (a) Stereonet showing the differential pore pressure needed to prevent compressive wellbore failures during drilling. The white points show the orientations of the wellbore cross-sections shown in parts (b–e). The colors correspond to the compressive rock strength needed to prevent failure in different parts of the rock surrounding the hole. (b) Vertical well. (c) Well inclined 45° at an azimuth of 280° (d) Horizontal well drilled toward S_{Hmax}. (e) Horizontal well drilled toward S_{hmin}.

142

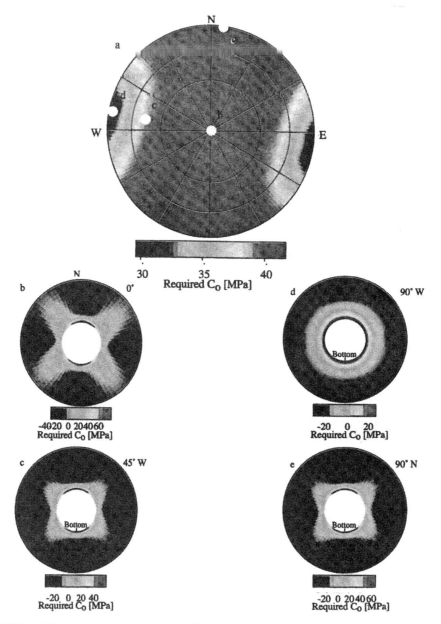

Fig. 9. (a) Stereonet showing the compressive rock strength needed to prevent wellbore failures after the completion of drilling. The white points show the orientations of the wellbore cross-sections shown in parts (b–e). The colors correspond to the compressive rock strength needed to prevent failure in different parts of the rock surrounding the hole. (b) Vertical well. (c) Well inclined 45° at an azimuth of 280°. (d) Horizontal well drilled toward S_{Hmax}. (e) Horizontal well drilled toward S_{hmin}.

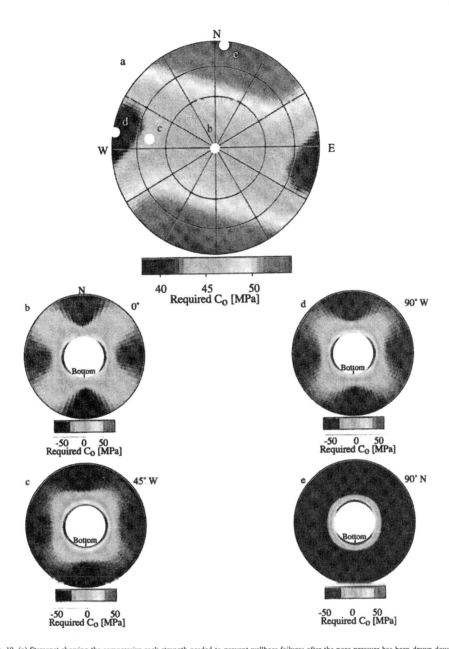

Fig. 10. (a) Stereonet showing the compressive rock strength needed to prevent wellbore failures after the pore pressure has been drawn down by 10 MPa and the horizontal stresses reduced by 6.5 MPa. The white points show the orientations of the wellbore cross-sections shown in parts (b–e). The colors correspond to the compressive rock strength needed to prevent failure in different parts of the rock surrounding the hole. (b) Vertical well. (c) Well inclined 45° at an azimuth of 280°. (d) Horizontal well drilled toward S_{Hmax}. (e) Horizontal well drilled toward S_{hmin}.

problems. Time was lost as a result of tight hole (i.e. hole pack-off, excessive overpull, or obstructions when running into the hole) and because of the extra time needed to ream the hole or circulate mud to clean cuttings from the bottom. Approximately 25% of the total downtime while drilling these holes was caused by hole-related, rather than equipment related, problems. Nearly 1200 cumulative hours of hole-related downtime, or approximately 10% of the total rig time, was lost on these holes. The time lost drilling these holes is almost equal to the time needed to drill well 5 to total depth of 3520 mTVD.

Compressive wellbore failures, if not controlled, may cause drilling problems like hole collapse, stuck pipe, pack-off, and obstructions when running into the hole, as well as problems while producing the reservoir such as sand production. Prevention of wellbore failure requires that the circumferential stresses around the wellbore be minimized. Fig. 8a shows a stereonet with the differential wellbore pressure ($\Delta P = P_m - P_p$) required to prevent compressive failures during drilling from growing beyond 90° in rock with 20 MPa uniaxial compressive strength. The stress state used in this analysis is the one found in the 10 S well at 2830 mTVD. Figs. 8b–e show cross-sections of wellbores for holes drilled in the directions shown in Fig. 8a, and assume a differential wellbore pressure of 6 MPa. The cross-sections are cut perpendicular to the axis of the wellbore and show the compressive rock strength needed to prevent the rock from failing and falling into the hole. A north arrow is shown in Fig. 8b for the vertical well, and the low side, or bottom, of the hole is shown if the well is inclined. Inclinations are shown next to each cross-section. As wells are increasingly inclined in the direction of the maximum horizontal stress (Figs. 8b–d), the compressive rock strength needed to prevent failure decreases. A well drilled in this stress field would experience problems if drilled vertically with an insufficient mud weight, and would experience fewer drilling problems as the well was inclined in the direction of the maximum horizontal stress. The most stable orientation is a wellbore drilled horizontally in the direction of the maximum horizontal stress because the nearly isotropic vertical and minimum horizontal stresses would not create large circumferential stress concentrations around the wellbore (Fig. 8d). If a well were drilled in the direction of the minimum horizontal stress it would encounter a stress field similar to that found in a well drilled vertically (Figs. 8b,e).

Well 10 S is deviated in the direction of the maximum horizontal stress, and it had the smallest amount of downtime of any well drilled in Visund. This well was also drilled with higher mud weights than were used in the previous eleven wells drilled in the field. The approach used in drilling this well is consistent with what we expect to be the best drilling strategy in Visund. The success of well 10 S compared to every well drilled before it serves to show that our strategy of increasing the mud weight and drilling in an optimal direction would have been effective in reducing drilling problems encountered in wells 1 through 9.

Fig. 9a shows the compressive rock strength needed to prevent breakouts from growing beyond 90° after the completion of drilling ($P = 0$), but before any pore pressure reduction has occurred. A wellbore drilled in the direction of the maximum horizontal stress would need compressive rock strengths as high as 30 MPa (Figs. 9a,d) in order to prevent significant wellbore failure. Because the uniaxial compressive rock strength in the reservoir sands is as low as 9–10 MPa in some sands (and as low as 15–25 MPa in others), we expect the onset of sand production problems in the weaker sands almost immediately after the initiation of production.

Fig. 10a shows the compressive rock strength needed to prevent breakouts from growing beyond 90° after the pore pressure in the reservoir has been drawn down by 10 MPa. We estimate the reduced horizontal stress by using a poroelastic model and a Poisson's ratio of 0.25, which reduces the horizontal stresses by 6.5 MPa [26]. The maximum horizontal stress is 65 MPa and the minimum horizontal stress is 46.7 MPa in this case. Even if the well is drilled in the optimal drilling direction we expect sand production to become significant as the reservoir is drawn down (Fig. 10d).

5. Conclusions

In this study we have shown that the full stress tensor can be reliably constrained using data that can be straightforwardly obtained as part of hydrocarbon field exploration and development. We demonstrate that in the Visund field, the maximum horizontal stress is significantly larger than the minimum horizontal and vertical stresses; and that our analysis is consistent with observations of recent deformation seen in studies of earthquakes and sub-seismic faults. We also show that the orientation of the stress tensor in this region is consistent both laterally and with depth.

Our analysis of wellbore stability illustrates that knowledge of stress magnitudes and orientations can be critical in designing successful exploration and production wells. By knowing the full stress tensor we are able to plan drilling strategies that minimize wellbore failure during drilling, and sand production while producing the reservoir. A significant reduction in the number of drilling problems and the amount of downtime of the rig was realized in well 10 S by modestly increasing the mud weight and by deviating the well in an optimally-stable direction.

Acknowledgements

We would like to thank Norsk Hydro for generously providing the data and financial support for this study.

Appendix A

Formation MicroImager data from taken from well 10 S are shown in Figs. A1–A4. These plots are "unwrapped" views of the wellbore with the azimuth from north shown along the top of each figure. The gray lines at an azimuth of approximately 270° show the pad 1 azimuth, and are used to orient the tool in the hole.

Appendix B

In wells 6, 8, 10 S, and 11 we observe sporadic occurrences of tensile fractures inclined with respect to the axis of the wellbore (Fig. 2). Because only well 10 S has a significant number of inclined fractures, we focus on this well. We split the inclined fractures into two sets, depending on the azimuth at which they formed (Fig. B1a). Fractures that formed at an azimuth of approximately 100° are gray, and those at approximately 300° are shown in black. There are fewer tensile fractures at 300° due to the keyseating in well 10 S, which tends to erode the tensile fractures. The azimuth of the data on this side of the hole is biased as a result, and does not show the expected average azimuth of 280°. Fig. B1b shows the inclination of the fractures as a function of depth. We measure the fracture inclination counter-clockwise from the downhole direction (insets of Fig. B1b). The fractures in this hole are typically inclined less than about 30° and more than about 150°. Fig. B1b shows that there is not a clear tendency for the fractures to be preferentially inclined in one direction. Fig. B1c shows the azimuth of the inclined

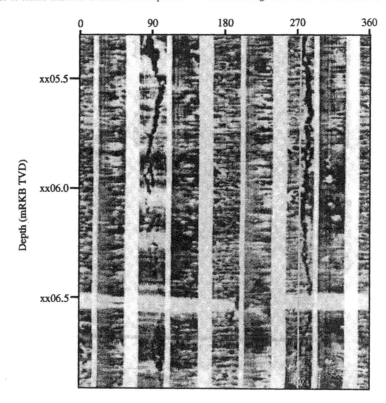

Fig. A1. FMI image of near-axial drilling-induced tensile wellbore failures. The tensile cracks appear at azimuths of approximately 90° and 270° in this plot. Note that the fractures are diametrically opposed, as we expect from theory.

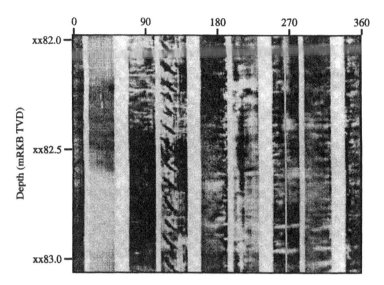

Fig. A2. FMI image of inclined drilling-induced tensile fractures. The tensile cracks appear at an azimuth of approximately 110° in this plot. No fractures are observed on the opposite side of the wellbore. This may be the result of poor image quality, reaming of the hole, or keyseating.

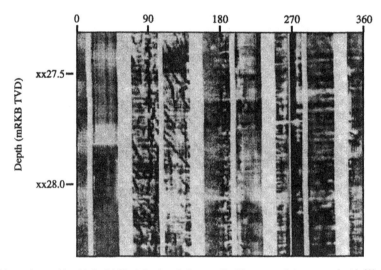

Fig. A3. FMI image of near-axial and inclined drilling-induced tensile fractures. The failures appear between approximately 80° and 110°. Some inclined fractures appear to grow into the near-axial fracture in this image. The dark band beneath the pad 1 azimuth is a keyseat, as the wellbore is deviated in this direction.

fractures plotted as a function of the inclination. The fractures generally fall into four groups. Each cluster represents fractures that have formed on the same side of the hole at similar inclinations. The fractures are rotated an average of approximately 25° from the axial orientation on both sides of the hole. The gray and black lines shown in the middle of the figure illustrate the simplified observations of the inclined tensile fractures in well 10 S. Inclined fractures formed at azimuths between 60° and 150° on one side of the hole, and between 270° and 340° on the opposite side. The inclinations of the tensile fractures range between near-axial fractures (0°) to 50° and from 115° back to near-axial fractures (180°).

We are able to reproduce the occurrence and orientation of the inclined tensile fractures shown in Fig. B1c using only minor perturbations to both the orientation of the stress tensor and the mud weight in well 10 S. Fig. B2 shows the expected minimum tangential (hoop) stress and corresponding fracture inclination (omega) as a function of the azimuth within the wellbore. The azimuth within the wellbore is defined as the angle measured clockwise from the bottom of an inclined well to a tensile or compressive failure.

In the first analysis we rotate S_{Hmax} 10° counter-clockwise (assuming a vertical principal stress), and increase ΔP by 3 MPa to a total of 9 MPa (Fig. B2a). Tensile fractures are expected to form when the minimum tangential stress falls below the tensile rock strength (zero stress) line. The gray shaded regions show the expected azimuths within the well where tensile fractures should form. Hatched areas along the omega axis show the possible inclinations for these fractures. Tensile fractures will form at any inclination between the two gray and the two black lines shown in the lower portion of the plot. These lines correspond to inclinations measured counter-clockwise from the downhole direction of approximately 155° and 25° for both the gray and black lines. The lines also correspond to azimuths within the well from about 165° to 185°, and from 345° to 5°. Because well 10 S is inclined to the west at an azimuth of 280°, fractures that form at azimuths within the well close to 0° and 180° (the low and high sides of the inclined hole) are

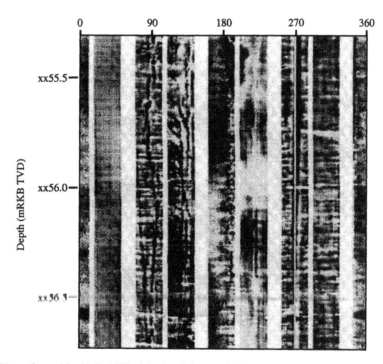

Fig. A4. FMI image of near-axial and inclined drilling-induced tensile fractures. The failures appear between approximately 90° and 120°. There may be a near-axial fracture at 120° that has inclined fractures growing toward the main axial fracture at 90°. A keyseat can be seen at 270°.

148

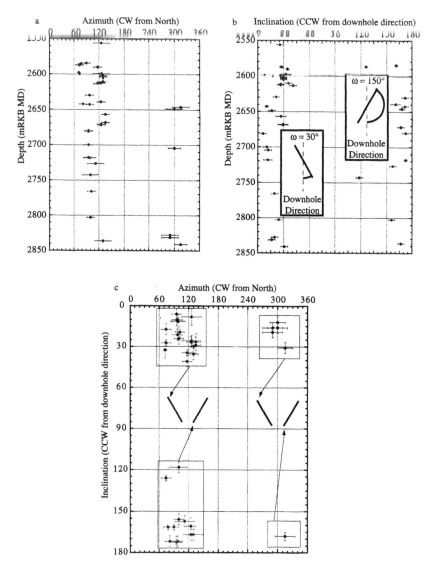

Fig. B1. (a) Azimuth of inclined fractures versus measured depth. The fractures are split into two groups depending on the side of the hole where they formed. (b) Inclination of fractures with respect to the downhole direction versus measured depth. The inclination of the fractures remains relatively small throughout the hole and has an average value of approximately 25° or 155°. (c) Azimuth of inclined fractures versus inclination measured counter-clockwise (ccw) from the downhole direction. The simplified observations are shown in the center of the plot.

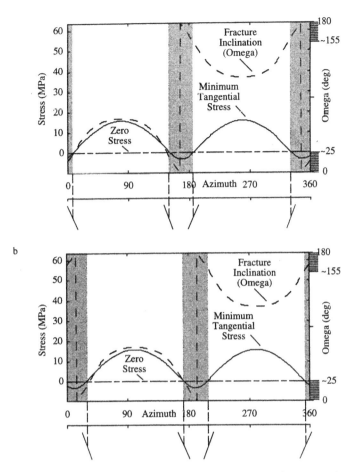

Fig. B2. (a) Minimum tangential stress (hoop stress) around the well for a stress state in which the maximum horizontal stress has been rotated 10° counter-clockwise. A slight increase in the differential fluid pressure of 3 MPa is used in this analysis. The gray and black lines below the plot show the expected range of orientations of the inclined tensile fractures. In this case, the fractures should be inclined less than 25° or more than 155° (gray and black lines respectively). (b) This plot is the same as in (a), except the maximum horizontal stress is rotated 10° clockwise.

in fact forming at azimuths measured from north of approximately 100° and 280°. Fig. B2b is the same as Fig. B2a, except the azimuth of S_{Hmax} is rotated 10° clockwise. This analysis provides similar results to those found from Fig. B2a. Thus, the stress state shown above for well 10 S predicts both the azimuth and inclination of the inclined tensile fractures, consistent with our observations from the FMI log.

The azimuth and inclination of the inclined tensile fractures can similarly be predicted by perturbing the vertical and minimum horizontal stresses. Therefore,

the overall stress tensor can be accurately described with approximately vertical and horizontal principal stresses, however the orientation of the stress tensor clearly deviates from this approximation by small amounts.

References

[1] Lindholm CD, Bungum H, Villagran M, Hicks E. Crustal stress and tectoncis in Norwegian regions determined from earthquake

focal mechanisms. In: Proceedings of the Workshop on Rock Stresses in the North Sea, Trondheim, Norway, 1995.

[2] Færseth RB, Sjøblom TS, Steel RJ, Liljedahl T, Sauar BE, Tjelland T. Sequence Stratigraphy on the Northwest European Margin. Amsterdam: Elsevier 1995.

[3] Müller B, Zoback ML, Fuchs K, Mastin L, Gregersen S, Pavoni N, Stephansson O, Ljunggren C. Regional patterns of tectonic stress in Europe. J Geophys Res 1992;97(B8):11,783–803.

[4] Gough DI, Bell J3. Stress orientations from borehole wall fractures with examples from Colorado, east Texas, and northern Canada. Can J Earth Sci 1981;19:1358–70.

[5] Zoback MD, Moos D, Mastin L, Anderson RN. Wellbore breakouts and in situ stress. J Geophys Res 1985;90:5523–30.

[6] Plumb RA, Hickman SH. Stress induced borehole elongation: a comparison between the four-arm dipmeter and the borehole televiewer in the Auburn geothermal well. J Geophys Res 1985;90:5513–21.

[7] Aadnoy BS. In situ stress direction from borehole fracture traces. J Pet Sci Eng 1990;4:143–53.

[8] Moos D, Zoback MD. Utilization of observations of well bore failure to constrain the orientation and magnitude of crustal stresses: application to continental, deep sea drilling project and ocean drilling program boreholes. J Geophys Res 1990;95(B6):9305–25.

[9] Brudy M, Zoback MD. Compressive and tensile failure of boreholes arbitrarily-inclined to principal stress axes: application to the KTB boreholes, Germany. International Journal of Rock Mechanics and Mining Sciences & Geomechanics Abstracts 1993;30(7):1035–8.

[10] Brudy M, Zoback MD. Drilling-induced tensile wall-fractures: implications for determination of in situ stress orientation and magnitude. Int. J. Rock Mech Min Sci & Geomech Abstr 1999; (in press).

[11] Zoback MD, Apel R, Baumgärtner J, Brudy M, Emmermann R, Engeser B, Fuchs K, Kessel W, Rischmüller H, Rummel F, Vernik L. Upper crustal strength inferred from stress measurements to 6 km depth in the KTB borehole. Nature 1993;365:633–5.

[12] Brudy M, Zoback MD, Fuchs K, Rummel F, Baumgaertner J. Estimation of the complete stress tensor to 8 km depth in the KTB scientific drill holes; implications for crustal strength. JGR 1997;102-8:18,453–75.

[13] Ekstrom MP, Dahan CA, Chen MY, Lloyd PM, Rossi DJ. Formation imaging with microelectrical scanning arrays. In: Transactions of the SPWLA Annual Logging Symposium, 27. 1986. p. BB1–BB21.

[14] Mardia KV. In: Statistics of directional data. New York: Academic Press, 1972. p. 357.

[15] Fejerskov M. Determination of in situ rock stresses related to petroleum activities on the Norwegian continental shelf, Doctoral Thesis: Department of Geology and Mineral Resources Engineering, Norwegian University of Science and Technology, 1996.

[16] Shamir G, Zoback MD. Stress orientation profile to 3.5 km depth near the San Andreas Fault at Cajon Pass, California. J Geophys Res 1992;97(4):5059–80.

[17] Barton CA, Zoback MD. Stress perturbations associated with active faults penetrated by boreholes; possible evidence for near-complete stress drop and a new technique for stress magnitude measurement. J Geophys Res 1994;99(5):9373–90.

[18] Paillet FL, Kim K. Character and distribution of borehole breakouts and their relationship to in situ stresses in deep Columbia River Basalts. J Geophys Res 1987;92(7):6223–34.

[19] Borgerud L, Svare E. In-situ stress field on the Norwegian Margin, 62°–67° north. In: Proceedings of the Workshop on Rock Stresses in the North Sea, Trondheim, Norway, 1995.

[20] Peska P, Zoback MD. Compressive and tensile failure of inclined well bores and determination of in situ stress and rock strength. J Geophys Res 1995;100(B7):12,791–811.

[21] Gaarenstroom L, Tromp RAJ, de Jong MC, Brandenburg AM. Overpressures in the Central North Sea: implications for trap integrity and drilling safety. In: Parker JR, editor. Geology of Northwest Europe: Proceedings of the 4th Conference, 1993. p. 1305–13.

[22] Byerlee JD. Friction of Rocks. Pure and Applied Geophysics 1978;116:615–29.

[23] Wiprut DJ, Zoback MD, Peska P, Hanssen TH. Constraining the full stress tensor from observations of drilling-induced tensile fractures and leak-off tests: application to borehole stability and sand production on the Norwegian margin. Int J Rock Mech Min Sci 1997;34(34) Paper No. 00365.

[24] Vernik L, Bruno M, Bovberg C. Empirical relations between compressive strength and porosity of siliciclastic rocks. Int J Rock Mech Min Sci & Geomech Abstr 1993;30(7):677–80.

[25] Jørgensen T, Bratli RK. In situ stress determination and evaluation at the Tampen Spur area. In: Proceedings of the Workshop on Rock Stresses in the North Sea, Trondheim, Norway, 1995.

[26] Engelder T, Fischer MP. Influence of poroelastic behavior on the magnitude of minimum horizontal stress, Sh, in overpressured parts of sedimentary basins. Geology 1994;22:949–52.

Estimating mechanical rock mass parameters relating to the Three Gorges Project permanent shiplock using an intelligent displacement back analysis method

Xia-Ting Feng[a],*, Zhiqiang Zhang[b], Qian Sheng[c]

[a] Institute of Rock and Soil Mechanics, The Chinese Academy of Sciences, Wuhan 430071, People's Republic of China
[b] College of Resources and Civil Engineering, Northeastern University, Shenyang 110006, People's Republic of China
[c] Yangtze River Science Institute, Wuhan 430070, People's Republic of China

Accepted 19 April 2000

Abstract

Establishing the mechanical rock mass parameters is one of the important tasks for the highwall stability analysis of the permanent shiplock at the Three Gorges Project in China. Existing back analysis methods are not sufficient to provide the necessary accuracy and to recognize non-linear relations. The new displacement back analysis method proposed in this paper is a combination of a neural network, an evolutionary calculation, and numerical analysis techniques. The non-linear relation involving displacement and mechanical parameters is adequately recognized by the neural network techniques. The neural networks learn using an evolutionary technique, with samples created by orthogonal design and tested with new cases given by event design. With the neural network model established, the mechanical parameters are recognized using a genetic algorithm over a large search space in the global range. The predicted displacement occurring for each excavation step from January 1998 to the end of excavation and their cumulative values for 5 later excavation steps are closely characterized by the new analysis technique. © 2000 Elsevier Science Ltd. All rights reserved.

1. Introduction

The Permanent Shiplock is one of the main parts of Three Gorges Project in China. It is one of the largest artificial navigation structures excavated in a rock mass in the world. The permanent shiplock is located on the right side of the Yangtze river. The ground level to the left of the project site increases gradually from about 60 m at the riverbank to 260 m at the Tanzi ridge. The ground then reduces from the ridge and then increases gradually to 260 m (Fig. 1). The permanent shiplock is constructed along an azimuth direction of 111° as a double channel with five stages (Fig. 1) and a total length of 161/m. The single shiplock room is excavated in granite and is 280 m in length, 34 m in width, and 5 m in depth for storing water. Both sides of the shiplock room are high and steep granite slopes. The slope angle

varies, depending on the weathering degree of the granite, e.g., it is 1 : 1 in a completely weathered zone, 1 : 0.5 in a moderately weathered zone, and 1 : 0.3 in a slightly weathered and fresh zone. The sidewall of the shiplock room is vertically cut with a height of 40–50 m. The deepest excavation is about 170 m deep.

The Section 17-17 is located at the head of the third shiplock room. Its slope is the highest of the permanent shiplock area. Also, the geological structures there are the most complex and there is a fault (F_{215}) in this area (Table 1). The design and stability analysis of this slope is critical in the construction of the permanent shiplock. For both analysis and design, proper recognition of the rock mass mechanical parameters is important.

Displacement back analysis is a common method to establish mechanical rock mass parameters [1–6]. There are three main types of displacement back analysis methods: inversive solving method, atlas method and direct (i.e., optimal) method [6]. Because of its special advantages, the optimal method is becoming more and more used in the geotechnical field. Therefore, many

*Corresponding author. Tel.: +86-27-87880913; fax: +86-27-87863386.
E-mail address: xtfeng@dell.whrsm.ac.cn (X.-T. Feng).

Reprinted from International Journal of Rock Mechanics and Mining Sciences 37 (7), 1039-1054 (2000)

152

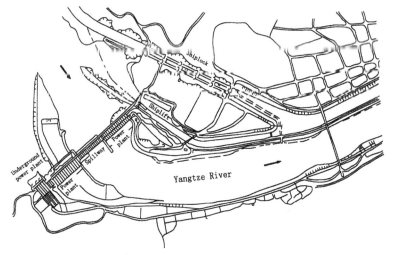

Fig. 1. Layout of the Three Gorges Project.

Table 1
Characteristics of fault F_{215}

No.	Distribution	Orientation	Characteristics
F_{215}	From south slope to the isolated rock mass, the length is larger than 300 m	N40–53°E/NW60–80°	Extension with dyke $\beta_{\mu1003}$. They are separated and joined at different places. The main fault face is plane and straight, and the width of fault is about 0.4–0.8 m. The geological structure is mainly fractured rocks with poor cementation

methods (for example, Levenber–Marquardt method [7], Gauss–Newton method [8], Bayesian method [9], Powell method, Rosenbork method) are proposed to obtain optimal values of parameters from measured displacement data. However, there are two problems associated with the optimal method. One is that the relation between the displacement and the mechanical parameters is highly non-linear, although neural networks can provide appropriate techniques to learn and represent this non-linear relation [10–13]. Another is that the search for the estimated parameter values is in a large space and is highly multi-modal. some existing techniques such as calculus-based and enumerative techniques are often insufficient to handle this problem. Thus, the two most efficient algorithms that are robust enough to search for a global minimum, given this highly non-linear back analysis in this highly multi-modal phase space (many local minima), are the genetic algorithm and the simulated annealing technique [14–17]. In this paper, the genetic algorithm [15,17]

was chosen for its biological and evolutionary appeal to find the set of unknown parameters that best matched the modeling prediction with the measured displacement data.

Therefore, a new displacement back analysis method is proposed in this paper. The paper presents the details of the techniques which incorporate a neural network and a genetic algorithm. The estimation of the mechanical rock mass parameters at the permanent shiplock at the Three Gorges Project is illustrated.

2. Genetic algorithms

A genetic algorithm operates on the Darwinian principle of "survival of the fittest". An initial population of size m is created from a random selection of the parameters in the parameter space. Each parameter set represents the individual's chromosomes. Each individual is assigned a fitness based on how well each individual's chromosomes allow it to perform in its environment. Through selection, crossover, and mutation operations, with the probabilities P_s, P_c, and P_m, respectively, the next generation is created. Fit individuals are selected for mating, whereas weak individuals die off. Mated parents create a child with a chromosome set that is some mix of the parent's chromosomes. For example, parent 1 has chromosomes HIJKL, whereas parent 2 has chromosomes ABCDE, one possible chromosome for the child is HICDE, where the position between the chromosomes I and C is the crossover point. There is a small probability that one or more of the child's chromosome will be mutated, e.g., the child ends up with chromosome HOCDE. The process of mating and child creation is continued until an entire population of size m is generated, with the hope that strong parents will create a fitter generation of children; in practice, the average fitness of the population tends to increase with each generation. The fitness of each of the children is determined, and the process of selection/crossover/mutation is repeated. Successive generations are created until very fit individuals are obtained. To assist the visualization of the genetic algorithm procedure a logic flow diagram is presented in Fig. 2.

3. Displacement back analysis based on evolutionary neural network

3.1. Determination of non-linear relation between displacement and parameters to be recognized

A relation between displacement and mechanical rock mass parameters can be described using a neural network $NN(n, h_1, \ldots, h_k, m)$ as

$$NN(n, h_1, \ldots, h_k, m) : R^n \to R^m,$$

$$D = NN(n, h_1, \ldots, h_k, m)(P), \tag{1}$$

$$P = (p_1, p_2, \ldots, p_n), \qquad D = (d_1, d_2, \ldots, d_m),$$

where the P_i are the ith parameters, $i = 1, 2, \ldots, n$. For example, $(P) = \{\sigma_x, \sigma_y, \tau_{xy}, E, \mu, c, \phi\}$, where σ_x, σ_y and τ_{xy} are components of the 2-D in situ stress, E, μ, c, and ϕ are deformation modulus, Poisson's ratio, cohesion, and friction angle, respectively d_j is the jth component of displacement, $j = 1, 2, \ldots, m$.

In this representation, based on neural network $NN(n, h_1, \ldots, h_k, m)$, $D = (d_1, d_2, \ldots, d_m)$ is the output of the neural network, and $P = (p_1, p_2, \ldots, p_n)$ is the input to the neural network, $n, h_1, \ldots, h_k, m$ are the number of input nodes, the number of nodes on

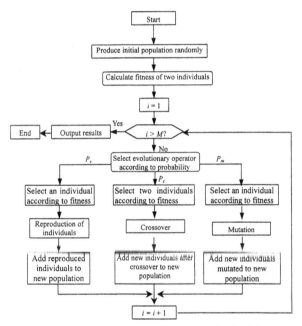

Fig. 2. Logic flow diagram for a typical genetic algorithm, M is number of evolutionary generation.

154

the first hidden layer, the number of nodes on the pth hidden layer, and the number of output nodes, respectively.

Now there is a problem to be solved: how do we obtained the model $NN(n, h_1, \ldots, h_k, m)$ as the global optimal value? A set of representative learning samples and the most reasonable network topology with its connection weights are required. Before the mechanical parameters of the rock mass are back recognized, we do not know the actual values of the parameters. So, how is a set of reasonable learning samples obtained? One way is to use the orthogonal design method to construct a set of parameters for a possible range. With the proposed parameter set, numerical analysis (e.g., the finite element method) for a given constitutive model can be conducted to calculate the displacement values at some characteristic points or lines in the rock mass. The current parameter set and its corresponding calculated displacement data could be used as a learning sample. Following this method, a set of learning samples can be generated to train the neural network.

The second problem is to establish what algorithm can be used to obtain the model $NN(n, h_1, \ldots, h_k, m)$ for the global optimal range. As we know, a popular algorithm is the Back propagation (BP) algorithm [12]. This algorithm does not over come "over-training" and local minima, i.e., the generalization ability of the network for prediction of new cases is not always improved as the progress of the learning process continues (i.e., there is not necessarily a decrease in the learning error). Thus, a new learning procedure is proposed in this paper.

In this new algorithm, the topology of the network is encoded in a binary string. The network may be fully connected as shown in Fig. 3. Not only nodes in neighboring layers but also nodes in different layers are all connected. Granularity bits define the level of the connections. The binary string will be evolved using genetic operations such as reproduction, crossover, and mutation to obtain the global optimum model. This

evolutionary process does not finish until the minimum error for the prediction of new cases is achieved.

To overcome "over-training" of the network, a set of testing samples is needed. This testing sample data set can be constructed using even-design method. The even-design of parameter combinations is developed based on the consideration that experimental points are even distributed over the investigated range. For an experiment on x factors with n value levels, it only needs an n number of experiments.

This algorithm can be described as follows:

Step 1: Determine initially the number of hidden layers, number of hidden nodes, range of granularity bits, range of connection bits, evolutionary generation of topology, learning iterations of connection weights, population size N_p, probability of single point mutation P_{jm}, probability of creep mutation P_{cm}, number of random seeds I_s. Create a binary string as shown in Fig. 3. Prepare a set of learning samples, (x_p, y_p) $(p = 1, 2, \ldots, N)$, using orthogonal design training for the coded neural network. Generate another sample set data, $(x_l, y_l)(l = 1, 2, \ldots, M)$, using the even design method to test the applicability of the model obtained.

Step 2: Generate randomly N_p groups of network topologies at the given range of granularity bits and use them as parent individuals. Generate random connection weights for every topology of the network.

Step 3: Modify the connection weights of the network $NN(n, h_1, \ldots, h_k, m)$ under the current individual topology using the delta rule. Predict the behavior of all learning samples according to the formula

$$\hat{y}_p(i) = NN(n, h_1, \ldots, h_k, m)x_p, \quad p = 1, 2, \ldots, N,$$

where i represents the ith learning cycle of the current individual.

Step 4: Use the individual $NN(n, h_1, \ldots, h_k, m)$ above to predict the output for testing samples as

$$\hat{y}_l(i) = NN(n, h_1, \ldots, h_k, m)x_l, \quad l = 1, 2, \ldots, M.$$

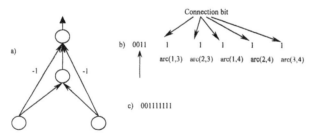

Fig. 3. Network coding scheme for a completed topology. (a) A layered neural network with the completed topology, i.e., it has full connections between nodes on different leayers. (b) The binary codes corresponding to (a). Each arc (connection) has only one binary bit. If connection is presented, its binary code is denoted 1. Otherwise, it is zero. Decimal values of the granularity are equal to bits of the connections.

Step 5: Evaluate the applicability of the current individual by calculating the prediction error for testing samples as

$$E_l(i) = |y_l - y_l|,$$

$$E(i) = \sqrt{\frac{1}{M} \sum_{l=1}^{M} E_l(t)} \qquad (2)$$

to obtain the best connection weights for the current topology.

Step 6: If all individuals have given their best predictions for the learning sample and testing samples, then go to Step 7. Otherwise, go to Step 3.

Step 7: If the evolutionary generation is over or the best individual (the network topology and its connection weights), i.e., the minimum prediction error, is achieved, then the evolutionary process ends. Otherwise, go to Step 8.

Step 8: Select randomly two individuals i_1 and i_2 from the parent generation. Carry out the crossover operation on the individuals i_1 and i_2 to generate a new chromosome. The new chromosome will pass on a single mutation and creep mutation at probability of P_{cm} and P_{jm} to generate a new network topology and granularity bits.

Step 9: Repeat Step 8 until generation of all N_p new individuals. They are used as offspring.

Step 10: Replace randomly an individual in the offspring generation using the best parent's individual.

Step 11: Regard the offspring as a parent and go to Step 3.

3.2. Objective function

To formulate the genetic evolutionary approach to this problem, as in any conventional approach to the displacement back analysis, one problem must be solved, i.e., what is the object function to be used? Here the objective function is defined as

$$F(P) = \sqrt{\frac{1}{M} \sum_{i=1}^{q} \sum_{j=1}^{w} (NN(n, h_1, \ldots, h_k, m)_{ij}(P) - u_{ij})^2}, \qquad (3)$$

where (P) is a set of parameters to be recognized, $NN(n, h_1, \ldots, h_k, xm)_{ij}(P)$ is the calculated displacement of the component j of the ith monitoring points, u_{ij} is the measured displacement, q is the number of monitoring points, w is the number of displacement component, $w \times q = m$.

3.3. Evolutionary neural network displacement back analysis algorithm

If the neural network model representing the non-linear relation between the displacement and a parameter is obtained, the model can be used to recognize parameters. The genetic algorithm is used to search for the best parameter system having the minimum error between the predicted displacements as predicted by the model and the actual measured displacement. This back analysis algorithm can be described as follows:

Step 1: Determine the range of parameters to be recognized and the genetic parameters.

Step 2: Generate randomly N_p group of parameters at their given range. Each individual represents an initial solution.

Step 3: Input a set of parameters to the model $NN(n, h_1, \ldots, h_k, m)$ obtained above to calculate the displacement values at given monitoring points.

Step 4: Use Eq. (3) to evaluate the fitness of the current individuals, i.e., the reasonability of the parameter set.

Step 5: If all individuals are evaluated, then go to Step 6. Otherwise, go to Step 3.

Step 6: If the given evolutionary generation is reached, or the best individuals (the parameter to be back recognized) are obtained, then the evolutionary process ends. Otherwise, go to Step 7.

Step 7: Select randomly two individuals i_1 and i_2 from the parent generation. Carry out the crossover operation on the individuals i_1 and i_2 to generate a new chromosome. The new chromosome will pass on a single mutation and creep mutation at probabilities of P_{cm} and P_{jm} to generate a new parameter set.

Step 8: Repeat Step 7 until all N_p new individuals are generated. They are used as offspring.

Step 9: Replace randomly an individual in the offspring generation using the best parent's individual.

Step 10: Take the offspring as parent and go to Step 3.

The entire flow for intelligent back analysis based on the evolutionary neural network is shown in Fig. 4.

4. Intelligent recognition of mechanical rock mass parameters for the Three Gorges Project permanent shiplock

4.1. Engineering geology

At the shiplock region, there is a core of the Huangling anticline and mainly plagioclase granite. It is a massive, jointed rock mass with steep faults and developed joints, of which, faults with an NNW strike, SW dip and 65–75° dip angle are the most developed. There are also faults with an NWW strike. According to the weathering degree of the granite, the rock mass is

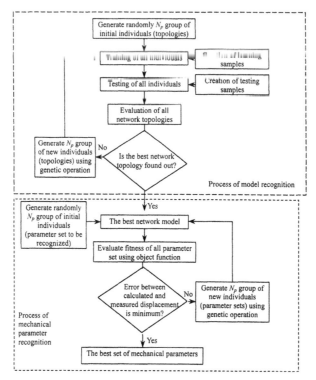

Fig. 4. Intelligent displacement back analysis method based on evolutionary neural network.

classified into completely, heavily, moderately, and slightly weathered, and fresh zones. The completely and heavily weathered zones are 15–30 m thick on the average, with 38 m thickness in one area. The average thickness of the moderately weathered zone is about 10 m. In the area around the second and the third shiplock rooms, the mountain extends obliquely from the Da ridge to the Tanzi ridge with the greatest height being 266.7 m.

There are mainly plagioclase granites isolated by schistose masses and dolerite dykes in this area. A zone of schistose mass exists on the east slope of the ridge and across the third shiplock room with a 340–360° strike, SW dip and 40–80° dip angle (mostly 50–60° dip angles), and at most a width of 12–35 m. The developed joints have the same strike as the strata. Quartz schist is the surrounding rock. The dolerite dykes have different orientations: (1) N50–70°E/NW60–85° for a length of about 200–1000 m and width of about 0.3–2 m; (2) N0–30°W/NE55–85° for a length mainly less than 100 m but sometimes larger than 500 m, and width about 1 m.

There are four weathered zones in the base rock mass, including completely, heavily, moderately, and slightly weathered zones. The thickness of the weathered zones changes at the junction of the natural slope and the river valley.

There are 77 faults investigated on the 1/1000 geological map at this shiplock area. There are 36 faults with a length of less than 50 m, 29 faults with a length of 50–100 m, and 12 faults with a length of 100 m. These faults are classified into four groups: (1) NNW group with a strike of 330–353°, dipping SW at most, with a dip angle of 63–84°; (2) NE–NEE group with a strike of 40–85°, dipping NW mainly with a dip of 60–82°; (3) NNW group with a strike of 0–35°, dipping NW mainly, with a dip angle of 65–85°; (4) NW–NWW group with strike of 270–330° dipping NE mainly.

4.2. Engineering geological zoning for Section 17-17

The section 17-17 is located at the head of the third shiplock room. The corresponding engineering

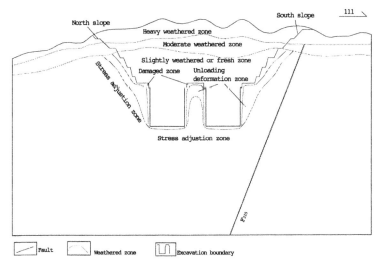

Fig. 5. Engineering geology zones of section 17-17 (scale: 1 : 500).

geological zones are shown in Fig. 5. They consist mainly of a hard and intact rock mass slightly weathered in some places. It is suitable for deep excavation of high and steep slopes. The mechanical properties of the rock mass do not vary at the slightly weathered and fresh zones. The upper portion of the strata has completely and heavily weathered thin strata and then it is a moderately weathered zone. Therefore, mechanical parameters for these natural rock mass zones, such as the slightly and non-weathered zone, the completely and heavily weathered zone, and the moderately weathered zone are considered to be recognized.

The in situ stress is disturbed due to excavation of the slope, which generates unloading and relaxation zones. Because of excavation and blasting, a damage zone forms at the boundary of the excavation. Therefore, there are two other zones related to engineering activities, unloading and damaged zones, to be considered in the back analysis of displacement.

4.3. Initial in situ stress field at Section 17-17

Some researchers have suggested regressive equations for calculating stress fields from in situ measurement in the region as follows:

$$\sigma_x = 4.3982 + 0.01168H \text{ (MPa)},$$

$$\sigma_y = 1.6628 + 0.03039H \text{ (MPa)},$$

$$\sigma_z = 4.7152 + 0.01027H \text{ (MPa)},$$

$$\tau_{xy} = -0.0472 + 0.00008H \text{ (MPa)},$$

$$\tau_{yz} = 0.7470 - 0.00046H \text{ (MPa)},$$

$$\tau_{zx} = 0.4048 + 0.00005H \text{ (MPa)}, \tag{4}$$

where H is depth below ground surface (m), and compressive stress is defined to be positive.

The following assumptions are made before back analysis:

(1) Since being significantly smaller than σ_x, σ_y, σ_z, the shear stresses τ_{xy}, τ_{yz}, τ_{zx}, are taken to be zero, i.e., the principal stresses are orientated vertically and horizontally.

(2) Eq. (4) is obtained from testing results in the slightly weathered and fresh granites. Coefficients in these equations representing tectonic stress are mainly used in this rock mass region. Heavily weathered granite zones lying above the slightly weathered and fresh granites has a lower Young's modulus. This is because the stress field here is generated by the self-weight rather than the tectonic stress. The tectonic stress field in the moderately weathered granite zone is determined from the tectonic stress values at the top boundary of the slightly weathered granite zone and the one at the bottom boundary of the strongly weathered granite zone, according to the depth below the ground surface.

(3) There are two stress factors in the formula for calculating σ_x, σ_y, σ_z. The first factor is related to the tectonic stress in the shiplock region. The second factor is related to the depth below surface, generated from the self-weight of the rock mass.

Because of the large influence of tectonic stress on the deformation of slope, the coefficients in the equations for σ_x, and σ_z are determined through back analysis. This also verifies regressive equations obtained by other researchers.

Therefore, we have the following equations for slightly weathered and fresh rock mass:

$$\sigma_x = a_x + 0.01168H,$$
$$\sigma_y = a_y + 0.03039H,$$
$$\sigma_z = 4.7152 + 0.01027H,$$
$$\tau_{xy} = 0 \quad \tau_{yz} = 0 \quad \tau_{zx} = 0. \tag{5}$$

and the following equations for the strongly weathered rock mass:

$$\sigma_x = \mu\gamma H,$$
$$\sigma_y = \gamma H,$$
$$\sigma_z = \mu H/(1-\mu),$$
$$\tau_{xy} = 0, \quad \tau_{yx} = 0, \quad \tau_{zx} = 0, \tag{6}$$

where γ is the unit weight of the rock mass $(\times 10^6 \, \text{N/m}^3)$, and a_x, a_y are coefficients to be back analyzed.

4.4. Excavation zoning of the slope and shiplock room and arrangement of monitoring points at Section 17-17

There are ten measuring points for monitoring displacement at the surface (Table 2). Eight inclined holes, and two multi-point displacement meters are installed in the section 17-17. Among these, TP/BM29GP02 was the headmost installed on December 1994.

The 12 excavation steps (shown in Fig. 6) are defined according to the actual construction of the slope and shiplock room as follows:

Step 1: Excavation from start to January 1995 to a height of 215 m above sea level.

Step 2: Excavation from February to June 1995 to a height of 200 m above sea level.

Step 3: Excavation from July to October 1995 to a height of 185 m above sea level.

Step 4: Excavation from October to November 1995 to a height of 170 m above sea level.

Step 5: Excavation from December 1995 to October 1996 to a height of 160 m above sea level.

Step 6: Excavation from November 1996 to April 1997 to a height of 155 m above sea level.

Table 2
Arrangement of displacement monitoring points for section 17-17 of the permanent shiplock

	Height above sea level (m)	Location	Date at initial monitoring
Monitoring point no.			
TP/BM10GP01	230	North slope	June 1995
TP/BM11GP01	200	North slope	January 1996
TP/BM71GP01	160	Top surface of stand wall at north slope	January 1998
TP/BM70GP01	160	Top surface at north wall of the isolated rock mass	August 1997
TP/BM97GP02	160	Top surface at south wall of the isolated rock mass	August 1997
TP/BM98GP02	160	Top surface of stand wall at south slope	August 1997
TP/BM26GP02	170	South slope	November 1996
TP/BM27GP02	200	South slope	November 1995
TP/BM28GP02	230	South slope	March 1995
TP/BM29GP02	245	South slope	December 1994

	Height above sea level (m)	Location	Depth (m)	Date for initial monitoring
Clinometer monitoring point no.				
IN04GP01	215	North slope	45.5	28 November 1995
IN08GP01	170	North slope	17.5	25 April 1996
IN13GP02	185	South slope	15.5	29 January 1996
IN06GP02	230	South slope	36.0	21 June 1995
IN11GP01	160	North slope	70.0	21 January 1998
IN01CZ32	160	North straight wall of the isolated rock mass	70.0	27 July 1997
IN02CZ32	160	South straight wall of the isolated rock mass	70.0	1 August 1997
IN16GP02	160	South slope	70.0	22 January 1998
Multi-point displacement meter monitoring point no.				
M03GP01	215	North slope	45.0	20 July 1995
M06GP02	230	South slope	38.0	23 April 1995

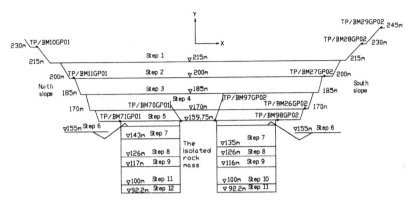

Fig. 6. Excavation steps and location of measurement points.

Table 3
Values of mechanical rock mass parameter at section 17-17

Zone		Moderately weathered zone E (GPa)	Damaged zone E (GPa)	Unloading deformation zone E (GPa)	Slightly weathered or fresh zone E (GPa)	Coefficient of geo-stress field	
						a_x	a_y
Values of parameters	1	6.0	8.0	15.0	25.0	3.0	0.8
to be back recognized	2	8.0	10.0	18.0	28.0	4.0	1.2
	3	10.0	12.0	20.0	30.0	5.0	1.6
	4	12.0	15.0	23.0	32.0	6.0	1.8
	5	15.0	18.0	25.0	35.0	7.0	2.0

Step 7: Excavation from May to December 1997 to a height of 160 m above sea level at south slope and a height of 143 m above sea level at north side.

Step 8: Excavation from January to June 1998 to a height of 126 m above sea level.

Step 9: Excavation from July to October 1998 to a height of 116 m above sea level at south slope and a height of 117 m above sea level at north side.

Step 10: Excavation from November 1998 to January 1999 to a height of 100 m above sea level at south slope and a height of 117 m above sea level at north side.

Step 11: Excavation in February 1999 to a height of 92.2 m above sea level at south slope and a height of 100 m above sea level at north side.

Step 12: Excavation in April 1999 to a height of 92.2 m above sea level at south slope and a height of 92.2 m above sea level at north side.

The interface between the heavily weathered zone and the moderately weathered zone, and the interface between the moderately weathered zone and slightly weathered or fresh zone are modified based on the borehole histograms for IN06GP02, IN07GP02, and IN04GP01 and their monitored data for displacement.

4.5. Intelligent recognition of mechanical rock mass parameters

4.5.1. Parameters to be back-analyzed in Section 17-17

The parameters to be back analyzed are the coefficients a_x, a_y for the geostress equation and the deformation moduli for four kinds of rock mass zones (moderately weathered zone, slightly weathered or fresh zone, unloading deformation zone and damaged zone (Table 3)). Data for deformation modulus, Poisson's ratio, and weight for the strongly weathered zone and the fault F_{215} were provided by the Yangtze River Water Conservancy Committee (Table 4).

The value ranges for these six parameters to be back analyzed are determined based on the field monitoring data as follows:

- Deformation modulus for moderately weathered zone E_1: 6–15 GPa.
- Deformation modulus for damaged zone E_2: 8–18 GPa.
- Deformation modulus for unloading deformation zone E_3: 15–25 GPa.
- Deformation modulus for slightly or non-weathered zone E_4: 25–35 GPa.

Table 4
Values of parameters determined from expertise and testing data

	Moderately weathered zone	Damaged zone	Unloading deformation zone	Slightly weathered or fresh zone	Heavily weathered zone	F_{215}
E (GPa)					0.3	3.0
$\tan g\phi$	1.3	1.4	1.5	1.7	0.7	1.0
c (MPa)	1.0	1.0	1.6	2.0	0.1	0.5
μ	0.24	0.22	0.23	0.23	0.22	0.35
R_t (MPa)	1.0	1.0	1.2	1.5	0.2	0.5
γ (kN/m^3)	26.5	27	27	27	25	26.5

- The geostress coefficients a_x: 3–7 and a_y: 0.8–2.0.
- The rock masses in all zones are considered to be plastic. Their cohesion c and friction angle ϕ are determined directly from engineering tests and previous monitoring data. Table 4 shows parameters determined from tests.

4.6. Establishment of neural network model

Five levels are determined within the value ranges of the parameters (as shown in Table 3). By combining five levels of parameter values, the 50 groups of parameters shown in Table 5 were produced through orthogonal design. A two-dimensional forward analysis using FLAC (3.3 version) was performed with each set of parameters. The cumulative displacement increment in the x-direction (i.e., perpendicular to the axis of the shiplock) at six monitoring points (i.e., TP/BM26GP02, TP/BM27GP02, TP/BM28GP02, TP/BM29GP02, TP/BM10GP01 and TP/BM11GP01) is calculated for this case (Table 5). These are deformation increments from excavation step 7 to step 12, i.e., for the period May 1997 to April 1999. Each group of parameters with its calculated displacements at these six monitoring points is used as a sample to train the neural network.

The Drucker–Prager model was used in the FLAC^{2D} analysis. The computational model scope was 1000 m in the x-direction, 500 m extended from the isolated rock mass, and 510 m in the y-direction. There are more than 15 000 nodes and 15 000 quadrilateral elements (Fig. 7). The excavation steps for the FLAC^{2D} analysis were determined according to the progress of the actual excavation and the time of installing monitoring points.

In the same way, the 11 groups of parameters shown in Table 6 were determined through the even design method already described. Displacements in the x-direction at six monitoring points were calculated using two-dimensional FLAC analysis (Table 6). Each sample of parameters with its calculated displacements at six monitoring points in this case is used as a sample (Table 6) to test the generation ability of the neural network model.

Before training of the neural network, the input data are normalized to an interval [0.2, 0.8] using the following formula:

$$x_i' = 0.2 + (0.8 - 0.2)\frac{x_i - x_{\min}}{x_{\max} - x_{\min}},$$

where x_i' are normalized data, x_i are original data, $i = 1, 2, \ldots, n$, n is the number of samples, and $x_{\min} = \min\{x_i\}$, $x_{\max} = \max\{x_i\}$.

Based on the search with the genetic algorithm, the best topology for the neural network model is NN(6, 34, 26, 6) which gives good outputs for both the testing and learning cases for a learning error of 0.000089. The error for testing samples is 0.015816. This neural network model can thus be used for a forward prediction of new cases by replacing the FLAC analysis.

4.7. Recognition of parameters for Section 17-17

The cumulated displacement increments from excavation step 8 to step 12 that were monitored from May 1997 to April 1999 at TP/BM26GP02, TP/BM27GP02, TP/BM28GP02, TP/BM29GP02, TP/BM10GP01 and TP/BM11GP01 are used for back analysis of the parameters.

The parameters determined by the method are shown in Table 7. With input of recognized parameters, displacements at six monitoring points are calculated by both forward analysis of FLAC and the trained neural network mode. Predictions of the cumulative displacement increments generated by excavation at all six monitoring points are compared with the actual measurements (as shown in Fig. 8 and Table 8).

Furthermore, the applicability of the parameters back analyzed from the two points is tested. One approach is to compare the prediction with measurement of cumulative displacement. Another is to compare the prediction with measurement of the response of the rock mass during the entire excavation process. For the former approach, displacements at TP/BM71GP01 installed at the top of the stand wall on the North slope and TP/BM98GP02 installed at the top of the stand wall on the South slope are investigated as new cases. With the input of parameters recognized, the cumulative displacement increment in the x-direction from excavation Step 8 to Step 12 (i.e., for the period of January

Table 5
Set of parameters determined using the orthogonal design method and the computational displacements at monitoring points[a]

Scheme	Parameters to be recognized						Computational displacements for monitoring points (mm)					
	Young's modulus E (GPa)				Coefficient of geostress field		TP/BM 10GP01	TP/BM 11GP01	TP/BM 26GP02	TP/BM 27GP02	TP/BM 28GP02	TP/BM 29GP02
	Wwz	Dz	Udz	Swf	a_x	a_y						
1	6.0	8.0	15.0	25.0	3.0	0.8	13.11	13.84	17.87	13.47	11.60	11.37
2	6.0	10.0	18.0	28.0	4.0	1.2	16.44	16.98	21.96	15.98	14.29	14.37
3	6.0	12.0	20.0	30.0	5.0	1.6	19.79	20.21	26.51	18.97	17.12	17.37
4	6.0	15.0	23.0	32.0	6.0	1.8	23.00	23.38	30.51	22.18	19.79	20.03
5	6.0	18.0	25.0	35.0	7.0	2.0	24.94	25.26	33.14	24.52	21.46	21.56
6	8.0	8.0	18.0	30.0	6.0	2.0	23.58	23.92	31.19	22.56	20.14	20.60
7	8.0	10.0	20.0	32.0	7.0	0.8	26.28	26.62	35.16	25.94	22.43	22.73
8	8.0	12.0	23.0	35.0	3.0	1.2	9.56	10.02	13.18	9.63	8.23	8.30
9	8.0	15.0	25.0	25.0	4.0	1.6	18.70	19.17	23.96	17.79	16.07	16.26
10	8.0	18.0	15.0	28.0	5.0	1.8	20.92	21.40	28.45	19.93	17.89	18.23
11	10.0	8.0	20.0	35.0	4.0	1.8	12.81	13.17	17.74	12.33	10.84	11.23
12	10.0	10.0	23.0	25.0	5.0	2.0	23.15	23.45	30.17	21.79	19.89	20.22
13	10.0	12.0	25.0	28.0	6.0	0.8	25.47	25.78	33.86	24.73	21.94	22.13
14	10.0	15.0	15.0	30.0	7.0	1.2	27.62	28.04	37.67	26.98	23.31	23.75
15	10.0	18.0	18.0	32.0	3.0	1.6	10.38	10.89	14.24	10.15	8.73	8.94
16	12.0	8.0	23.0	28.0	7.0	1.6	29.31	29.51	38.13	28.24	24.94	25.44
17	12.0	10.0	25.0	30.0	3.0	1.8	10.87	11.37	14.66	10.59	9.17	9.40
18	12.0	12.0	15.0	32.0	4.0	2.0	13.80	14.26	19.34	13.39	11.66	12.02
19	12.0	15.0	18.0	35.0	5.0	0.8	16.31	16.73	23.62	16.51	14.11	14.33
20	12.0	18.0	20.0	25.0	6.0	1.2	28.17	28.47	37.48	26.95	24.17	24.39
21	15.0	8.0	25.0	32.0	5.0	1.2	17.49	17.88	24.09	17.17	15.11	15.48
22	15.0	10.0	15.0	35.0	6.0	1.6	19.39	19.82	27.55	19.17	16.42	16.95
23	15.0	12.0	18.0	25.0	7.0	1.8	32.12	32.48	42.73	30.70	27.36	27.77
24	15.0	15.0	20.0	28.0	3.0	2.0	11.45	12.05	15.58	11.22	9.66	9.85
25	15.0	18.0	23.0	30.0	4.0	0.8	14.85	15.40	20.58	14.98	13.00	13.00
26	6.0	8.0	15.0	32.0	7.0	1.8	26.28	26.63	34.85	25.67	22.25	22.69
27	6.0	10.0	18.0	35.0	3.0	2.0	9.56	9.98	13.05	9.19	7.97	8.20
28	6.0	12.0	20.0	25.0	4.0	0.8	18.65	19.29	24.48	18.36	16.39	16.22
29	6.0	15.0	23.0	28.0	5.0	1.2	21.51	21.99	28.41	20.76	18.75	18.80
30	6.0	18.0	25.0	30.0	6.0	1.6	24.75	25.16	32.44	23.81	21.39	21.50
31	8.0	8.0	18.0	25.0	5.0	1.6	23.00	23.43	30.34	21.76	19.86	20.20
32	8.0	10.0	20.0	28.0	6.0	1.8	25.50	25.79	33.48	24.26	21.80	22.19
33	8.0	12.0	23.0	30.0	7.0	2.0	28.35	28.54	36.92	27.24	24.13	24.51
34	8.0	15.0	25.0	32.0	3.0	0.8	10.58	11.14	14.62	11.21	9.46	9.21
35	8.0	18.0	15.0	35.0	4.0	1.2	13.13	13.62	18.59	13.01	11.23	11.41
36	10.0	8.0	20.0	30.0	3.0	1.2	10.77	11.32	14.79	10.65	9.20	9.37
37	10.0	10.0	23.0	32.0	4.0	1.6	14.17	14.55	19.12	13.62	12.06	12.41
38	10.0	12.0	25.0	35.0	5.0	1.8	16.69	16.98	22.79	16.05	14.29	14.72
39	10.0	15.0	15.0	25.0	6.0	2.0	28.10	28.49	37.46	26.54	23.92	24.30
40	10.0	18.0	18.0	28.0	7.0	0.8	29.90	30.24	40.13	29.02	25.44	25.70
41	12.0	8.0	23.0	35.0	6.0	0.8	19.95	20.27	27.42	19.93	17.05	17.42
42	12.0	10.0	25.0	25.0	7.0	1.2	32.97	33.19	42.63	31.60	28.19	28.55
43	12.0	12.0	15.0	28.0	3.0	1.6	11.41	12.01	15.71	11.13	9.60	9.82
44	12.0	15.0	18.0	30.0	4.0	1.8	14.92	15.39	20.46	14.44	12.68	13.00
45	12.0	18.0	20.0	32.0	5.0	2.0	18.10	18.42	24.97	17.36	15.41	15.82
46	15.0	8.0	25.0	28.0	4.0	2.0	15.59	16.07	21.00	15.07	13.34	13.66
47	15.0	10.0	15.0	30.0	5.0	0.8	18.26	18.87	26.34	18.28	15.79	16.08
48	15.0	12.0	18.0	32.0	6.0	1.2	21.38	21.77	30.00	21.02	18.23	18.65
49	15.0	15.0	20.0	32.0	7.0	1.6	23.37	23.66	32.48	23.08	19.76	20.26
50	15.0	18.0	23.0	25.0	3.0	1.8	12.96	13.59	17.79	12.74	11.00	11.16

[a] Wwz: moderately weathered zone, Dz: damaged zone, Udz: unloading deformation zone, Swf: slightly weathered or fresh zone.

1998 to April 1999 at TP/BM71GP01 and TP/BM98GP02) is calculated. Their calculated values are about 26 and 24 mm, respectively, compared with measurements of 23.4 and 18 mm (as shown in Fig. 8 and Table 8). Error is incurred from the deformation complexity of the stand wall.

Finally, verification is carried out on the variation tendency of cumulative displacement increment in the

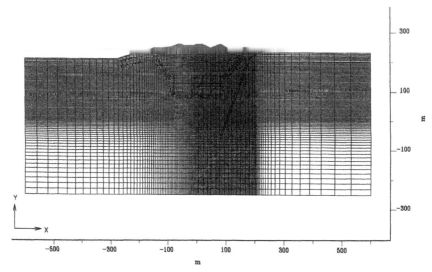

Fig. 7. Mesh for FLAC calculation for section 17-17.

Table 6
Set of parameters determined using the even design method and their computational displacement at monitoring points

Scheme	Parameters to be recognized						Computational displacements of monitoring points (mm)					
	Young's modulus E (GPa)				Coefficient of geostress field		TP/BM 10GP01	TP/BM 11GP01	TP/BM 26GP02	TP/BM 27GP02	TP/BM 28GP02	TP/BM 29GP02
	Wwz	Dz	Udz	Swf	a_x	a_y						
1	6.0	9.0	17.0	29.0	5.4	1.9	22.00	22.41	29.24	20.90	18.91	19.29
2	6.9	11.0	20.0	34.0	3.8	1.8	12.82	13.20	17.35	12.28	10.85	11.16
3	7.8	13.0	23.0	28.0	6.6	1.6	28.64	28.87	37.25	27.37	24.50	24.78
4	8.7	15.0	15.0	33.0	5.0	1.5	17.55	17.99	24.72	17.07	15.02	15.39
5	9.6	17.0	18.0	27.0	3.4	1.4	14.12	14.70	18.83	13.64	12.01	12.19
6	10.5	8.0	21.0	32.0	6.2	1.3	22.70	23.02	30.45	22.18	19.36	19.79
7	11.4	10.0	24.0	26.0	4.6	1.2	21.10	20.55	26.70	19.38	17.39	17.62
8	12.3	12.0	16.0	31.0	3.0	1.0	10.30	10.89	14.54	10.39	8.79	8.93
9	13.2	14.0	19.0	25.0	5.8	0.9	26.64	27.04	36.09	27.79	22.97	23.16
10	14.1	16.0	22.0	30.0	4.2	0.8	15.68	16.20	21.77	15.71	13.68	13.72
11	15.0	18.0	25.0	35.0	7.0	2.0	23.70	23.89	32.39	23.23	20.13	20.59

Table 7
Rock mechanical parameters determined for section 17-17

Young's modulus (GPa)				Coefficients in geo-stress field	
Slightly weathered or fresh zone	Unloading deformation zone	Damaged zone	Moderately weathered zone	a_x	a_y
32.1	18.95	9.683	7.515	4.793	1.599

x-direction at TP/BM26GP02, TP/BM27GP02, TP/BM28GP02, TP/BM29GP02, TP/BM10GP01, TP/BM11GP01, TP/BM98GP02 and TP/BM71GP01 for the progress of excavation (Fig. 9). It is obvious that the tendency is for the predictions for all eight monitoring points to be close to the corresponding measurements.

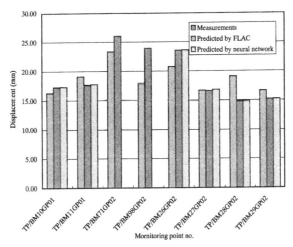

Fig. 8. Comparison of accumulative displacement increment at *x*-direction (i.e., vertical to axial of shiplock) measured actually with predictions of both FLAC and neural network model with input of mechanical parameters determined by intelligent back-analysis method.

Table 8
Comparison of measured displacement with predictions from FLAC and neural network model with input of mechanical parameters determined by the intelligent back analysis method

	Displacement at monitoring point (mm)								Average absolute error
	TP/BM 10GP01	TP/BM 11GP01	TP/BM 71GP01	TP/BM 98GP02	TP/BM 26GP02	TP/BM 27GP02	TP/BM 28GP02	TP/BM 29GP02	
Measurement	16.32	19.11	23.38	17.91	20.76	16.71	19.10	16.71	
Predictions from FLAC^{2D}	17.30	17.68	26.01	23.94	23.59	16.62	14.88	15.25	2.46
Predictions of neural network model	17.36	17.79	–	–	23.64	16.85	14.91	15.29	1.83

5. Conclusion and discussions

Due to the complexity of the rock masses and the geological conditions, it is difficult to create true physical or numerical models for back recognizing parameters. Neural networks have a strong capability for self-learning and non-linear representation. They provide attractive tools to represent the non-linear relations between mechanical parameters and deformation behavior (i.e., displacement) of a rock mass affected by excavation. The algorithm for evolving network topology and its connection weights and learning with the minimum prediction error leads to the creation of a global optimal model. The search using the genetic algorithm in the global space enables the optimal parameter set to be established.

The parameters recognized are the macro- and equivalent behavior corresponding to the integration of the rock mass, geological structures and excavation factors. They can be used not only to obtain results corresponding to the measured displacements, but also to give reasonable prediction for the macro-mechanical behavior of excavations at a later period and for similar conditions using the same forward analysis.

The following two points support the above idea:

(1) Results of the forward analysis indicate that the plastic zone on the surface of the straight wall of the isolated rock mass expands with an increase of excavation depth. By contrast, the slopes at both sides have smaller plastic zones (Fig. 10). Results analyzed from the stress contours indicate that there is a large stress release at the bottom of the model because of the higher initial stress field. This leads to the unloading on the surface of the isolated rock mass and the upper part of the straight wall as

164

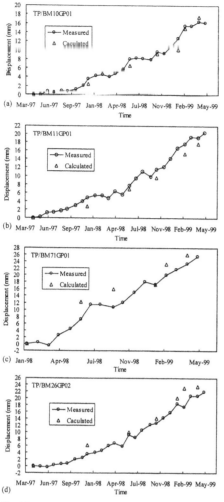

(a) Mar-97 Jun-97 Sep-97 Jan-98 Apr-98 Jul-98 Nov-98 Feb-99 May-99
Time

(b) Mar-97 Jun-97 Sep-97 Jan-98 Apr-98 Jul-98 Nov-98 Feb-99 May-99
Time

(c) Jan-98   Apr-98   Jul-98   Nov-98   Feb-99   May-99
Time

(d) Mar-97 Jun-97 Sep-97 Jan-98 Apr-98 Jul-98 Nov-98 Feb-99 May-99
Time

Fig. 9(a)–(g). Comparison of the measured variation tendency of displacement at x-direction as excavation progress with forward analysis of FLAC with input of mechanical parameters determined by intelligent back analysis method during excavation process.

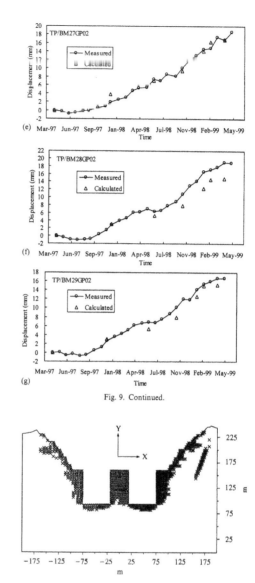

(e) Mar-97 Jun-97 Sep-97 Jan-98 Apr-98 Jul-98 Nov-98 Feb-99 May-99
Time

(f) Mar-97 Jun-97 Sep-97 Jan-98 Apr-98 Jul-98 Nov-98 Feb-99 May-99
Time

(g) Mar-97 Jun-97 Sep-97 Jan-98 Apr-98 Jul-98 Nov-98 Feb-99 May-99
Time

Fig. 9. Continued.

Fig. 10. Distribution of plastic region at final excavation step.

excavation depth increases. Therefore, its influence on the failure on the isolated rock mass is larger than that on the slope. This result is consistent with the actual measurements.

(2) On the other hand, it is seen from the calculated displacement vectors from the forward analysis that the top of the isolated rock mass generally inclines

to the north side (Fig. 11). This result is also consistent with the field measurements. For example, some results indicate that the isolated rock mass has been unloaded in three dimensions after the excavation of the shiplock room and experiencing a

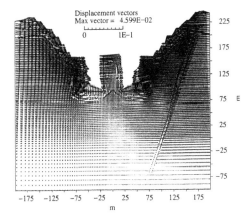

Fig. 11. Vectors of accumulative displacement increment for excavation step 8 to step 12.

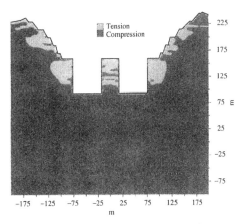

Fig. 12. Distribution of tension and compression zone after excavation of Step 12.

tensile state from a change of the compressive state. In addition, cracks experienced tensile shear, changed from compressive shear (Fig. 12). It is possible to have opening and elongation of cracks.

The results from the field investigation indicate that concrete constructed on the top surface of the isolated rock mass in the third shiplock room was cracking. The maximum opening displacement is 20 mm.

Special attention should be paid to safety of the isolated mass of section 17-17. Therefore, efficient monitoring measurements must be adopted to control excess deformation of the isolated mass in order to guarantee long-term safety of the navigation buildings. Further results will be reported on the long-term deformation behavior of the isolated mass and its control, as well as how cracks are open and elongated toward the deep rock mass, together with the effectiveness of rockbolt and anchor ropes to control these openings and elongations.

Acknowledgements

The results reported in this paper were supported by the National Natural Science Foundation, National Development Plan Key Basic Research, and Hundred Talents Program of CAS to Xia-Ting Feng. The financial support and advice of Professors Yongjia Wang, Shiwei Bai and Tingjie Li are gratefully acknowledged. Special acknowledgement should be given to Mr. Ping Xu and Mis. Xiuli Ding for their help on the use of FLAC, and to Prof. J.A. Hudson and Dr. Jian Zhao for their English improvements and recommendations.

References

[1] Kirsten HAD. Determination of rock mass elastic moduli by back analysis of deformation measurement. Proceedings of the Symposium on Exploration for Rock Engineering, Johannesburg, 1976.
[2] Gioda G, Jurina L. Numerical identification of soil structure interaction pressures. Int J Numer Anal Meth Geomech 1981;5:33–56.
[3] Sakurai S, Takeuchi K. Back analysis of measured displacements of tunnels. Rock Mech Rock Engng 1983;16:173–80.
[4] Sakurai S, Dees Wasmongkol N, Shinji M. Back analysis for determining material characteristics in cut slopes. Proceedings of the International Symposium on ECRF, vol. 11, Beijing, 1986. p. 770–76.
[5] Sakurai S. Interpretation of the results of displacement measurements in cut slopes. Proceedings of the Second International Symposium on FMGM87, vol. 2, Kobe, 1987. p. 528-40.
[6] Feng ZL, Lewis RW. Optimal estimation of in-situ ground stress from displacement measurements. Int J Numer Anal Meth Geomech 1987;11:397–408.
[7] Okabe T. Inverse of drilling-induced tensile fracture data obtained from a single inclined borehole. Int J rock Mech Min Sci 1998;35(6):747–58.
[8] William WGY. Aquifer parameter identification with optimum dimension in parameterization. Water Resour Res 1981;17(3):664–72.
[9] Cividini A. Parameter estimation of a static geotechnical model using a Bayes' approach. Int J Rock Mech Min Sci 1938;20(5):215–26.
[10] Kohonen T. An introduction to neutral computing. Neural Networks 1998;1:3–16.
[11] Grossberg S. Nonlinear neural networks: principles, mechanisms and architectures. Neural Networks 1988;1:17–61.

166

[12] Rumelhart DE, Hinton GE, Williams RJ. Learning internal representations by err propagation. In: Rumulhart DE, McClelland JL, editors. Parallel distribution processing. Cambridge, MA: MIT Press, 1986.

[13] Kartam N, Flood I, Garret JH. Artificial neural networks for civil engineers: fundamentals and applications. Virginia: American Society for Civil Engineers, 1997. p. 216.

[14] Holland JH. Adaptation in natural and artificial systems. Ann Arbor, MI: University of Michigan Press, 1975.

[15] Goldberg DE. Genetic algorithms in search, optimization and machine learning. vol. 77. Reading, MA: Addison Wesley, 1989. p. 106–22.

[16] Holland JH. Adaptation in natural and aritificial systems. Cambridge, MA: MIT Press, 1992.

[17] Goldberg DE, Deb K, Clark JH. Genetic algorithms, noise, and the sizing of populations, Complex Systems. Complex System Pub 1992;6:19,333–62.

The strength of hard-rock pillars

C.D. Martin[a],*, W.G. Maybee[b]

[a] *Department of Civil & Environmental Engineering, University of Alberta, Edmonton, Alberta, Canada T6G 2G7*
[b] *Geomechanics Research Centre, Laurentian University, Fraser Building F217, Ramsey Lake Road Sudbury, ON, Sudbury, Canada P3E 2C6*

Accepted 14 April 2000

Abstract

Observations of pillar failures in Canadian hard-rock mines indicate that the dominant mode of failure is progressive slabbing and spalling. Empirical formulas developed for the stability of hard-rock pillars suggest that the pillar strength is directly related to the pillar width-to-height ratio and that failure is seldom observed in pillars where the width-to-height ratio is greater than 2. Two-dimensional finite element analyses using conventional Hoek−Brown parameters for typical hard-rock pillars (Geological Strength Index of 40, 60 and 80) predicted rib-pillar failure envelopes that did not agree with the empirical pillar-failure envelopes. It is suggested that the conventional Hoek−Brown failure envelopes over predict the strength of hard-rock pillars because the failure process is fundamentally controlled by a cohesion-loss process in which the frictional strength component is not mobilized. Two-dimensional elastic analyses were carried out using the Hoek-Brown brittle parameters which only relies on the cohesive strength of the rock mass. The predicted pillar strength curves were generally found to be in agreement with the observed empirical failure envelopes. © 2000 Elsevier Science Ltd. All rights reserved.

1. Introduction

Pillars can be defined as the in situ rock between two or more underground openings. Hence, all underground mining methods utilize pillars, either temporary or permanent, to safely extract the ore. In coal mines rectangular pillars are often designed in regular arrays such that should a single pillar inadvertently fail the load could be transferred to adjacent pillars causing these to be overloaded. This successive overloading process can lead to an unstable progressive "domino" effect whereby large areas of the mine can collapse. This type of failure occurred in 1960 and resulted in the collapse of 900 pillars in the Coalbrook coal mine in South Africa and the loss of 437 lives. Recently, Salamon [1] summarized the extensive research into coal-pillar design that followed the Coalbrook disaster. The key element that has been used since 1960 for the successful design of coal pillars is "back-calculation", an approach that has been used extensively in geotechnical engineering [2]. This approach has led to the development of empirical

*Corresponding author. Tel.: +1-780-492-2332; fax: +1-780-492-8198.

E-mail address: dmartin@civil.ualberta.ca (C.D. Martin).

pillar strength formulas but can only be implemented by observing and documenting failed pillars.

The design of hard-rock pillars has not received the same research attention as coal pillar design. This is partly because fewer mines operate at depths sufficient to induce the stresses required to cause hard rocks to fail, and in hard-rock mining pillar and mining geometries are irregular making it difficult to establish actual loads. Nonetheless as mining depth increases the potential for the failure of hard-rock pillars also increases. This paper focuses on the strength of hard-rock pillars, particularly rib-pillars, and presents a stability criterion that can be used to establish hard-rock pillar geometries.

2. Empirical pillar strength formulas

Following the Coalbrook disaster, a major coal-pillar research program was initiated in South Africa. One of the main objectives of this research was to establish the in situ strength of coal pillars. Using the back-calculation approach Salamon and Munro [9] analyzed 125 case histories involving coal-pillar collapse and proposed that the coal-pillar strength could be

Reprinted from *International Journal of Rock Mechanics and Mining Sciences* **37 (8)**, 1239-1246 (2000)

Table 1
Summary of empirical strength formula for hard-rock pillars where the pillar width and height is in metres.

Reference	Pillar strength formulas (MPa)	σ_c (MPa)	Rock mass	No. of pillars
[3]	$133\frac{W^{0.5}}{H^{0.75}}$	230	Quartzites	28
[4]	$65\frac{W^{0.46}}{H^{0.66}}$	94	Metasediments	57
[5]	$35.4(0.778 + 0.222\frac{W}{H})$	100	Limestone	14
[6]	$0.42u_c\frac{W}{H}$		Canadian Shield	23
[7]	$74(0.778 + 0.222\frac{W}{H})$	240	Limestone/Skarn	9
[8]	$0.44\sigma_c(0.68 + 0.52\kappa)$	—	Hard rocks	178[a]

[a] Database compiled from published sources including those listed in this table.

adequately determined using the power formula

$$\sigma_p = K\frac{W^{\alpha}}{H^{\beta}}, \qquad (1)$$

where σ_p (MPa) is the pillar strength, K (MPa) is the strength of a unit volume of coal, and W and H are the pillar width and height in metres, respectively. The notion that the strength of a rock mass is to a large part controlled by the geometry of the specimen, i.e., the width-to-height ratio, has since been confirmed by extensive laboratory studies, e.g., [10]. The data from the 125 case studies gave the following values for the parameters in Eq. (1): $K = 7.176$ MPa, $\alpha = 0.46$ and $\beta = 0.66$. According to Madden [11] and Salamon [1] Eq. (1) has been applied extensively to the design of pillar layouts in South Africa since its introduction in 1967. While it is tempting to apply Eq. (1) to other pillar designs, it must be remembered that Eq. (1) was developed for room and pillar mining of horizontal coal seams and that the value of K is only typical for South African coal.

One of the earliest investigations into the design of hard-rock pillars was carried out by Hedley and Grant [3]. They analyzed 28 rib-pillars (3 crushed, 2 partially failed, and 23 stable) in massive quartzites and conglomerates in the Elliot Lake room and pillar uranium mines. These pillars were formed with the long-axis of the rib-pillar parallel to the dip direction of the quartzites. They concluded that Eq. (1) could adequately predict these hard-rock pillar failures but that the parameters needed to be modified to

$$\sigma_p = K\frac{W^{0.5}}{H^{0.75}}, \qquad (2)$$

where the units are the same as Eq. (1). The value of K in Eq. (2) was initially set as 179 MPa but later reduced to 133 MPa [12].

Since 1972 there have been several additional attempts to establish hard-rock pillar strength formulas, using the "back-calculation" approach (Table 1). Inspection of Table 1 reveals that the best-fit formulas for the observed pillar failures take the form of either a power- or linear-type equation, and that these equations

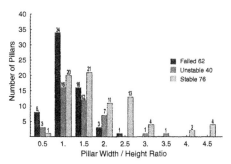

Fig. 1. Summary of the failed, transitional and stable pillars used to establish the pillar strength formulas in Table 1.

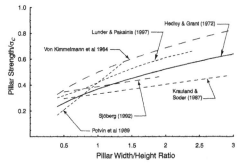

Fig. 2. Comparison of the empirical pillar strength formulas in Table 1.

have been used to predict the pillar strength for a wide range of pillar shapes and rock mass strengths as indicated by the unconfined compressive strength (94–240 MPa). Fig. 1 provides a summary of the "failed", "unstable" and "stable" classes of pillars that were used to establish the pillar strength formulas in Table 1. Fig. 2 shows the predicted pillar strength from the various formulas using a pillar height of 5 m. The pillar strengths in Fig. 2 have been normalized to the laboratory uniaxial compressive strength (σ_c). As shown in Fig. 2 the formulas predict very similar strengths, particularly for the pillar W/H ratio between 0.5 and 2.5, the range over which all of the pillar failures occur (see Fig. 1).

The stress magnitudes used to establish the pillar strengths formulas in Table 1, which were determined using either the tributary area method, or two and three dimensional elastic analyses, represent either the average maximum pillar stress or the maximum stress at the center of the pillar. In all cases, except the formula presented by Lunder and Pakalnis [8], the pillar-strength formulas ignore the effect of σ_3 and rely on a simple stress to strength ratio based on the maximum pillar stress and the uniaxial compressive strength. While Lunder and Pakalnis [8] attempt to include the effect of σ_3 through their parameter κ (see Table 1) their formula predicts similar strengths to the other formulas in Table 1 (Fig. 2). Hence the effect of σ_3 is essentially ignored by the empirical formulas to match the observed failures. This is similar to tunnel stability observations in South African mines where the stability is expressed as a simple stress to strength (σ_1/σ_c) ratio [13].

The elastic stress distribution in pillars is a function of the pillar geometry. These distributions can readily be determined through numerical computer programs. Lunder and Pakalnis [8] examined the stress distribution in hard-rock pillars in Canadian mines and proposed that the average confinement in a pillar could be expressed in terms of the ratio of σ_3/σ_1. They then expressed this ratio in terms of the pillar width and pillar height as

$$\frac{\sigma_3}{\sigma_1} = 0.46\left[\log\left(\frac{W}{H} + 0.75\right)\right]^{\frac{1.4}{(W/H)}}. \qquad (3)$$

Fig. 3 illustrates Eq. (3) and shows that the confinement in pillars increases significantly beyond a pillar W/H

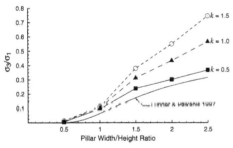

Fig. 3. The increase in confinement at the center of the pillar as a function of k, the ratio of the far-field maximum horizontal stress and vertical stress. The predicted effect of confinement using Eq. (3) is also shown.

ratio of 1. Recently, Maybee [14] showed however, that the rate of increase is a function of k, the ratio of the far-field horizontal stress σ_1 and σ_3 (Fig. 3). Fig. 3 shows that beyond a pillar W/H ratio of 1 the effect of k is significant but for pillar W/H ratios less than 1 the effect of k can be ignored.

The strength of a rock mass is usually described in terms of a constant cohesive component and a normal-stress or confinement dependent component. Hence for pillars with W/H ratios greater than 1, the strength should increase as the confining stress increases. In the next section the Hoek–Brown failure criterion is used to investigate the effect of confinement on pillar strength.

3. Pillar and rock mass strength

One of the most widely used empirical failure criteria is the Hoek–Brown criterion [13]. Since its introduction in 1980 the criterion has been modified several times, most recently in 1997 [15]. The generalized form of the criterion for jointed rock masses is defined by

$$\sigma_1 = \sigma_3 + \sigma_{ci}\left(m_b\frac{\sigma_3}{\sigma_{ci}} + s\right)^a, \qquad (4)$$

where σ_1 and σ_3 are the maximum and minimum effective stresses at failure, respectively, m_b is the value of the Hoek–Brown constant m for the rock mass, and s and a are constants which depend upon the characteristics of the rock mass, and σ_{ci} is the uniaxial compressive strength of the intact rock pieces. For hard-rock masses, Hoek and Brown [15] recommend a value of 0.5 for a. In order to use the Hoek–Brown criterion for estimating the strength and deformability of jointed rock masses, "three properties" of the rock mass have to be estimated. These are: (1) the uniaxial compressive strength σ_{ci} of the intact rock pieces in the rock mass; (2) the Hoek–Brown constant m_i for these intact rock pieces; and (3) the Geological Strength Index (GSI) for the rock mass. The GSI was introduced by Hoek and Brown [15] to provide a system for estimating the reduction in the rock mass strength for different geological conditions. The GSI can be related to either of the commonly used rock-mass classification systems, e.g., the modified rock-mass quality index Q' defined as

$$Q' = \frac{RQD}{J_n} \times \frac{J_r}{J_a}, \qquad (5)$$

where RQD is the rock quality designation, J_n is the joint set number, J_r is the joint roughness number, J_a is the joint alteration number or the rock mass rating RMR. Hoek and Brown [15] suggested that GSI can be related to Q' by

$$GSI = 9\ln Q' + 44 \qquad (6)$$

and to RMR by

$$GSI = RMR_{89} - 5, \tag{7}$$

where RMR_{89} has the Groundwater rating set to 15 and the Adjustment for Joint Orientation set to zero. The parameters m_b and s can be derived from GSI by the following:

$$m_b = m_i \exp\left(\frac{GSI - 100}{28}\right), \tag{8}$$

$$s = \exp\left(\frac{GSI - 100}{9}\right). \tag{9}$$

The Elliot Lake uranium orebody was actively mined from the early 1950s through to the mid-1990s. The shallow ($10-15°$) dipping tabular deposit was characterized by uranium bearing conglomerates separated by massive quartzite beds $3-30$ m thick [3,16]. Mining was carried out using room-and-pillar and stope-and-pillar methods with long (76 m) narrow rib-pillars formed in the dip direction. The rock mass quality of the pillars ranged from good to very good ($Q = 10-100$)(C. Pritchard, pers. comm.). Seismic surveys carried out across various pillars indicated that at the core of stable pillars the P-wave velocity averaged about 6 km/s while at the edge of the pillars the P-wave velocity dropped to 5.5 km/s [16]. Barton and Grimstad [17] proposed the following correlation between seismic compressional wave velocity and rock mass quality Q for non-porous rocks:

$$Q = 10^{\frac{V_p - 3500}{1000}}, \tag{10}$$

where V_p is the P-wave velocity in m/s. This relationship is shown in Fig. 4 along with the results from the pillar velocity surveys. These results also support the notion that the pillars were excavated in a very good-quality

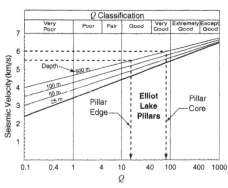

Fig. 4. Estimation of the rock mass quality from pillar seismic surveys.

Table 2
Parameters used in the Phase2 modelling to estimate the strength of the Elliot Lake pillars, assuming an elastic brittle response.

Parameter	Description/value
Rock-type	Quartzite, Conglomerate
Insitu stress	$\sigma_1 = 2\sigma_3$ and $\sigma_2 = 1.66\sigma_3$
	$\sigma_3 = 0.028$ MPa/m
Intact rock strength	$\sigma_{ci} = 230$ MPa
Geological Strength Index	$GSI = 80$
Hoek–Brown constants	$m_i = 22$
	$m_b = 10.7$
	$s = 0.108$
	$m_r = 1$
	$s_r = 0.001$

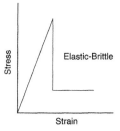

Fig. 5. Illustration of the suggested post-failure characteristic for a very good quality hard-rock mass.

rock mass. These descriptions and measurements indicate that the rock mass strength can be characterized by a GSI value of 80 (very good category), which was used to establish the parameters required for the Hoek–Brown failure criterion (Table 2).

Hoek and Brown [15] suggested that for good-quality rock masses the progressive spalling and slabbing nature of the failure process should be treated in an elastic–brittle manner as shown in Fig. 5. This failure process involves significant dilation, and provided there is support to the broken pieces, it is assumed that the failed rock behaves as a cohesionless frictional material. The post-peak Hoek–Brown parameters (m_r, s_r) provided in the Table 2, reflect this assumption.

The original Elliot Lake pillar-database used by Hedley and Grant [3] to establish Eq. (2) is shown in Fig. 6. Pritchard and Hedley [18] described the progressive spalling and slabbing nature of the failure process of the pillars at these mines and highlighted the difficulty of determining when a pillar had failed. Hedley and Grant [3] classified their pillars as "crushed", "partial failure" and "stable" to reflect the progressive nature of hard-rock pillar failures, and used elastic analyses to determine the loads on the pillars. An example of a "crushed" pillar is given in Fig. 7. Hence, the elastic loads for the "partial failure" or "crushed"

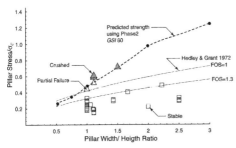

Fig. 6. Comparison of the predicted rib-pillar strength using Eq. (2) and the observed rib-pillar behavior in the Elliot Lake uranium mines. Data from Hedley and Grant [3].

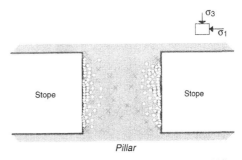

Fig. 8. Example of the output from Phase2 showing complete yielding of a pillar with a W/H ratio of 1. The × represents shear failure and the ° represents tensile failure.

Fig. 7. Photo of a "crushed" pillar in massive quartzite, courtesy of Mr. C. Pritchard.

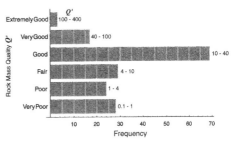

Fig. 9. Distribution of the rock mass quality Q' in Canadian hard-rock mines, data from Potvin et al. [6].

pillars shown in Fig. 6 are not the actual loads because once failure initiates the loads are redistributed internally within the pillar and/or to adjacent pillars. Numerical analyses were carried out to determine if the pillar strengths predicted using the Hoek–Brown failure criterion with the parameters in Table 2 were similar to the strength predicted by Eq. (2) in Fig. 6.

The numerical analyses were carried out using the two-dimensional finite element program Phase2.[1] This program has the capability to incorporate the elastic–brittle post peak response using the Hoek–Brown parameters. A pillar was considered to have failed when the elements across the pillar had yielded (Fig. 8). This was considered similar to the "crushed" conditions in Fig. 6. However, in order to compare the stress to strength ratio from the numerical program to the data in Fig. 6, the elastic stresses had to be determined for the

elastic–brittle failure conditions. These results are presented in Fig. 6 and agree with the failed observations for the pillar W/H ratio from 0.5 to 1.5. Beyond a pillar W/H ratio of 1.5 the elastic–brittle response appears to over predict the pillar strength compared to Eq. (2).

4. Pillar stability criterion and *GSI*

The Phase2 modelling for the Elliot Lake case study used the **GSI** for a very good-quality rock mass. In the hard-rock mines of the Canadian Shield experience suggests that the *GSI* will vary significantly. Potvin et al. [6] collected 177 case studies from Canadian hard rock mines and found that Q' ranged from 0.1 to 120 (Fig. 9). The *GSI* values, using Eq. (6) and Potvin et al. [6] database are shown in Fig. 10. The *GSI* values range from 31 (fair) to 87 (very good) with a mean value of 67, suggesting that the *GSI* values of 40 (fair), 60 (good) and 80 (very good) would represent the range of typical strength conditions for Canadian hard-rock mines. The

---

[1] Available from RocScience Inc. 31 Balsam Ave., Toronto, Ontario, Canada M4E 3B5; Internet:www.rocscience.com

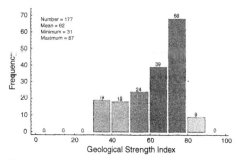

Fig. 10. Distribution of the Geological Strength Index (*GSI*) in Canadian hard-rock mines.

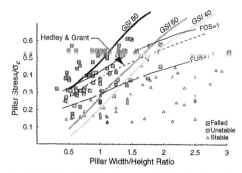

Fig. 11. The Pillar Stability Graph developed by Lunder and Pakalnis [8] compared to the pillar strength equation proposed by Hedley and Grant [3] and the Phase2 modeling results indicated by *GSI* values of 40, 60 and 80.

Table 3
GSI and Hoek−Brown strength parameters used in the Phase2 modeling.

	GSI		
	80	60	40
σ_{ci} (MPa)	230	230	230
m_i	22	22	22
m_b	10.77	5.27	2.58
s	0.108	0.0117	0.0013
Residual			
m_r	1	1	1
s_r	0.001	0.001	0.001

corresponding Hoek−Brown parameters for these strength conditions, using Eq. (8) and (9), are given in Table 3. Experience suggests that $m_i = 22$ and $\sigma_{ci} = 230$ MPa are typical values for the hard-rocks found in many Canadian underground mines.

The most extensive database of hard-rock pillar failures was compiled by Lunder and Pakalnis [8] who analyzed 178 case histories from hard-rock mines, 98 of which were located in the Canadian Shield (Fig. 11). Many of these pillars were rib or sill pillars from steeply dipping ore bodies. Lunder and Pakalnis proposed that the pillar strength could be adequately expressed by two factor of safety (*FOS*) lines. Pillars with a *FOS* < 1 fail while those with a *FOS* > 1.4 are stable. The region between 1 < *FOS* < 1.4 is referred to as unstable and pillars in this region are prone to spalling and slabbing but have not completely failed, similar to the "partial failure" used by Hedley and Grant [3]. It should be noted that of the pillars investigated 76 were classed as stable; 62 were classed as failed; and 40 were classed as unstable (see Fig. 1). For comparison purposes, the Hedley and Grant pillar strength equation is also shown in Fig. 11.

Phase2 numerical analyses were carried out using the same procedure discussed in the Elliot Lake case study

to develop pillar stability lines based on the rock mass strength. The Hoek−Brown parameters for *GSI* 40, 60 and 80 given in Table 3 were considered to be representative of the variation of rock mass strength found in Canadian hard-rock mines. The results from this Phase2 modelling are also shown in Fig. 11. While the Hedley and Grant pillar strength equation is in good agreement with the stability lines proposed by Lunder and Pakalnis [8], Phase2 modelling results using the Hoek−Brown failure criterion for *GSI* 40, 60 and 80 do not follow the trends of the stability lines proposed by Hedley and Grant [3] or Lunder and Pakalnis [8]. In particular, the slope of the stability lines predicted by the empirical formulas for $W/H < 1$ tend to be somewhat flat and for $W/H = 0.5$ the empirical formulas indicate that the pillar strength ranges from 0.2 to $0.35\sigma_c$. The slope of *GSI* lines, however, tend to be steeper and for $W/H = 0.5$ indicate a pillar strength ranging from 0.05 to $0.33\sigma_c$. The generally steeper slope of the *GSI* lines reflect the effect of increasing confinement, e.g., see Fig. 3, on the rock mass strength while the observed failure lines appear to be less dependent on confinement. This noticeable trend would suggest that the confining stress-dependent frictional strength component contributes less to the overall pillar strength than the conventional Hoek−Brown failure envelop predicts. However, only with additional case studies can the overall importance of the noted difference in these trend lines be fully assessed.

5. Pillar failure and cohesion loss

The failure of hard-rock pillars involves spalling, i.e., slabbing and fracturing, which leads to the progressive deterioration of the pillar strength. Pritchard and

Hedley [18] noted that in the early (pre-peak strength) stages of pillar failure at Elliot Lake stress-induced spalling, dominated the failure process while in the latter stages (post-peak strength), after spalling had created the typical hour-glass shape, slip along structural features such as bedding planes and joints played a more significant role in the failure process. These observations are in keeping with the laboratory findings of Hudson et al. [10] and Martin and Chandler [19], who demonstrated that the development of the shear failure plane occurs after the peak strength is reached. Martin [20] proposed that this pre-peak stress-induced spalling/fracturing-type failure is fundamentally a cohesion-loss process and Martin et al. [21] suggested that in order to capture this process in numerical models the Hoek–Brown parameters needed to be modified. They proposed that this spalling- or brittle-type failure could adequately be captured using elastic models and the following Hoek–Brown brittle parameters:

$$m_b = 0 \quad \text{and} \quad s = 0.11.$$

The fundamental assumption in using these brittle parameters is that the failure process is dominated by cohesion loss associated with rock mass fracturing, and that the confining stress-dependent frictional strength component can be ignored when considering near surface failure processes. Hence, it is not applicable to conditions where the frictional strength component can be mobilized and dominates the behavior of the rock mass.

A series of elastic numerical analyses were carried out using the boundary element program Examine2D[1] and the Hoek–Brown brittle parameters to evaluate pillar stability over the range of pillar W/H ratios from 0.5 to 3. The analyses were carried out using a constant k ratio of 1.5 and the results are presented as solid lines in Fig. 12 for both a factor of safety (FOS) equal to 1 and 1.4. A pillar was considered to have failed when the core of the pillar had a $FOS = 1$. A similar approach was used to establish when the pillar reached a $FOS = 1.4$. Fig. 12 shows a better agreement between the FOS lines predicted using the Hoek–Brown brittle parameters, and the FOS lines empirically developed by Lunder and Pakalnis [8] and Hedley and Grant [3] for pillar W/H ratio less than 1.5. More importantly in contrast to the failure envelopes developed using the Geological Strength Index and the traditional Hoek–Brown parameters (see Fig. 11), the slope of the failure envelope using the Hoek–Brown brittle parameters is in closer agreement with the empirical failure envelopes, particularly for pillar W/H ratios from 0.5 to 1.5. Also note that for the pillar W/H ratio less than 1, the strength is essentially constant, reflecting the low confinement for these slender pillars.

Fig. 13 shows the comparison of all the empirical formulas listed in Table 1 with the numerical results

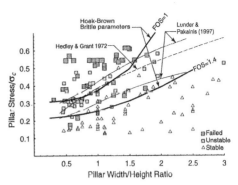

Fig. 12. Comparison of the pillar stability graph and the Phase2 modeling results using the Hoek–Brown brittle parameters.

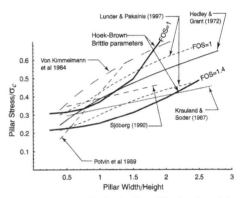

Fig. 13. Comparison of hard-rock pillar stability formulas and the elastic two-dimensional modeling results using the Hoek–Brown brittle parameters.

using the Hoek–Brown brittle parameters. There is general agreement with the empirical formulas and the predicted pillar strength for $W/H < 1.5$ where the majority of the pillar failure occur. Beyond a pillar $W/H > 2$ the Hoek–Brown brittle parameters suggest that the strength increases significantly which is in contrast to the empirical formulas. Because at pillar $W/H > 2$ the confinement at the core of the pillar is increasing significantly the use of Hoek–Brown brittle parameters will be less appropriate. It should be noted that the pillar-failure database shows that there are only a few pillar failures for pillar $W/H > 2$ (Fig. 1). Hence, the empirical pillar strength formulas should be limited to pillar $W/H < 2$.

6. Conclusions

Observations [8] of 178 pillar case studies in hard-rock mines indicate that the nearly all failures occur when the pillar W/H ratio is less than 2.5 and that the dominant mode of failure is progressive slabbing and spalling which eventually leads to an hour-glass shape. The Lunder and Pakalnis pillar stability graph documents over 98 rib-pillar observations in Canadian hard-rock mines and is based on the calculated average maximum stress in the pillar, the uniaxial strength of the intact rock and the pillar W/H ratio. Their findings are in keeping with other pillar formulas developed for hard-rock pillars and suggest that for slender pillars ($W/H < 1$) failure initiates at approximately $\frac{1}{3}$ of the laboratory uniaxial compressive strength. For squat pillars ($W/H > 1.5$) these empirical formulas only predict an increase in pillar strength to approximately $\frac{1}{2}-\frac{2}{3}$ of the laboratory uniaxial compressive strength, despite asignificant increase in confinement at the core of the pillar.

The conventional Hoek–Brown failure envelope is based on a cohesive strength component and a confining stress-dependent frictional component. In a confined state, such as pillar W/H ratios greater than 1, the frictional strength component increases significantly. Two-dimensional finite element analyses using conventional Hoek–Brown parameters for typical hard-rock rib-pillars (GSI of 40, 60 and 80) predicted pillar-failure envelopes that did not agree with the observed empirical failure envelopes. It is suggested that the conventional Hoek–Brown failure envelopes over predict the strength of the hard-rock pillars because the failure process is fundamentally controlled by a cohesion-loss process and for practical proposes the frictional strength component can be ignored at pillar width-to-height ratios less than 1.5.

Two-dimensional elastic analyses were carried out using the Hoek–Brown brittle parameters ($m_b = 0, s = 0.11$). The predicted rib-pillar strength curves were generally found to be in agreement with the observed empirical failure envelopes. It should be noted however, that the Hoek–Brown brittle parameters are not applicable to conditions where the frictional component of the rock-mass strength can be mobilized and dominates the behavior of the rock mass.

Acknowledgements

This work was supported by the Natural Sciences and Engineering Research Council of Canada (NSERC) and through collaboration with the hard-rock mining industry in Northern Ontario.

References

[1] Salamon M. Strength of coal pillars from back-calculation. In: Amadei B, Kranz RL, Scott GA, Smeallie PH, editors. Proceedings of 37th US Rock Mechanics Symposium, Vail, volume 1. Rotterdam: A.A. Balkema, 1999. p. 29–36.

[2] Sakurai S. Back analysis in rock engineering. In: Hudson JA, editor. Comprehensive rock engineering–excavation, support and monitoring, vol 4. Oxford: Pergamon Press, 1993. p. 543–69.

[3] Hedley DGF, Grant F. Stope-and-pillar design for the Elliot Lake Uranium Mines. Bull Can Inst Min Metall 1972;65:37–44.

[4] Von Kimmelmann MR, Hyde B, Madgwick RJ. The use of computer applications at BCL Limited in planning pillar extraction and design of mining layouts. In: Brown ET, Hudson JA, editors. Proceedings of ISRM Symposium: Design and Performance of Underground Excavations. London: British Geotechnical Society, 1984. p. 53–63.

[5] Krauland N, Soder PE. Determining pillar strength from pillar failure observations. Eng Min J. 1987;8:34–40.

[6] Potvin Y, Hudyma MR, Miller HDS. Design guidelines for open stope support. Bull Can Min Metall 1989;82:53–62.

[7] Sjöberg J. Failure modes and pillar behaviour in the Zinkgruvan mine. In: Tillerson JA, Wawersik WR, editors. Proceedings of 33rd U.S. Rock Mechanics Symposium, Sante Fe. Rotterdam: A.A. Balkema, 1992. p. 491–500.

[8] Lunder PJ, Pakalnis R. Determination of the strength of hard-rock mine pillars. Bull Can Inst Min Metall 1997;90:51–5.

[9] Salamon MDG, Munro AH. A study of the strength of coal pillars. J S Afr Inst Min Metall 1967;68:55–67.

[10] Hudson JA, Brown ET, Fairhurst C. Shape of the complete stress-strain curve for rock. In: Cording E, editor. Proceedings of 13th US Symposium on Rock Mechanics, Urbana. New York: American Society of Civil Engineers, 1972. p. 773–95.

[11] Madden BJ. A re-assement of coal-pillar design. J S Afr Inst Min Metall 1991;91:27–36.

[12] Hedley DGF, Roxburgh JW, Muppalaneni SN. A case history of rockbursts at Elliot Lake. In: Proceedings of 2nd International Conference on Stability in Underground Mining, Lexington. New York: American Institute of Mining, Metallurgical and Petroleum Engineers, Inc., 1984. p. 210–34.

[13] Hoek E, Brown ET. Underground excavations in rock. London: The Institution of Mining and Metallurgy, 1980.

[14] Maybee WG. Pillar design in hard brittle rocks. Master's thesis, School of Engineering, Laurentian University, Sudbury, ON, Canada, 1999.

[15] Hoek E, Brown ET. Practical estimates of rock mass strength. Int J Rock Mech Min Sci. 1997;34:1165–86.

[16] Coates DF, Gyenge M. Incremental design in rock mechanics. Monograph, vol. 880. Ottawa: Canadian Government Publishing Centre, 1981.

[17] Barton N, Grimstad E. The Q-System following twenty years of application in NWT support selection. Felsbau, 1994;12:428–36.

[18] Pritchard CJ, Hedley DGF. Progressive pillar failure and rockbursting at Denison Mine. In: Young RP, editor. Proceedings of 3rd International Symposium on Rockbursts and Seismicity in Mines, Kingston. Rotterdam: A.A. Balkema, 1993. p. 111–6.

[19] Martin CD, Chandler NA. The progressive fracture of Lac du Bonnet granite. Int J Rock Mech Min Sci Geomech Abstr 1994;31:643–59.

[20] Martin CD. Seventeenth Canadian Geotechnical Colloquium: the effect of cohesion loss and stress path on brittle rock strength. Can Geotech J 1997;34:698–725.

[21] Martin CD, Kaiser PK, McCreath DR. Hoek-Brown parameters for predicting the depth of brittle failure around tunnels. Can Geotech J 1999;36:136–51.

Papers from

Tunnelling and Underground Space Technology

Tunnels in Opalinus Clayshale — A Review of Case Histories and New Developments

H. H. Einstein

Abstract —*Numerous tunnels have been built and new tunnels are planned in the Swiss Jura Mountains, and most of them intersect Opalinus Clayshale. This paper reviews several tunnel case histories which show that Opalinus Clayshale can be quite problematic by swelling during construction and also during operation, if the invert cover is not strong enough and if water can penetrate into the shale. Modern testing methods and a newly developed behavioral model based on these tests are then presented. They clearly show that Opalinus Clayshale when unloaded as can occur around the tunnel perimeter swells which in turn can induce creep and even failure. Considering this behavior in design and construction will reduce swelling and creep. Modern, TBM-based tunnel construction which utilizes prefabricated liner elements satisfies these requirements. In addition, the paper describes project management procedures that allow one to include the risks associated with swelling when assessing and awarding bids.* © 2000 Published by Elsevier Science Ltd. All rights reserved.

Sommaire —*Un grand nombre de tunnels ont été et sont en train d'être construits dans le Jura Suisse. La plupart de ces tunnels traversent la marne aalénienne. Dans cet article on présente plusieurs cas de tunnels qui montrent que la marne aalénienne peut être problématique et gonfler pendant la construction et l'exploitation, si la couverture du radier n'est pas suffisamment épaisse et si l'eau peut infiltrer le rocher. Des méthodes expérimentales modernes et un modéle de comportement, qui a été développé récemment en se basant sur ces expériences, sont ensuite présentés. Ils montrent clairement que la marne aalénienne gonfle, quand elle est déchargée comme cela peut se produire autour d'un tunnel, ce qui à son tour produit fluage et même rupture. Si l'on considère ce comportement dans le dimensionnement et la construction de tunnels, on peut réduire l'intensité du gonflement et du fluage. Les méthodes modernes de construction de tunnels par tunneliers et éléments de soutènement préfabriqués satisfont à ces exigences. En outre des procédés de gestion de projet sont décrits qui permettent de considérer les risques liés au goflement dans l'évaluation et l'adjudication des offres.*

Preamble: *A substantially abbreviated version of this paper was presented at the Session on Retrospective Case Histories at the 37th U.S. Rock Mechanics Symposium in Vail, June 1999 and is included in the Proceedings of that Symposium. Since the old and the new cases were only summarized in that paper and since the engineering community can benefit from a more detailed description, the following more encompassing paper has been written.*

1. Introduction

A number of railroad and highway tunnels have been built through the Swiss Jura mountains (Fig. 1). Some of these railroad tunnels are over 100 years old, while new highway and railroad tunnels are being built at present or will be built in the near future. This history and present state of tunnelling is ideal for making comparisons between past and present design and construction.

The Jura mountains consist of two distinct parts, the "Chain Jura" and the "Plateau Jura" (Fig. 2). Figure 3, a longitudinal geologic cross section along the Hauenstein Base Tunnel, shows the clear distinction between the "Chain Jura" and the "Plateau Jura" going from south to north. Several of the geologic formations associated with the Jura

mountains cause significant problems in tunneling mostly through swelling and softening or a combination of both during tunnel construction and tunnel lifetime. Best known for causing problems are the clay sulfate rocks in the Keuper and the Opalinus Clayshale of the lower Dogger (lower Aalénien). This paper deals with the latter. An extensive collection of case studies on the clay-sulfate rocks in the Keuper will be published by the ISRM Commission on Swelling Rocks in early 2000.

Most of the tunnels which were built and are being built in the Jura Mountains intersect the Opalinus Clayshale. This provides a number of case studies in which the past performance of tunnels was studied in detail and over long periods of time. On the other hand, the new tunnels are being built with new construction methods, notably Tunnel Boring Machines (TBMs), and new testing and modelling methods are applied in designing them, a fact that involves extensive research, for instance, at the Ecole Polytechnique Fédérale - Lausanne (Switzerland) (see e.g. EPFL 1990) and two Ph.D. theses at M.I.T. (Bellwald 1991, Aristorenas 1992). Finally, the Opalinus Clayshale of the Plateau Jura is considered as a possible host formation for the Swiss high-level radioactive waste repository (e.g. see NAGRA 1988; Einstein et al. 1996, Bobet et al. 1999).

2. Opalinus Clayshale

The Opalinus Clayshale, as indicated in Figure 3, is a part of the Lower Dogger (Lower Aalénien) formation of Jurassic age. It was deposited in a shallow sea with limited

Present address: Dr. Herbert H. Einstein, Massachusetts Institute of Technology, Room 1-342, Cambridge, Massachusetts 02139, U.S.A.

Reprinted from *Tunnelling and Underground Space Technology* **15 (1)**, 13-29 (2000)

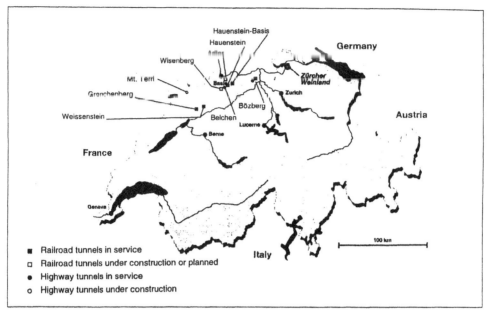

Figure 1. Tunnels through the Jura mountains in Switzerland.

circulation (see Nüesch, 1991). Layer thicknesses, if undisturbed, are 100 m or less, but folding and faulting have led to greater thicknesses in some locations. The mineralogy, as listed in Table 1, indicates medium-activity clay; and the grain size distribution in Figure 4 indicates, on first sight, an equal distribution of silt- and clay-size particles. However, Nüesch (1991) found that a substantial portion of the grain-size fraction greater than 2 microns are actually clay aggregates which break down when sheared. Both the

mineralogy and grain size distribution can vary with the facies. Opalinus Clayshale, looked at from a soil mechanics point of view, and as will be shown in Section 4, has the characteristics of an overconsolidated clay.

In addition to being encountered in many tunnels, Opalinus Clayshale also outcrops in many areas and often involves slope failures. On the other hand, Opalinus Clayshale is a resource for many clay-based products, particularly bricks. All this led to numerous laboratory and some field experi-

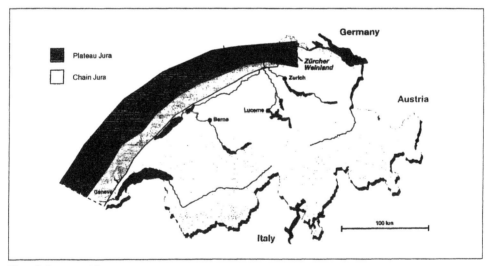

Figure 2. Jura mountains.

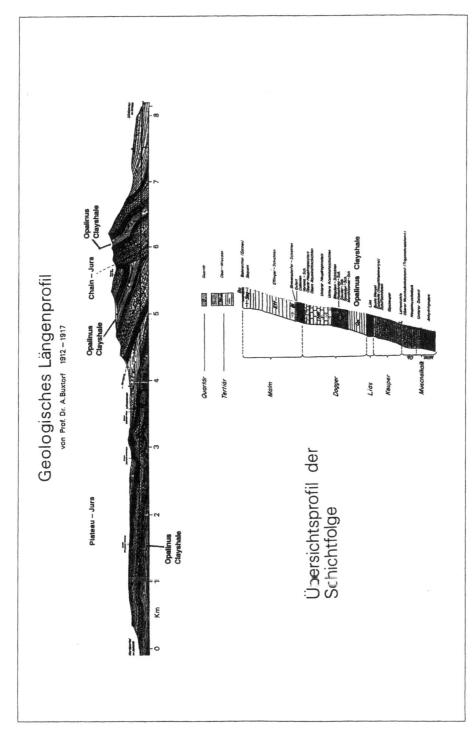

Figure 3. Longitudinal cross-section—Hauenstein Base Tunnel. (from Ettertin 1988)

ments, including the already mentioned Ph.D. thesis by N.esch (1991). Most of this experimental research has been reviewed and summarized in Einstein et al. 1995. Typical characteristics from these earlier studies are:

Table 1. Mineralogy of Opalinus Clayshale.

Mineral	
Mixed layer clays	20%
Illite	15%
Chlorite	10%
Kaolinite	10%
Quartz	30%
Calcite	10%
Feldspar, Pyrite	5%

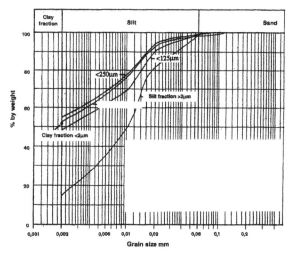

Figure 4. Grain size distribution — Opalinus Clayshale (from Nüesch 1991).

From Laboratory Tests

Unit weight	γ = 24.5 ± 0.2 kN/m³
Unconfined compressive strength	σ_c = 3 to 24 MPa*
Unconfined tensile strength	σ_t = 1 MPa
Modulus of elasticity	E = 1000–20,000 MPa*
Poisson's ratio	ν = 0.35
Porosity	n = 9–11%

* In general, higher values are for tests with the load parallel to the bedding planes, and lower values are for tests with the load perpendicular to the bedding planes.

A number of triaxial tests were also run in the past, but either on dry material or without pore pressure measurements:

- The Mohr envelopes are nonlinear. Consequently friction angles ranging between 40° and 28° and cohesion ranging between 0 and 1 MPa have been measured for confining pressures between 0 and 2.5 MPa (EPFL 1989, 1990). Nüesch (1991), who ran experiments mostly on air-dried specimens but at the much higher confining pressures between 50 and 400 MPa to model geologic processes, observed friction angles and cohesion values ranging from $\phi \cong 13.5°$, $c \cong 30$ MPa (σ_1 perpendicular to bedding) to $\phi \cong 14.5°$, $c \cong 34$ MPa (σ_1 parallel to bedding).

- Swell pressures, using the ISRM (1989) procedure were also measured and values between 1 and 2.2 MPa were observed; swell deformation measured according to ISRM (1989) ranged between 5% and 10%.

- Finally, triaxial creep tests showed secondary creep strain over a five-week period of about 0.2% (Descoeudres and Kharchafi 1993).

Field experiments in the Pilot Tunnel for the Mt. Terri Highway Tunnel (see Fig. 1) also produced a number of interesting results:

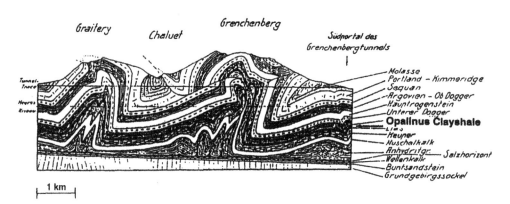

Figure 5. Geologic profile — Grenchenberg tunnel (from Buxdorf and Troesch in "Schlussbericht," 1916).

Table 2. Grenchenberg Tunnel: typical cross-sections (profiles) (from Steiner and Metzger 1988).

Length	%	Profile Type	Abutment Thickness	Arch Thickness	Invert Thickness
m			c m	c m	c m
3327	38.8	I	35	35	-
32	0.4	I a	45	35	-
56	0.7	I b	35	43	-
1489	17.4	II	45	45	-
20	0.2	Iia	60	45	-
40	0.5	Iib	45	60	-
1973	23.0	Iic	45	45	30
260	3.0	Iia/c	60	45	30
136	1.6	Iib/c	45	60	30
268	3.2	III a normal	60	60	50
784	9.2	III a/cl	60	60	50
80	0.9	III a/cl reinf	60	60	50
88	1.0	IIIb	60	60	50
12	0.1	Special	60	60	60
8566	100				

- Moduli of elasticity obtained with plate load tests produced E between 400 and 1,000 MPa while those backfigured from observed tunnel convergence (EPFL 1989) are in the range of 5,000 to 15,000 MPa. The plate load tests appear to be affected by construction disturbance while the backfigured values are in the range of the laboratory values.

What one can conclude from all these experimental results is the expected variability of a geologic material, and that the material characteristics are time-dependent with swell and creep phenomena, but also that running tests on a clayey material without pore pressure measurements is not entirely adequate.

Table 3. Grenchenberg Tunnel—reconstruction of invert (from Steiner and Metzger, 1988).

	1911–15 m	Replaced m	Still Exists m
Profile III a reinf	120		76
Replaced 1949 Concrete Blocks		120	
Profile IIIa	122		46
Replaced 1922/24 Nat. Bld. Stone		16	
Replaced 1949 Concrete Blocks		60	
Profile IIa/c	154		51.5
Replaced 1949 Concrete Blocks		102.5	
Profile IIc	47		18
Replaced 1922/24 Nat Bld. Stone		28	
Total	443	251.5	191.5

3. Performance-Observations in Existing Tunnels

The many railroad and highway tunnels built across the Jura Mountains (Fig. 1) provide a unique opportunity to assess the performance of Opalinus Clayshale. In this section some of these are discussed; more details can be found in Einstein et al. 1996; and Steiner and Metzger 1988.

3.1 Grenchenberg Rail Tunnel

This tunnel was built 1911–16. Opalinus Clayshale occurs at depths of 700 to 800 m below the surface, as shown in the profile (Fig. 5) developed by A. Buxdorf (Schlussbericht 1916). The profile clearly shows that all the formations, including the Opalinus Clayshale, are folded and shifted, and the final construction report (Schlussbericht 1916) mentioned "pressure zones" observed during tunneling, but these are in the Molasse and the strongly disturbed Keuper, not in the Opalinus Clayshale. The Opalinus Clayshale occurs five times for a total length of 416 m (longest section 195 m).

The construction conditions in the Opalinus Clayshale were "good", i.e. the material was dry; consequently, the advance rates of the invert drift were medium to high (7 to 11 m per day, compared to the average of 5.55 m/day).

Nevertheless, the support requirements were relatively high, as seen in Table 2 and in Figures 6a and 6c. As can be seen from this information, this means that either an invert cover (Sohlverkleidung) of 30 cm or invert arch (Sohlgewölbe) of 50 cm was used in the Opalinus Clayshale, i.e. some of the most substantial support in this tunnel was used there.

What is most interesting is that about half of the invert arches and covers had to be replaced (Figures 6b, 6d and Table 3), some of them actually only about 10 years after construction (Steiner and Metzger 1988). The reason seems to have been a "softening" of the rock, probably caused by a combination of water intake and dynamic effects from railroad traffic. No swelling has been reported, but it cannot be ruled out.

3.2 Hauenstein Base Rail Tunnel

The Opalinus Clayshale in this tunnel occurs roughly from km 8.5 to 8.8 (see Fig. 3) at a depth of 300 to 350 m. During construction (1912–16), a max. 10 cm invert heave was observed in Opalinus Clayshale, as illustrated in Figure 7. No inward abutment movement occurred. How-

182

ever, over time due to the combination of water and dynamic (train traffic) effects, the Opalinus Clayshale in the invert deteriorated. During the reconstruction 1980–86, an invert arch of 80 cm thickness was, therefore, placed in this tunnel section (Fig. 8).

Several years after this reconstruction of the tunnel, so-called "flat jack stress re-establishment" measurements were made in the tunnel liner, usually in the sidewalls at springline elevation and between the springline and the crown. With these measurements it is possible to measure the stresses in the tunnel liner, and from this one can estimate swelling pressures or other equivalent stresses acting on the liner. These measurements, conducted in July 1988, showed stresses in the liner in the Opalinus Clayshale between 2.9 and 0.9 MPa, from which swell pressures between 0.36 and 0.09 MPa were calculated (IG Wisenbergtunnel 1990a).

3.3 Bözberg Rail Tunnel

Opalinus Clayshale occurs from km 0.18 to 0.62 at a depth of roughly 200 m. The tunnel was built 1871–75, and according to Beck and Golta (1972), invert heave and in-

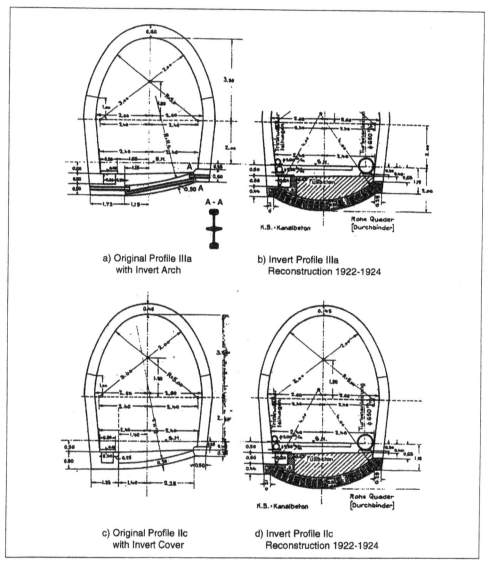

a) Original Profile IIIa
with Invert Arch

b) Invert Profile IIIa
Reconstruction 1922-1924

c) Original Profile IIc
with Invert Cover

d) Invert Profile IIc
Reconstruction 1922-1924

Figure 6. Typical sections (profiles), Grenchenberg tunnel (from Steiner and Metzger, 1988 and "Schlussbericht," 1916). a, c: Original profiles. b, d: Reconstructed profiles with natural building stone (Quader) and concrete (Füllbeton).

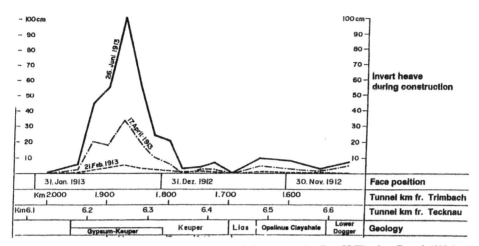

Figure 7. Hauenstein Base Tunnel— observed invert heave during construction (from IG Wisenberg Tunnel, 1990a).

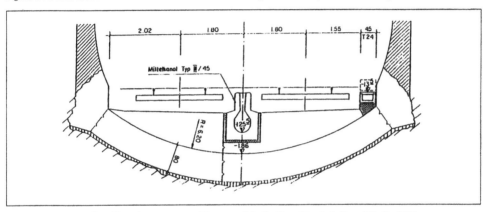

Figure 8. Hauenstein Base Tunnel— reconstructed invert arch in Opalinus Clayshale (from Etterlin 1988).

ward movement of the abutments started to occur early on. Bridel and Beck (1970) mention that some sections of the drainage channel in Opalinus Clayshale (and in the Lias and Anhydrite) had already had to be replaced by 1885.

Continuous measurements were conducted in the period 1923–54. In the Opalinus Clayshale section, the maximum invert heave during this period was about 10 cm and the inward movement of the lower abutments was about 13 cm (Fig. 9). The latter is mostly due to indirect effects of invert heave (see e.g. in Beck and Golta, 1972). Also, Bridel and Beck (1970) state that the main cause for the deterioration of the rock was the malfunctioning drainage channel which had an uneven longitudinal profile and in which water was standing. Consequently, an invert arch was built during the reconstruction in the mid-'60s, and this not only in the Opalinus Clayshale section but also in the Marl and Anhydrite sections. As of the date of Beck's and Golta's article (1972), no further movements had been observed.

3.4 Belchen Highway Tunnel

In this tunnel, which was built between 1963 and 1970, Opalinus Clayshale occurs from km 1.7 to km 2.0 at depths between 330 and 330 m below the surface, and from km 2.1 to 2.15 at a depth of 200 m (Fig. 10). According to Grob (1972, (see also Fig. 11), significant heave of the drainage channel (up to 50 cm) in Opalinus Clayshale occurred during construction, but once the invert arch was constructed, no further heave occurred. Measurements of the swell pressures on the invert arch and of stresses in the invert arch produced the results shown in Table 4; earlier measurements are shown in Figure 11. The stresses in the liner are quite significant although still substantially below the concrete compressive strength, which is probably in the order of 40 to 45 MPa.

During the recent past, some repairs have been made, notably of destroyed or damaged parts of the invert arch. According to the engineer (Chiaverio 1994), damage usually occurs if there is a combination of low structural

184

quality (e.g. low quality concrete) and of significant swell pressures. Most of these problem sections are in anhydritic rock. Nevertheless, one such combination occurred in Opalinus Clayshale, where it caused a shear failure of the invert arch.

3.5 Mont Terri Highway Tunnel

This tunnel, which is presently under construction, has a section in Opalinus Clayshale. Extensive field and laboratory tests were and are conducted in this context. The results of some of them are discussed in Section 6.

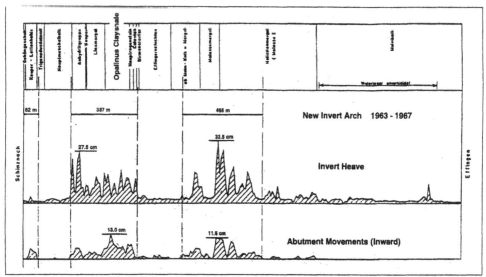

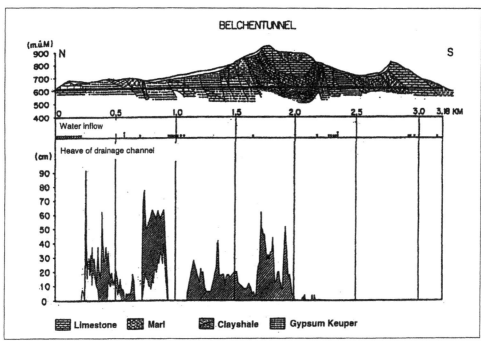

Figure 9. Bözberg rail tunnel. Invert heave and abutment movements, 1923–1945 (from Beck and Golta 1972).

Figure 10. Heave in the Belchen Tunnel (from Grob 1972).

3.6 Conclusions from Existing Tunnels

The information obtained from existing tunnels allows one to draw some definite conclusions, while other aspects are not as clear:

·Opalinus Clayshale in the tunnel invert deteriorates if it is exposed or if it has insufficient cover. "Insufficient cover" relates not only to allowing water to seep into the rock, but also to the strength and stiffness of the tunnel support. The "deterioration" ranges from softening to swelling which can be quite substantial, namely, 10 to 20 cm, with an extreme of 50 cm. Indications of swelling are also given by stresses measured in the tunnel supports in cases where substantial invert arches exist, i.e. where the tunnel support formed a closed "ring". The swelling pressures are between .1 and .3 MPa, producing liner stresses between 1 and 3 MPa, which can be easily managed by normal good quality concrete.

While all this indicates that Opalinus Clayshale may be dealt with in a satisfactory manner, it is much less clear how to incorporate this in a rational modern design also considering new construction methods. This uncertainty is caused to some extent by the fact that "satisfactory performance" was achieved only after repairs had been made and that there do not seem to be clear relations between the characteristics measured in the laboratory and the field on the one hand and tunnel performance on the other.

4. Behavior of Opalinus Clayshale Near a Tunnel — New Experiments and Models

4.1 Experiments

To get a better approach to tunnel design and construction in the swelling rocks of the Jura mountain, several Swiss entities funded research at MIT and elsewhere. Specifically, the Swiss Federal Office for Road Construction, funded the Ph.D. research by Bellwald (1991); the Swiss Federal Railroads funded research consisting of laboratory experiments and analytic modelling in the con-

text of the Wisenberg Tunnel Project, which led to another MIT Ph.D. thesis (Aristorenas 1992); and NAGRA (Swiss Cooperative for Radioactive Waste Storage) funded further modelling work, as well as a feasibility study (Einstein et al. 1996).

The experimental part of this research was conducted following the most up-to-date geotechnical techniques and, correspondingly, led to a state-of-the-art material model. The reason for doing all of this is that, as mentioned above, a rational relation between material characteristics and the performance of a tunnel during construction and its lifetime had to be established. This can only be done if the tests on a ground element duplicate the behavior it undergoes.

Figure 12 shows typical stress paths of ground elements in the invert and the springline of a circular tunnel with an in situ stress ratio $K_o = 1.5$. What is most interesting is not only that swelling in conjunction with the relief of negative pore pressures occurs, but that, for the invert element, the stress path is close to or actually touches the failure envelope. *This concept is probably one of the most significant statements in this paper:* It is possible that shear failure can occur during swelling. Even if no actual failure occurs, the stress conditions may favor creep. In other words, not only volumetric deformation but shear deformation may occur, and both can be time-dependent. Clearly, this behavior needs to be considered by appropriate experiments and by models built on these experiments.

Bellwald (1991) and Aristorenas (1992) conducted tests on a variety of clay shales from the Jura, including Opalinus Clayshale. The Opalinus Clayshale specimens came from two borings for the planned Wisenberg tunnel (Fig. 1) and from an exploratory boring for a clay pit near the German-Swiss border. The mineralogy was tested by the Clay-Mineralogy Laboratory at ETH-Zürich. In general the mineralogy was as shown in Table 1, but given that the specimens came from different borings and different depths, it is not surprising that there is quite a bit of variation.

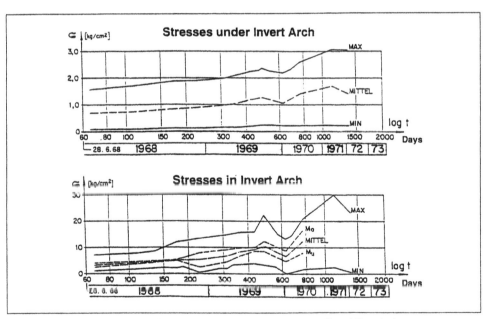

Figure 11. Stress measurement in the Opalinus Clayshale sections of the Belchen Tunnel (from Grob 1972).

Table 4. Belchen Tunnel—pressure in and on invert arch in Opalinus Clayshale (from Steiner and Metzger 1988).

	Mean kg/cm²	Minimum kg/cm²	Maximum kg/cm²
Measurement 1980:			
– Swell pressures on invert arch	3.00	0.80	0.00
– Stresses in concrete	22.90	4.20	46.50
Measurement 12.9.84:			
– Swell pressures on invert arch	3.41	1.20	7.00
– Stresses in concrete	39.30	17.20	70.50
Measurement 11:12.86:			
– Swell pressures on invert arch	3.03	1.10	6.40
– Stresses in concrete	34.40	13.20	82.50

The specimens were received in the form of 79-mm-diameter bore cores wrapped in aluminum foil and sealed with wax; they were kept in a humid room until used. Twelve triaxial tests were performed on Opalinus Clayshale, of which four (4) were drained tests, and the others undrained. The specimens for the undrained tests have the same diameter as the bore core and were thus simply cut to a length of twice the core diameter. For the drained tests, smaller specimens of 3.0 cm diameter and 7 cm length were produced by coring them under air cooling. (Smaller specimens are required in these experiments to reduce the testing time, which otherwise would be excessive and introduce experimental errors.)

In the triaxial test, cell pressures and pore pressures were applied by a combination "piston-stepping motor" device. This is now standard in all modern soil mechanics triaxial apparati, but to this author's knowledge, it was first built for this series of experiments. The cell- and pore-control devices allow one not only to measure and, if desired, to change the pressures, but also to measure volume changes. In the undrained tests, another innovative feature, a lateral inductive stress measurement device, was used (see Fig. 13). All this was connected to a computerized control system allowing one to apply any desired stress path. The descriptions in Table 5 provide examples of the variety of test conditions/stress paths that can be applied. The different testing conditions listed in Table 5 represent a systematic set:

- Isotropic and K_o-consolidation (K_o = no lateral strain).
- Undrained extension, undrained compression and drained compression loading.
- All tests subjected to pure shear during loading.

The stress paths for undrained and drained tests are shown in Figure 14a and 14b, respectively. In all cases, the specimen was first saturated through backpressuring (pore pressure of 0.5 or 1.0 MPa), which requires special consideration of the stiff rock matrix and compliance of the equipment. After saturation, the specimens were consolidated, followed by pure shear loading in compression (undrained and drained) and in extension (undrained only). The reason for choosing pure shear loading which produces a vertical total stress path in the q-p diagram is that this is the in-situ stress condition around a

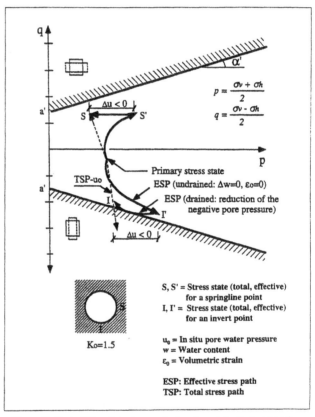

Figure 12. Typical Clayshale stress paths for elements in the invert arch (crown) and springline of a circular tunnel (K_o = 1.5).

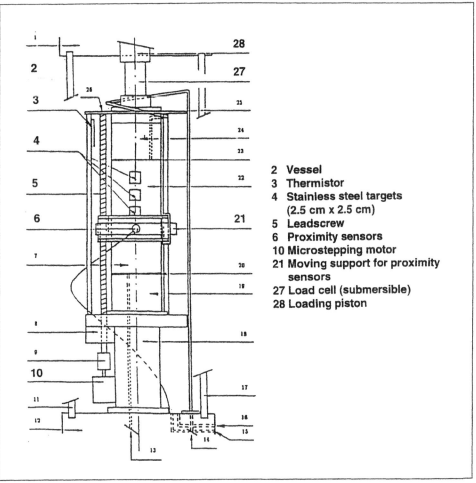

2 **Vessel**
3 **Thermistor**
4 **Stainless steel targets**
   (2.5 cm x 2.5 cm)
5 **Leadscrew**
6 **Proximity sensors**
10 **Microstepping motor**
21 **Moving support for proximity**
   sensors
27 **Load cell (submersible)**
28 **Loading piston**

Figure 13. Details of proximity measurement device in triaxial cell (from Bellwald 1991).

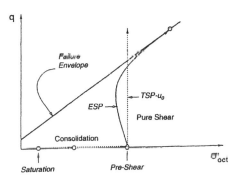

Figure 14a. Typical stress path for undrained tests (from Aristostenas 1992).

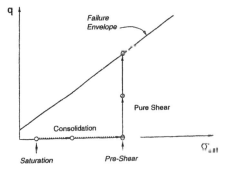

Figure 14b. Typical stress path for drained tests (from Aristostenas 1992).

Table 5. Experimental condition for tests on Opalinus Clayshale (from Aristorenas 1992).

Test	Shale Type	Test Code	Test Description
20-1q	Opalinus	CADU	Anisotropically Consolidated, Drained Unloading at constant q
23-1 (UC)	Opalinus	CIUC(PS)	Isotropically Consolidated, Sheared Undrained in Compression under Pure Shear
23-1 (DC)	Opalinus	CIDC	Isotropically Consolidated, Sheared Drained in Compression
23-5 (UC)	Opalinus	CIUC(PS)	Isotropically Consolidated, Sheared Undrained in Compression under Pure Shear
23-5 (DC)	Opalinus	CIDC	Isotropically Consolidated, Sheared Drained in Compression
23-6 (UC)	Opalinus	CIUC(PS)	Isotropically Consolidated, Sheared Undrained in Compression under Pure Shear
26B-4 (UE)	Opalinus	CK$_o$UE(PS)	K$_o$-Consolidated, Sheared Undrained Extension under Pure Shear
26B-6(UE)	Opalinus	CIUE(PS)	Isotropically Consolidated, Sheared Undrained in Extension under Pure Shear
26B-6 (DC)	Opalinus	CIDC(PS)	Isotropically Consolidated, Sheared Drained in Compression under Pure Shear
26B-8 (UE)	Opalinus	CIUE(PS)	Isotropically Consolidated, Sheared Undrained in Extension under Pure Shear
26B-8 (DC)	Opalinus	CIDC(PS)	Isotropically Consolidated, Sheared Drained in Compression under Pure Shear
ERZ4 (UE)	Opalinus	CK$_o$UE(PS)	K$_o$-Consolidated, Sheared Undrained Extension under Pure Shear

tunnel with equal horizontal and vertical stress (K = 1) and reasonably close also for K = 0.5 (horizontal stress 1/2 vertical stress).

Another advantage of pure shear tests is that the octahedral volumetric stress remains the same throughout the test and all observed volume/pore pressure changes are solely due to shearing. In undrained tests, negative excess pore pressure produce an effective stress path deviating to the right of the vertical 'TSP-u$_o$' path and vice-versa for positive excessive pore pressures. Note that for extension the same graph as in Figure 14a applies, simply flipped around the σ$_{oct}$ axis.

The observed shear strengths at failure are summarized in Figure 15 and indicate a cohesion of 1.25 MPa and friction angle of 31.5° for compression tests, while the extension tests produced no cohesion and a friction angle of 34.1°. The stress paths in most undrained compression tests and in all extension tests indicate that negative excess pore pressures are produced, i.e. dilation occurs. Shear stress (q)-shear strain (γ) curves have a typical hyperbolic shape as many clays do, a fact that was used in the analytical model to be discussed below. From the triaxial tests it was also possible to get some idea on the overconsolidation ratio (see Ladd and Foot 1974), i.e. the maximum pressure that once acted on the sample; it ranges from 4 to 14 for the Wisenberg specimens.

Aristorenas (1992) also devoted considerable effort to investigating the anisotropic/dilatant behavior and clearly found—and this is very important—that the Opalinus Clayshale is anisotropic and dilatant; i.e. it shears under isotropic loading and volume change occurs under pure shear.

Permeability backfigured from the triaxial tests produced values of 10^{-12} to 10^{-13} m/sec. (These low permeabilities made it necessary to run the drained tests over extended periods of time; one load stage lasted for several weeks to over a month.)

Another important result is that both shear- and volumetric creep occur and that they are both linearly related to the logarithm of time. Also, creep shows increase with shear stress, i.e. the closer to failure, the larger the creep strains. Volumetric and deviatoric (shear) creep seem to be related to each other, which may indicate that creep strain may be related to material anisotropy. (After the end of consolidation, no time dependent volume change should occur; however, if shearing induces volume change through anisotropy/dilatancy, then it is possible that volumetric creep occurs.)

4.2 Analytical Model

An analytical model for clayshales was developed by Aristorenas (1992); it has been calibrated with the data from the experiments on Opalinus Clayshale. Rather than describing the detailed mathematical formulation which can be found in Aristorenas (1992) and Bobet et al. (1999), it is more informative to describe and discuss the model parameters and their experimental determination first, then to briefly show some applications to tunnel analysis.

The model has nine (9) parameters to describe deformability (5 parameters), strength (2), permeability (1) and creep (1) characteristics.

Deformation parameters

The compressibility C$_{Be}$ is the slope of the σ'$_{oct}$- ε$_{vol}$ (effective octahedral stress-volume change) curve measured in isotropic consolidation tests.

The "shear stress-shear strain" behavior measured in triaxial tests can be expressed in the hyperbolic form:

$$\frac{q}{\sigma'_{co}} = \frac{\gamma}{a + b\gamma} \qquad \text{(Ladd and Foot 1974).}$$

where q = 1/2 (σ$_1$ - σ$_3$) and γ = |ε$_1$ - ε$_3$|, σ'$_{co}$ = effective octahedral consolidation stress. Hence, parameters a and b characterize shear deformation with 'a' defining the compliance and 'b' the non-linearity.

As was mentioned earlier, the type of triaxial test has to

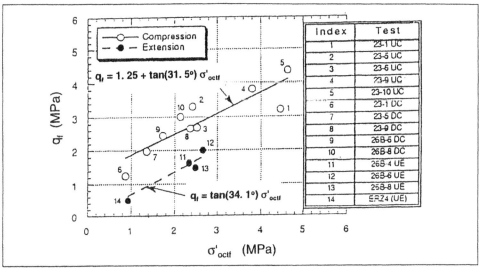

Figure 15. Results of triaxial tests conducted at MIT. Stress states at failure (from Aristostenas 1992).

be chosen to represent the appropriate stress path of the ground element considered.

In an anisotropic material such as shale, the shear strains and volumetric strains are coupled. Specifically, in the pure shear experiments conducted on Opalinus Clayshale shear induced volume changes express themselves as positive (contraction) or negative (dilatation) excess pore pressures in undrained tests. In drained tests, one obtains $q/\sigma'_{on} - \varepsilon_{vol}$ curves which are expressed as:

$$\varepsilon_{vol} = m_1 \left(\frac{q}{\sigma'_{co}}\right) + m_2 \left(\frac{q}{\sigma'_{co}}\right)^2$$

where the first and second term, respectively, describe the linear and non-linear anisotropic behavior.

Strength Parameters

These are simply the cohesion c' and friction angle φ', determined in triaxial tests.

Permeability

In essence, this can be determined from consolidation tests or from the consolidation phase prior to triaxial shear testing. It is also possible—and this was done in the experiments on Opalinus Clayshale—to derive permeability in drained triaxial testing in which shear stress/effective octahedral stress increments were applied and the volume change over time was measured after each increment; however, determining permeability with this procedure is difficult if the increment is shear dominated.

Creep

Information on creep, the time-dependent deformation under constant effective stresses, was obtained by measuring the time-dependent de-

formation after the end of primary consolidation in drained tests. The volumetric and shear creep increase linearly with time and the slopes of these curves, the parameters $m_\varepsilon = d\varepsilon_{vol}/d$ (log t) and $m_\gamma = d\gamma/d$ (log t) are the volumetric and shear creep parameters. The slopes increase with the magnitude of the applied normalized shear stress q/σ'_{oct}. This confirms what has been said in conjunction with Figure 12, namely, that creep increases with the proximity to failure. More interestingly, in the tests in Opalinus Clayshale, m_ε and m_γ are linearly related to each other (Fig. 16). The dependence or independence of shear – and volumetric creep is a major open question in soil and rock mechanics and shall not be addressed here. For the purposes

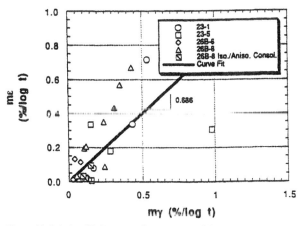

Figure 16. Relationship between volumetric m_ε and deviatoric m_γ creep strain rates. (symbols and numbers refer to different test types; see Table 5 and Aristorenas 1992).

of the clayshale model, the linear relationship between m_ε and m_γ allows one to work with only one of the creep parameters.

4.3 Application of Model to Tunnel Analysis

Numerous tunnel cases were studied (Aristorenas 1992; Bobet et al. 1999; Einstein et al. 1996) in which the Opalinus Clayshale model was implemented in the finite element approach ABAQUS (Hibbit 1993). Both drained and undrained cases were studied, as well as different delays of support installation. The material properties listed in Table 6 represent typical values obtained from the experiments mentioned earlier. The modelling is done by setting the pore pressures at the tunnel perimeter and the effective normal stresses to the tunnel perimeter to zero. Short-term and long-term effects are then modelled, the latter considering both consolidation (swelling) and creep.

Figure 17 shows the stress paths of elements in the invert (crown) and springline of an inclined tunnel of 5-m radius at the stress conditions indicated. The stress paths represent excavation and consolidation (swelling) after excavation. Figure 18 shows the effect of consolidation alone and of consolidation and creep on time dependent deformation. These results show that the stress paths as proposed in Figure 12 are actually ocurring and that the time dependent effects, notably creep can be quite substantial.

What the analysis can also show quite well is the interaction of two mechanisms influencing pore pressures development, namely: 1) the development of negative excess pore pressure due to unloading; and 2) the reduction of the initial hydrostatic pore pressure due to a zero pore pressure condition at the tunnel perimeter. This combination produces some unexpected fluctuation of pore pressures in the ground.

Another significant set of results from this analysis is the effect of the cross-sectional shape. Circular tunnels do reduce the shear stress levels and, thus, zones which fail (plastify), a fact that is widely known. The pore pressures in the ground around the tunnel initially differ between a flat and circular invert; under the flat invert, the pore pressures are lower than under the circular invert. They reach a steady state with a gradient toward the tunnel perimeter after two years.

In the analytical study of tunnel performance, particularly for NAGRA (Einstein et al. 1996; Bobet et al. 1999) tunnels ranging from 2.4 m to 5.5 m and

6.6 m diameter with 0.2- and 0.3-m-thick concrete liners were examined at depths of 400 m, 700 m and 1000 m and at horizontal stress ratios of 0.5, 1.0 and 2.0. The most interesting results are shown in Table 7.

As shown in Table 7, the liner stresses for $K_o = 1$ are greater than for $K_o = 2$. This is caused by the anisotropy of the shale for which deformability is greater in the vertical

Table 6. Opalinus Clayshale—typical parameter values.

Elastic Shear Compliance:	a	=	0.5%
Plastic Shear Compliance:	b	=	0.5%
Elastic Bulk Compliance:	$C_{B\varepsilon}$	=	3.0%
Elastic Coupling Compliance:	m1	=	0.4%
Plastic Coupling Compliance:	m2	=	-0.5%
Permeability:	k	=	5.0×10^{-8} cm/sec
Friction Angle:	ϕ'	=	30°
Cohesion:	c'	=	0.8 MPa

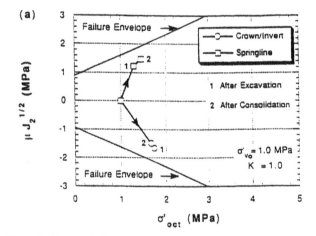

(a)

Figure 17. Stress paths for elements in the crown (invert) and springline for a circular tunnel with

σ'_{vo} = 1.0 MPa and K = 1.0:

J_2 = Second deviatoric stress invariant (proportional to shear stress q);

μ = factor indicating compression (+1) or extension (-1)

(from Aristorenas 1992).

Table 7. Maximum tangential stresses in concrete liner at depth of 700 m in MPa. D = Diameter; t = Liner Thickness

			Tunnel	
			D = 2.4 m, t = 0.2	D 6.6 m, t = 0.3 m
Drained Conditions				
- Delayed Support Installation	K_o =	1.0/0.5/2.0	20*/35/10	23/40/22
- Immediate Support Installation	K_o =	1.0	82	
Undrained Conditions				
- Immediate Support Installation	K_o =	1.0	88	

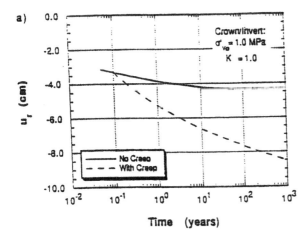

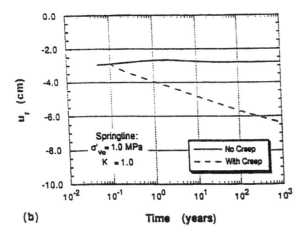

Figure 18. Time-dependent radial displacements of the perimeter of a circular tunnel in Opalinus Clayshale for consolidation alone (solid line) and consolidation and creep (dashed line); for σ'ₒₒ = 1.0 MPa and K = 1.0 (from Aristorenas 1992).

"support delay," and the portion of displacements acting on the liner and the induced stresses is smaller than for lower permeabilities.

5. New Construction Methods

While the Belchen Tunnel (built 1963–70) and all previous tunnels were constructed using the drill-and-blast method, the newer tunnels (Bözberg Highway Tunnel, Mt. Terri Tunnel, Wisenberg planned) are TBM tunnels (see Fig. 1 for location of these tunnels). This has a number of important implications, which are illustrated to some extent in the cross-section of the Wisenberg tunnel (Fig. 19):

- The cross-section is circular.
- The support is a closed ring.
- The construction procedure produces smooth surfaces.
- The exterior liner consists of prefabricated elements that are cast in factory conditions with high quality control. Also, the installation of prefabricated elements is a controlled manufacturing process.
- The invert is protected by liner elements at a relatively short distance behind the face. As a consequence, construction water and water emanating from the rock are not running on the exposed rock.
- The final liners are watertight, and well-aligned drainage channels prevent ponding of water.

If one compares this with the empirical evidence discussed in Section 3, it is evident that the installation of a good quality invert arch and the control or even elimination of water contact with the exposed rock, which were mentioned as two major requirements, are fulfilled. Given that the liner is installed shortly after excavation, these favorable conditions do exist both in the short and long term. As was mentioned in Section 4, the circular cross-section has the favorable effect of reducing the shear stress magnitudes and the size of plastic zones compared to flat inverts while the pore pressure related differences at least with the present state of knowledge are not that significant.

Combining the new experiments and analytical models with the new construction procedure provides further insight:

The analysis shows that unsupported Opalinus Clayshale can fail upon being excavated. Most importantly, it shows the ground elements and stress paths which are particularly problematic. Similarly, for lined openings, it is possible to consider ground structure interaction with all influencing factors. Time-dependent deformation or, if prevented, time-dependent stress increase, can be substantial.

Regarding consolidation, it was shown that the support delay effect is reduced and liner stresses become greater for lower permeabilities. Creep, which was analysed only for unsupported tunnels, can more than double the consolidation related deformation. Since it was also shown that creep increases with the level of the shear stresses ('proximity to

than in the horizontal direction, producing bending moments also for uniform stress conditions (Kₒ = 1). (The location of the max. liner stresses clearly differs for the different Kₒ).

Support delay is the time between excavation and support installation and corresponds to one tunnel shift. Having no delay leads to the expected very substantial increase in liner stresses.

The undrained case, in which the full hydrostatic pressure acts on the liner, produced the highest stresses. The values in Table 7 are for a permeability of K = 5 × 10⁻¹⁰ m/sec (drained case); if the permeability is reduced to 10⁻¹² m/sec, the liner stresses increase (e.g. the 20 MPa marked with an asterisk increases to 30 MPa). With higher permeability, more of the displacements take place during the

failure'), one can confidently assume that the prevention of such a near failure stress state through appropriate liner installation is important.

6. New Developments in Project and Construction Management

In this domain, many new tools and procedures made possible by information technology have been implemented in the last 20 to 30 years. Only one, which has particular relevance regarding time-dependent behavior, will be discussed here. Uncertainties in structural performance and in the construction process can be captured and included in a risk analysis, which in turn provides the basis for making decisions ranging from the initial go ahead to details of the construction process.

Risk analysis can also be used to decide to which extent it is worthwhile to include preventive measures in the initial design and construction to avoid later repairs. This was done in conjunction with the Adler Tunnel (see Fig. 1) with regard to swelling effects (see Einstein et al. 1994). Although the specific issue was swelling of clay-sulfate rock, any long-term behavior and, thus, swelling and creep of Opalinus Clayshale, could be treated analogously.

Specifically, in the Adler Tunnel, the long-term risk

R = Probability of cause × Probability cause will produce damage × Cost of damage

was assessed for the damage caused by swelling and by sulfatic and chloridic water. The cost of damage was based on the recent repairs in a number of existing tunnels. The probabilities (probability of cause, probability that cause will produce damage) were estimated for different construction procedures and liner configurations proposed by the bidding contractors. This resulted in different long-term risks for the different design-construction combinations. In addition, the so-called construction risk representing problems during construction and which also differ for the different methods were calculated. The present value of the long-term risk, the construction risk and the bid price were then summed (see Fig. 20). Details of this analysis can be found in Chiaverio et al. (1995) and (Einstein et al. 1994).

Quite importantly, regarding long-term effects, damage caused after 30 years has only a negligible present value with the assumed interest rates of 5 and 6%; this is not surprising but should be considered in decision making.

7. Conclusions

The empirical insight gained from the performance of old tunnels is that access of water should be limited both during and after construction and that an invert arch made of good quality concrete should be constructed to obtain satisfactory performance of tunnels in Opalinus Clayshale. The new TBM-based construction procedures which utilize prefabricated liners and provide a "clean" environment can satisfy these requirements. New experimentation and modelling approaches not only confirm what can cause short-term and long-term problems, but also make it possible to design tunnels such that they perform adequately during construction and during their lifetime. With appropriate risk analysis procedures, it is even possible to estimate to what extent it is worthwhile to prevent long-term damage.

While this indicates that the combination of modern design-, analysis- and construction-approaches allow one to build "better tunnels," it still leaves some questions unanswered, two of which are of particular interest to this author:

1. The experimentation and the model were adequate for the Opalinus Clayshale. It would be interesting to know how applicable they are to other clayshales.

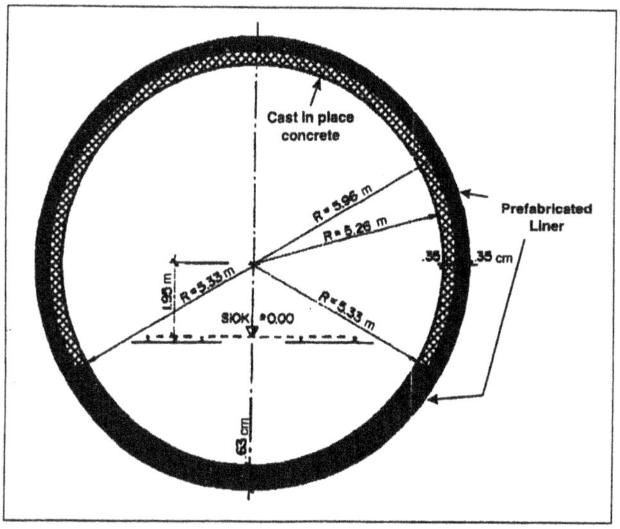

Figure 19. Wisenberg tunnel—typical cross-section in Opalinus Clayshale (from IG Wisenberg 1990b).

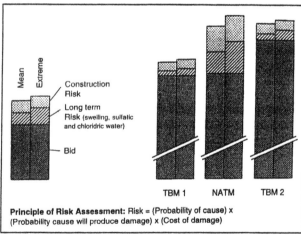

Figure 20. Risk analysis — Adler Tunnel.

2. It seems that the main driver toward better tunnels, the use of TBM's and prefabricated elements, was not the need to prevent the short- and long-term problems associated with Opalinus Clayshale but more productive and safer construction operation. New experimentation and analytical models had only a subsidiary function. The question is: Would the better knowledge of material behavior as provided by the experiments and models have by itself led to the better design?

Acknowledgements

The author would like to acknowledge the work of his former students and colleagues, W. Steiner, P. Bellwald, G. Aristorenas, A. Bobet and T. Meyer, who contributed to the research and reports underlying this paper. The research was funded by the Swiss Federal Office for Road Construction, the Swiss Federal Railroads and the Swiss National Cooperative for the Disposal of Radioactive Waste (NAGRA) and this support is gratefully acknowledged.

Literature References

Aristorenas, G. 1992. Time Dependent Behavior of Tunnels Excavated in Shale, Ph.D. Thesis, MIT, 1992.

Beck, A.; Golta, A. 1972. Tunnelsanierungen der Schweizerischen Bundesbahnen, *Schweizerische Bauzeitung* **90**(36).

Bellwald, P. 1991. A Contribution to the Design of Tunnels in Argillaceous Rocks, Ph.D. Thesis, MIT, 1991.

Berner Alpenbahn Gesellschaft Bern-Lötschberg-Simplon, 1916. *Schlussbericht* über den Bau der normalspurigen Hauptlinie Münster-Lengnau, Bern.

Bobet, A.; Aristorenas, G.; Einstein, H.H. 1999. Feasibility Analysis for a Radioactive Waste Repository Tunnel. *Tunnelling and Underground Space Technology* **13**(4), 409–426.

Bridel, G. and Beck, A. 1970. Zwei Tunnelsanierungen bei den SBB. *Eisenbahntechnische Rundschau*, Heft 1, 2.

Chiaverio, F.; Einstein, H.H. Köppel, U. 1995. Riskoanalyse beim Adlertunnel, *Schweiz. Ing. u. Arch.*, Jan. 19, 1995.

Descoeudres, F.; Kharchafi, M. 1993. Projet CERS No. 2672.1/2387, Rapport Intermédiaire sur la partie expérimentale en laboratoire —Conception et analyse numérique d'ouvrages souterrains, Laboratoire de Mécanique des Roches, EPFL.

Ecole Polytechnique Fédérale de Lausanne, (EPFL) 1989. N-16 Transjurane, Galérie de reconnaissance-Mont Terri, Essais de charge Niche TM 888 Etude demandée par le Service des Ponts et Chaussées de la République et Canton du Jura, Rapport EPFL R169.

Ecole Polytechnique Fédérale de Lausanne, (EPFL) 1990. N-16 Transjurane, Galérie de reconnaissance-Mont Terri, Essais de laboratoire, Forages TGR1-TGR2 et niches TM447-TM888. 1990. Etude demandée par le Service des Ponts et Chaussées de la République et Canton du Jura, Rapport EPFL R169.

Einstein, H.H.; Chiaverio, F.; Köppel, U. 1994. Risk Analysis for the Adler Tunnel, *Tunnels and Tunnelling*, November, 1994.

Einstein, H.H.; Bobet, A.; Aristorenas, G. 1995. Feasibility Study Opalinuston, NAGRA Interner Bericht NIB 95-61, A, B, C.

Etterlin, A. ed. 1988. Rekonstruktion Hauenstein-Basistunnel, Hasler+Holtz.

Grob, II. 1972. Schwelldruck im Belchentunnel. Proc Int'l. Symp. on Underground Construction, Lucern.

Hibbit, K. Sorensen, Inc. 1993. ABAQUS. User's Manual, Version 5.3.

IG Wisenberg-Tunnel. 1990a. Bahn 2000-Wisenberg Tunnel. Schlitzpressenmessungen in der Auskleidung des Hauenstein-Basistunnels, Bericht.

IG Wisenberg-Tunnel. 1990b. Bahn 2000-Wisenberg Tunnel. Profiltypen.

International Society for Rock Mechanics (ISRM). 1989. *Suggested Methods for Laboratory Testing of Argillaceous Swelling Rock.*

Ladd, C.C.; Foot, R. 1974. New Design Practice for Stability of Soft Clays. *ASCE Journal* of the Geotech. Eng. Div., Vol. 100, No. GT7.

NAGRA, 1988. Sedimentstudie Endlager für hochaktive Abfälle, Bau and Betrieb, *NAGRA. Interner Bericht*, 88-26.

Nüesch, R. 1991. Felsmechanische Resultate aus Untersuchungen an Opalinuston, *NAGRA Interner Bericht*, NB 89-17. (also ETH Doctoral Dissertation No. 9349).

SBB Kreis II. 1987. Hauenstein Basistunnel Rekonstruktion, Typenplan Sohlengewölbe.

Schlussbericht. 1916. Berner Alpenbahngesellschaft Schlussbericht an das Schweiz. Post u. Eisenbahndept. über den Bau der normalspurigen Hauptlinie Münster Lengnau, Selbstverlag.

Steiner, W. Metzger, R. 1988. IG Wisenberg-Tunnel, Bahn 2000-Wisenbergtunnel. *Erfahrungen aus Tunneln in quellendem Gestein.*

Application of the Convergence-Confinement Method of Tunnel Design to Rock Masses That Satisfy the Hoek-Brown Failure Criterion

C. Carranza-Torres and C. Fairhurst

Abstract —This paper discusses the practical application of the Convergence-Confinement Method of tunnel design to rock-masses that satisfy the Hoek-Brown failure criterion. The strength of intact rock and jointed rock-masses, as defined by the Hoek-Brown criterion, and the basis of the Convergence-Confinement method are reviewed. Equations that allow the construction of the three basic components of the Convergence-Confinement method, i) the Longitudinal Deformation Profile (LDP), ii) the Ground Reaction Curve (GRC) and iii) the Support Characteristic Curve (SCC) are given. A practical case of support design for a circular tunnel is discussed and solved using the Convergence Confinement method. A spreadsheet summarizing the implementation of the method is also included. Reference values of typical rock properties and geometrical and mechanical properties for typical support systems are presented in tables and charts. © 2000 Published by Elsevier Science Ltd. All rights reserved.

1. Introduction

Estimation of the support required to stabilize a tunnel excavation, especially in the vicinity of the face, is essentially a four-dimensional problem. Time-dependent weakening of the rock compounds the three-dimensional redistribution of forces around the excavation, and the nature of the rock is uncertain until it is exposed in the face. Labasse (1949) describes the situation as follows:

> First, the types of supports to be used must be limited to one or two in order not to disrupt the material supply operations underground. This standardization makes precise calculation of a support for each cross-section useless.

> Further, the need to install the support immediately after excavation does not allow time to make calculations and fabricate the support. In order to arrive at a precise determination it would be necessary, in fact, to study each cross-section separately because it would differ from neighboring cross-sections with respect to the rock layers encountered, their dip and their deposition. It would be necessary to take a test specimen from each layer, determine its properties and the influence of these properties on neighboring layers. This would require a series of experiments and mathematical analyses whose solution, assuming that a solution is possible, would take up precious time during which the excavation would certainly have collapsed.

Given these constraints, it is valuable to have a general albeit simplified appreciation of the nature of the interplay between the (variable) rock-mass and the installed support, and the effect of variation in assumed rock properties on the support loads.

The 'Convergence-Confinement' method is such a tool. Although the term was developed in the 1960's and 70s (see, for example, AFTES, French Association for Underground Works 1978), the method has been known at least since the paper by Fenner (1938). Application of the Convergence-Confinement method, as is discussed later in detail, requires a knowledge of the deformation characteristics of the ground and of the support.

Estimation of the mechanical response of a jointed rock-mass is one of the fundamental problems in rock mechanics. The Hoek-Brown criterion (Hoek and Brown 1980) for rock failure is widely used as an attempt to address the problem. The presence of joints and associated in situ geological effects (e.g., weathering and inhomogeneities) can considerably reduce the mechanical strength and stiffness of the rock-mass compared to the corresponding properties of intact specimens taken from the mass. The Hoek-Brown criterion 'adjusts' the strength properties of intact rock cores measured in triaxial tests in order to estimate the reduced strength that the rock-mass will exhibit in the field scale.

The following sections review the practical implementation of the Convergence-Confinement method to rock-masses that can be described by the Hoek-Brown failure criterion. Because of its importance in defining the strength and deformability properties of rock-masses, a detailed discussion of the Hoek-Brown failure criterion is presented as Appendix A to this paper.

Present addresses: Dr. Carlos Carranza-Torres, Consulting Engineer, Itasca Consulting Group Inc., 708 South Third St., Suite 310, Minneapolis, MN 55415, U.S.A. (cct@itascacg.com); Dr. Charles Fairhurst, Professor Emeritus, Department of Civil Engineering, University of Minnesota, Minneapolis, MN, U.S.A. (fair001@tc.umn.edu)

Reprinted from *Tunnelling and Underground Space Technology* **15** (2), 187-213 (2000)

2. The Convergence-Confinement Method of Tunnel Design

As noted in the Introduction, the Convergence-Confinement method is a procedure that allows the load imposed on a support installed behind the face of a tunnel to be estimated. When a section of support is installed in the immediate vicinity of the tunnel face, it does not carry the full load to which it will be subjected eventually. A part of the load that is redistributed around the excavation is carried by the face itself. As the tunnel and face advance (i.e., away from the installed support), this 'face effect' decreases and the support must carry a greater proportion of the load that the face had carried earlier. When the face has moved well away from the support in question, it carries effectively, the full design load.

The problem is illustrated in Figure 1a. A cylindrical tunnel of radius R is driven (e.g., by the conventional drill-and-blast method) through a rock-mass that is assumed to be subject initially to a uniform (i.e., hydrostatic) stress field. A circular support is installed at a section A—A' located a distance L from the face of the tunnel (the support is assumed to be of unit length in the direction of the tunnel axis). The objective of the analysis is to determine the load that the rock-mass will transmit to the support at section A—A', from the time of installation —indicated in Figure 1a— until the time when the face has moved ahead, sufficiently far that the 'face effect' has disappeared.

The variables involved in the analysis are shown in Figure 1b, which is a cross-section of the excavation at the position A—A' (the support has been 'removed' for clarity in this figure). The stress σ_o represents the hydrostatic far-field stress acting on the rock-mass. The radius R_{pl} indicates the extent of the 'failure' (or plastic) zone that develops around the tunnel (a discussion of the development of this failure region will be presented in Section 3).

To simplify the problem, it is assumed that all deformations occur in a plane perpendicular to the axis of the tunnel (i.e., the problem is two-dimensional plane strain) and that the radial displacement u_r and the pressure p_i—the latter representing the reaction of the support on the walls of the tunnel—are uniform at the section. Figure 1c shows a cross-section of a circular annular support of thickness t_c and external radius R installed at the section A—A'. The uniform pressure p_s represents the load transmitted by the rock-mass to the support; the radial displacement u_r represents the displacement induced by the load p_s. For compatibility of deformations at the rock support interface, the radial displacement of the support must equal the radial displacement of the rock wall u_r indicated in Figure 1b.

The basis of the Convergence-Confinement method is illustrated in the sequences (a) through (c) in Figure 2. The situation at the initial time t_o, when the lining is installed at section A—A', is represented in the upper sketch (Figure 2a). At this instant, the section is located at a distance L from the face and the ground has converged radially by the amount u_r^0. It is assumed that, provided the face does not advance, the rock-mass transmits no load to the support — i.e., $p_s^0 = 0$ at this stage. (Time-dependent weakening, with associated deformation, is not considered in this analysis.)

As the tunnel advances to the right, the ground and the support (at section A—A') deform together and the support receives part of the load that the face had been carrying previously. Figure 2b shows the situation at a time t when the section is located at the distance L_i from the face; at that moment, the ground has converged the amount $u_r^i > u_r^0$ and the rock-mass transmits the pressure p_s^i to the support.

Once the face of the tunnel has moved ahead far enough (Fig. 2c), the ground-support system at the section A—A' is in equilibrium and the support carries the final (or design) load p_s^D. At this time t_D, the effect of the face has disappeared and the support and ground have converged together by the final amount u_r^D.

As can be seen from Figure 2, determination of the load transferred to the support requires an analysis of the interaction of the load-deformation characteristics of the elements comprising the system, (i) the tunnel as it moves forward; (ii) the section of excavation perpendicular to the tunnel axis; and (iii) the support installed at that section.

The three basic components of the Convergence-Confinement method are, therefore, (i) the Longitudinal Deformation Profile (LDP); (ii) the Ground Reaction Curve (GRC); and (iii) the Support Characteristic Curve (SCC).

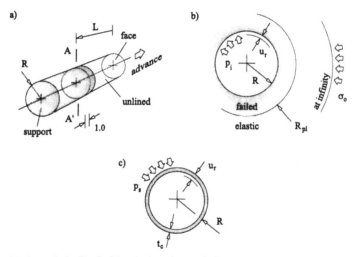

Figure 1. a) Cylindrical tunnel of radius R driven in the rock-mass. b) Cross-section of the rock-mass at section A—A'. c) Cross-section of the circular support installed at section A—A'.

The LDP is the graphical representation of the radial displacement that occurs along the axis of an *unsupported* cylindrical excavation —for sections located ahead of and behind the face. The upper diagram in Figure 3 represents such a profile. The horizontal axis indicates the distance x from the section analyzed to the tunnel face; the vertical axis indicates the corresponding radial displacement u_r (the right part of the diagram is included for use later in relating the LDP to the GRC and the SCC). The diagram indicates that at some distance *behind* the tunnel face the effect of the face is negligibly small, so that beyond this distance the unlined tunnel section has converged by the final amount u_r^M. Similarly, at some distance *ahead* of the face, the advancing tunnel has no influence on the rock-mass and the radial displacement is zero.

Considering now the section of unlined tunnel represented in Figure 1b, the GRC is defined as the relationship between the decreasing internal pressure p_i and the increasing radial displacement of the wall u_r. The relationship depends on the mechanical properties of the rock-mass and can be obtained from elasto-plastic solutions of rock deformation around an excavation (Section 3 discusses the construction of the GRC). The GRC is shown as the curve *OEM* in the lower diagram of Figure 3, extending from point O —where the internal pressure p_i is equal to the initial stress σ_o, to point M corresponding to the case where the internal pressure is equal to zero (i.e., the tunnel is unsupported) and the maximum closure (i.e., the radial displacement) u_r^M is the maximum possible. Point E defines the internal pressure p_i^{cr} and corresponding closure at which the elastic limit of the rock is reached (at the tunnel wall) —if the internal pressure falls below this value, a failed region of extent R_{pl} develops around the tunnel, as shown in Figure 1b.

The SCC is similarly defined as the relationship between the increasing pressure p_s on the support (shown in Fig. 1c) and the increasing radial displacement u_r of the support. This relationship depends on the geometrical and mechanical characteristics of the support (Section 4 discusses the construction of the SCC). The SCC is shown as the curve *KR* in the lower diagram of Figure 3. Point K corresponds to a support pressure equal to zero (i.e., when the support is first installed) and point R to the pressure p_s^{max} that produces failure of the support.

Interpretation of the interaction between the LDP, GRC and SCC allows us to define the pressure p_s that the ground transmits to the support as the face advances. To illustrate the procedure, consider again the sequences *a)* through *c)* illustrated in Figure 2. Installation of the support at section $A—A'$ at time t_0 in Figure 2a, corresponds in the LDP of Figure 3 as point I of coordinates $x = L$ and $u_r = u_r^0$. Point J on the right side of the diagram has a horizontal coordinate $u_r = u_r^0$ and defines point K of the SCC in the diagram below. As long as the face does not move, stability is maintained solely by the ability of the face to carry the load redistributed by excavation. Thus, the vertical segment KN in the lower diagram of Figure 3 corresponds to the pressure taken by the face at time t_0. (Again, note that time dependent weakening of the rock-mass is not considered here.)

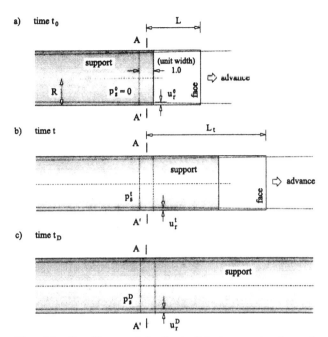

Figure 2. Loading of the support at section A—A' due to progressive advance of the tunnel face.

As the face advances in Figure 2, both the support and excavation deform by the same amount —with the pressure p_s on the support increasing and the confining effect p_i on the periphery of the tunnel decreasing.

At time t_D in Figure 2c, when the supporting effect of the face has disappeared completely, the system reaches equilibrium at point D in the lower diagram of Figure 3—i.e. at the intersection of the GRC and the SCC. The pressure p_s^D defined by point D then represents the final pressure (or design load) that the rock transmits to the support.

Inspection of the LDP, GRC and SCC in Figure 3 leads to two conclusions of practical interest:

i) the support will not be subject to a radial pressure larger than p_s^L —defined by point L in the lower diagram. This pressure would be achieved only in the hypothetical case of an infinitely rigid support installed at the face itself —i.e., the SCC would be a vertical one, starting from point H; and

ii) a support will take no load if placed beyond point M, since the maximum possible convergence has occurred already.

These two cases correspond to the two limiting cases of load that the rock-mass can transmit to the support. In general, as is seen from the LDP, GRC and SCC in Figure 3, the further that the support is installed from the tunnel face, the lower the final load p_s^D on the support (assuming, again, that no time-dependent weakening or disintegration of the rock-mass occurs).

3. Construction of the Ground Reaction Curve

The Ground Reaction Curve (GRC) shown in Figure 3 can be constructed from the elasto-plastic solution of a

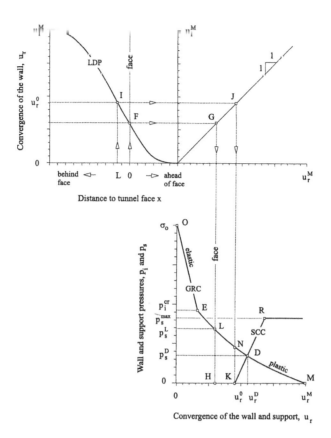

Figure 3. Schematic representation of the Longitudinal Deformation Profile (LDP), Ground Reaction Curve (GRC) and Support Characteristic Curve (SCC).

circular opening subject to uniform (i.e., hydrostatic) far-field stresses and uniform internal pressure (see Fig. 1b). Several solutions of this type, based on the Hoek-Brown failure criterion for the rock, have been published in the past. Some of these solutions include approximations in the equations of deformations to simplify the problem —see for example Brown (1983); others resort directly to numerical treatment to obtain a relationship between internal pressure and radial deformation (for example, Wang 1996).

In the present work, an analytical solution derived by Carranza-Torres and Fairhurst (1999) will be considered. The solution is based on the 'general' form of the Hoek-Brown criterion proposed by Londe (1988) —the reader is referred to Appendix A for a detailed review of the Hoek-Brown failure criterion and the coefficients characterizing the strength and deformability of the rock-mass, that will be used in the equations presented this section.

Consider the section of a cylindrical tunnel of radius R subject to uniform far-field stress σ_o and internal pressure p_i shown in Figure 1b. The rock-mass is assumed to satisfy the Hoek-Brown failure criterion defined by equation (A-2); the variables characterizing the strength of the rock-mass

are the unconfined compressive strength σ_{ci}, the intact rock parameter m_i, and the rock-mass parameters m_b and s discussed in Appendix A (as mentioned there, the analysis assumes the parameter a to be 0.5).

The uniform internal pressure p_i and far-field stress σ_o can be 'scaled' according to the transformation (A-8), to give the scaled internal pressure P_i and far-field stress S_o, respectively,

$$P_i = \frac{p_i}{m_b \, \sigma_{ci}} + \frac{s}{m_b^2} \tag{1}$$

$$S_o = \frac{\sigma_o}{m_b \, \sigma_{ci}} + \frac{s}{m_b^2} \tag{2}$$

The pressure p_i^{cr}, defined by point E in the GRC of Figure 3, marks the transition from elastic to plastic behavior of the rock-mass —i.e., for an internal pressure $p_i \geq p_i^{cr}$, the rock remains elastic, and for $p_i < p_i^{cr}$, a plastic region of radius R_{pl} develops around the tunnel (see Fig, 1b).

The scaled critical (internal) pressure P_i^{cr} for which the elastic limit is achieved is given by the following expression:

$$P_i^{cr} = \frac{1}{16}\left[1 - \sqrt{1 + 16\,S_o}\,\right]^2 \tag{3}$$

The actual (i.e., non-scaled) critical pressure is found from the inverse of equation (1),

$$p_i^{cr} = \left[P_i^{cr} - \frac{s}{m_b^2}\right]m_b\,\sigma_{ci} \tag{4}$$

Provided $p_i \geq p_i^{cr}$, the relationship between the radial displacements u_r^{el} and internal pressure p_i in the elastic part of the GRC (i.e., segment OE in Fig. 3) is given by the equation,

$$u_r^{el} = \frac{\sigma_o - p_i}{2G_{rm}}R \tag{5}$$

where G_{rm} is the shear modulus of the rock-mass defined by equation (A-10).

For values of internal pressure $p_i < p_i^{cr}$, the extent of the plastic region R_{pl} that develops around the tunnel is

$$R_{pl} = R\,\exp\left[2\left(\sqrt{P_i^{cr}} - \sqrt{P_i}\right)\right] \tag{6}$$

To define the plastic part of the Ground Reaction Curve (i.e., the curve EM in the GRC of Figure 3), a flow rule for the material is needed. The flow rule defines the relationship between the strains that produce distortion and those that produce volumetric changes, as plastic deformation occurs in the material—see, for example, Atkinson (1993). In underground excavation practice, the flow rule is usually assumed to be linear, with the magnitude of volumetric change characterized by a 'dilation' angle ψ, such that, if $\psi = 0°$, the material undergoes no change in volume during plastic deformation; if $\psi > 0°$, the volume increases during plastic deformation.

In the solution described here, the flow rule will be characterized by a dilation coefficient K_ψ that is computed from the dilation angle, ψ, according to the expression $K_\psi = [1 + \sin\psi]/[1 - \sin\psi]$. Note, for example, that for $\psi = 0°$, the dilation coefficient is $K_\psi = 1$ and for $\psi = 30°$, the coefficient is $K_\psi = 3$.

With the flow rule characterized by the dilation coefficient K_ψ the plastic part of the GRC —i.e., the segment EM in Figure 3—is given by

$$\frac{u_r^{pl}}{R}\frac{2G_{rm}}{\sigma_o - p_i^{cr}} = \frac{K_\psi - 1}{K_\psi + 1} + \frac{2}{K_\psi + 1}\left(\frac{R_{pl}}{R}\right)^{K_\psi + 1} \tag{7}$$

$$+ \frac{1 - 2v}{4(S_o - P_i^{cr})}\left[\ln\left(\frac{R_{pl}}{R}\right)\right]^2$$

$$- \left[\frac{1 - 2v}{K_\psi + 1}\frac{\sqrt{P_i^{cr}}}{S_o - P_i^{cr}} + \frac{1 - v}{2}\frac{K_\psi - 1}{(K_\psi + 1)^2}\frac{1}{S_o - P_i^{cr}}\right]$$

$$\times \left[(K_\psi + 1)\ln\left(\frac{R_{pl}}{R}\right) - \left(\frac{R_{pl}}{R}\right)^{K_\psi + 1} + 1\right]$$

where v is Poisson's ratio for the rock-mass.

Hoek and Brown (1997) suggest that in some cases the assumption of no plastic volume-change for the rock-mass may be more appropriate. For the case of non dilating rock-masses, characterized by the coefficient $K_\psi = 1$, equation (7) becomes

$$\frac{u_r^{pl}}{R}\frac{2G_{rm}}{\sigma_o - p_i^{cr}} = \left[\frac{1 - 2v}{2}\frac{\sqrt{P_i^{cr}}}{S_o - P_i^{cr}} + 1\right]\left(\frac{R_{pl}}{R}\right)^2 \tag{8}$$

$$+ \frac{1 - 2v}{4(S_o - P_i^{cr})}\left[\ln\left(\frac{R_{pl}}{R}\right)\right]^2$$

$$- \frac{1 - 2v}{2}\frac{\sqrt{P_i^{cr}}}{S_o - P_i^{cr}}\left[2\ln\left(\frac{R_{pl}}{R}\right) + 1\right]$$

To illustrate the construction of Ground Reaction Curves using equations (1) through (8), let us consider the case shown in Figure 4 of an *unsupported* section A—A' of a tunnel, radius R, located at a distance of $10\,R$ behind the face of the tunnel (as will be discussed in Section 5), a distance of $10\,R$ is large enough that the face has no further effect on the section).

We wish to determine the radial convergence and the extent of the plastic region as the internal pressure p_i is reduced from the (initial) in-situ stress value σ_o towards zero.

In this particular example, the radius of the tunnel is $R = 1$ m, the initial stress field is $\sigma_o = 7.5$ MPa and the properties of the intact rock are $\sigma_{ci} = 20$ MPa and $m_i = 15$. Rock-masses of decreasing quality, characterized by GSI values equal to 50, 40 and 30, are considered. The parameters defining the strength and deformability of the rock-mass according to the Hoek-Brown criterion are computed from equations (A-3) through (A-5) and (A-9) through (A-10) are listed in Figure 4a.

The ground reaction curves constructed using expressions (1) through (8) for *GSI* values equal to 50, 40 and 30 are shown in Figure 4b. The dashed-line curves represent the corresponding extent R_{pl} of the failure zone (the values of R_{pl} are read on the vertical axis on the right side of the diagram). Points A, B and C represent the condition at which the elastic limit of the rock-mass is reached. Note that these points are associated with a failure region of extent $R_{pl} = R = 1$ m (i.e., a failure zone that is about to start to develop around the tunnel).

To validate the analytical results presented in Figure 4, a numerical analysis was carried out with the finite difference code FLAC³D (Itasca Consulting Group 1997). The models were set up and solved for values of internal pressures $p_i = 0.5$, 1.0 and 1.5 MPa. The values of radial displacement obtained from these models, represented as open squares in Figure 4b, are in good agreement with the analytical results.

4. Construction of Support Characteristic Curves

The Support Characteristic Curve (SCC) shown in Figure 3 can be constructed from the elastic relationship between the applied stress p_s and the resulting closure u_r for a section of the support of unit length in the direction of the tunnel (see Fig. 1c).

If the elastic stiffness of the support is denoted by K_s, the elastic part of the SCC —i.e., segment KR in Figure 3, can be computed from the expression,

$$p_s = K_s\,u_r \tag{9}$$

Note that from equation (9), the unit of the stiffness K_s is pressure divided by length (e.g., MPa/m if the stresses are expressed in MPa and the displacements in meters).

The plastic part of the SCC in Figure 3—i.e., the horizontal segment starting at point R, is defined by the maximum pressure p_s^{max} that the support can accept before collapse.

The following subsections present the equations needed to compute the maximum pressure p_s^{max} and the elastic stiffness K_s for three different support systems:

  i)   shotcrete or concrete rings,
  ii) blocked steel sets and
  iii) ungrouted bolts and cables.

These equations have been adapted from Hoek and Brown (1980) and Brady and Brown (1985). The reader is referred to the original source for a detailed description of each of these support systems.

4.1 Shotcrete or Concrete Rings

Considering the closed ring of shotcrete or concrete represented in Figure 5a, the maximum pressure provided by the support is

a)

$R = 1$ m
$\sigma_o = 7.5$ MPa
$\sigma_{ci} = 20$ MPa
$m_i - 15$
$\nu = 0.25$
$\psi = 30°$

GSI	m_b	s	G_{rm} (GPa)
50	2.5	3.9×10^{-3}	1.8
40	1.8	1.3×10^{-3}	1.0
30	1.2	0.4×10^{-3}	0.6

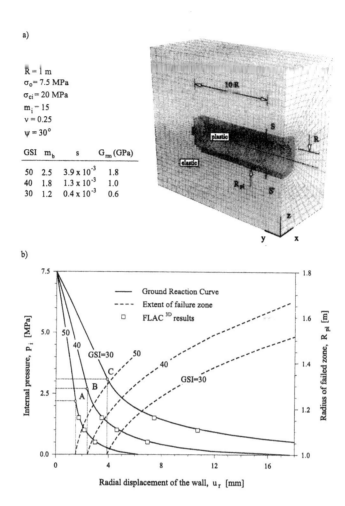

b)

Figure 4. a) Analysis of the convergence and the extent of the plastic zone for a section of tunnel located well behind the tunnel face. b) Ground Reaction Curves and extent of failure curves for section A-A' in the model—for GSI values of 50, 40 and 30. The points A, B and C represent the elastic limit in each GRC. The open squares in the diagram correspond to results obtained with FLAC ³D.

Table 1. Values of σ_{cc} and E_c for dry and wet shotcrete mixtures after 1 and 28 days (after Singh and Bortz 1975).

Type of mixture	σ_{cc} [MPa]	E_c [MPa]
Dry (1 day)	20.3	$13.6 \times 10^3 - 23.4 \times 10^3$
(28 days)	29.6	$17.8 \times 10^3 - 23.1 \times 10^3$
Wet (1 day)	$18.9 - 20.3$	$12.3 \times 10^3 - 28.0 \times 10^3$
(28 days)	$33.3 - 39.4$	$23.8 \times 10^3 - 35.9 \times 10^3$

Cement type III. Maximum aggregate size 13 mm. Mix design –expressed as a percentage of total bulk weight: *i)* Dry mixture: 17.9% cement; 29.9% coarse aggregate; 52.2% sand. *ii)* Wet mixture: 16.7% cement; 27.9% coarse aggregate; 48.7% sand; 6.7% water.

Table 2. Values of σ_{cc} and E_c for concrete mixtures used in the construction industry (after Leonhardt 1973).

Designation	σ_{cc} [MPa]	E_c [MPa]
Bn 150	14.7	25.5×10^3
Bn 250	24.5	29.4×10^3
Bn 350	34.3	33.3×10^3
Bn 450	44.1	36.3×10^3
Bn 550	53.9	38.2×10^3

Properties after 28 days, obtained from tests on cubic samples of 200 mm side. The strength of concrete at the early age of 7 days is approximately 80% of the σ_{cc} values listed above.

$$p_s^{\max} = \frac{\sigma_{cc}}{2}\left[1 - \frac{(R-t_c)^2}{R^2}\right] \quad (10)$$

The elastic stiffness is

$$K_s = \frac{E_c}{(1-v_c)R}\frac{R^2-(R-t_c)^2}{(1-2v_c)R^2+(R-t_c)^2} \quad (11)$$

where

σ_{cc} is the unconfined compressive strength of the shotcrete or concrete [MPa];

E_c is Young's Modulus for the shotcrete or concrete [MPa];

v_c is Poisson's ratio for the shotcrete or concrete [dimensionless];

t_c is the thickness of the ring [m];

R is the external radius of the support [m] (taken to be the same as the radius of the tunnel)

Typical values for σ_{cc} and E_c for *dry* and *wet* shotcrete mixtures[1] are given in Table 1. Poisson's ratio for the shotcrete is usually assumed to be $v_c = 0.25$. The thickness t_c of the shotcrete depends on the roughness of the surface

---

[1] The distinction between *dry* and *wet* mixtures comes from the moment at which water is added to the cement/sand mixture. In the former, sand and cement are mixed dry and water is added at the nozzle of the shotcrete equipment. In the latter, sand, cement and water are mixed at the same time and the mixture pumped to the nozzle for application.

Table 3. Values of Young's modulus E_s and yield strength σ_{ys} for different steel types (adapted from Gieck 1977).

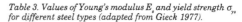

Designation	E_s [MPa]	σ_{ys} [MPa]
St 37–11	210×10^3	$80-120$
St 50–11	210×10^3	$100-150$
GS 38	220×10^3	$80-100$

The ranges of admissible stress σ_{ys} listed above are for static-compressive loads. This assumes a safety coefficient of 1.75 with respect to the yield strength.

after blasting and scaling. When the shotcrete is applied as a temporary support system, the thickness usually varies between 50 and 100 mm.

For pre-cast or cast-in-place concrete support, the parameters σ_{cc} and E_c depend mainly on the type of cement and aggregate used in the mixture. Table 2, adapted from Leonhardt (1973), lists values of σ_{cc} and E_c for typical concrete mixtures used in the construction industry. Poisson's ratio v_c for concrete varies between 0.15 and 0.25; the value $v_c = 0.2$ is normally used in practice (Leonhardt 1973). The thickness t_c for pre-cast or cast-in-place support is usually larger than that for shotcrete, partly because structural steel reinforcement is commonly used (structural steel reinforcement requires a sufficient cover of concrete to protect the steel from corrosion).

4.2 Blocked Steel Set

Considering steel sets spaced a unit length apart in the direction of the tunnel axis and tightened against the rock by wood blocks that are equally spaced circumferentially— as shown in Figure 5b— the maximum pressure that the system can sustain is

$$p_s^{\max} = \frac{3}{2}\frac{\sigma_{ys}}{S\ R\ \theta}\frac{A_s I_s}{3I_s+DA_s[R-(t_B+0.5D)](1-\cos\theta)} \quad (12)$$

The elastic stiffness is

$$\frac{1}{K_s} = \frac{S\ R^2}{E_s A_s}+\frac{S\ R^4}{E_s I_s}\left[\frac{\theta(\theta+\sin\theta\cos\theta)}{2\sin^2\theta}-1\right]+\frac{2S\ \theta\ t_B\ R}{E_B B^2} \quad (13)$$

where

B is the flange width of the steel set and the side length of the square block [m]

D is the depth of the steel section [m]

A_s is the cross-sectional area of the section [m²]

I_s is the moment of inertia of the section [m⁴]

E_s is Young's modulus for the steel [MPa]

σ_{ys} is the yield strength of the steel [MPa]

S is the steel set spacing along the tunnel axis [m]

q is half the angle between blocking points [radians]

t_B is the thickness of the block [m]

E_B is Young's modulus for the block material [MPa]

R is the tunnel radius [m]

a)

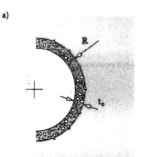

b)

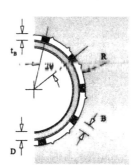

Figure 5. Schematic representation of sections of a) shotcrete or concrete rings and b) blocked steel sets (adapted from Brady and Brown 1983).

Table 4. Values of D, B, A$_s$ and I$_s$ for typical steel sections (adapted from CONSTRADO 1988).

Universal Beams

Section	D [mm]	B [mm]	A$_s$ [m^2]	I$_s$ [m^4]	t [mm]	T [mm]
457 × 152	461	153	9.50×10^{-3}	324.35×10^{-6}	9.9	17.0
406 × 140	402	142	5.90×10^{-3}	156.47×10^{-6}	6.9	11.2
356 × 127	353	126	4.94×10^{-3}	100.87×10^{-6}	6.5	10.7
305 × 127	304	124	4.75×10^{-3}	71.62×10^{-6}	7.2	10.7
254 × 102	260	102	3.62×10^{-3}	40.08×10^{-6}	6.4	10.0
203 × 133	203	133	3.23×10^{-3}	23.56×10^{-6}	5.8	7.8

Joists

Section	D [mm]	B [mm]	A$_s$ [m^2]	I$_s$ [m^4]	t [mm]	T [mm]
203 × 102	203	102	3.23×10^{-3}	22.94×10^{-6}	5.8	10.4
152 × 89	152	89	2.18×10^{-3}	8.81×10^{-6}	4.9	8.3
127 × 76	127	76	1.70×10^{-3}	4.76×10^{-6}	4.5	7.6
102 × 64	102	64	1.23×10^{-3}	2.18×10^{-6}	4.1	6.6
89 × 89	89	89	2.94×10^{-3}	3.07×10^{-6}	9.5	9.9
76 × 76	76	76	1.63×10^{-3}	1.59×10^{-6}	5.1	8.4

Universal beams, in contrast to *joists*, have flanges of the same thickness throughout. The figure on the right shows a typical joist section with non-parallel flanges (the thickness T is measured at the mid-distance on the flange). Note that the values of moment of inertia I_s listed above are with respect to the axis x–x indicated in the figure. This considers that the shortest side (the flange of width B) is in contact with the wood block placed between the rock surface and the steel section (see Figure 5b).

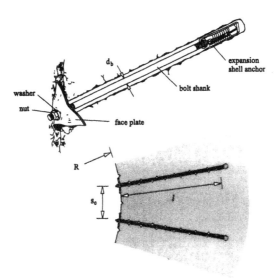

Figure 6. Representation of an ungrouted mechanical-anchored bolt (adapted from Stillborg 1994 and Hoek and Brown 1980).

For the case of blocked steel sets spaced at intervals d_s other than unity, both the maximum stress p_s^{max} given by equation (12) and the stiffness K_s given by equation (13) should be divided by $(d_s/1.0 \text{ m})$.

Values of Young's modulus E_s and yield strength σ_{ys} for different types of steel are listed in Table 3.

Values of D, B, A_s and I_s for typical sections of steel are given in Table 4.

The angle θ in equations (12) and (13) can be computed as $\theta = \pi/n_B$, where n_B is the total number of equally spaced blocks installed in the cross-section.

Young's Modulus for the wood block depends on the type of wood used and on the tightness of the block at installation. For hard woods (e.g., ash, maple, oak) Young's Modulus is typically $E_b = 10 \times 10^3$ MPa and for conifers (e.g., pine, cypress, cedar) it is $E_b = 7 \times 10^3$ MPa (Derucher and Korfiatis 1988). In order to take into account the tightness of the block at installation, Hoek and Brown (1980) suggest the values $E_b = 10 \times 10^3$ MPa for stiff blocking and $E_b = 500$ MPa for soft blocking.

4.3 Ungrouted Bolts and Cables

The sketches in Figure 6 represent mechanically anchored bolts installed in the rock-mass surrounding a circular tunnel of radius R. Assuming that the bolts are equally spaced in the circumferential direction, the maximum support pressure provided by this support system is

$$p_s^{max} = \frac{T_{bf}}{s_c \, s_l} \qquad (14)$$

The stiffness is

$$\frac{1}{K_s} = s_c \, s_l \left[\frac{4\,l}{\pi\,d_b^2\,E_s} + Q \right] \qquad (15)$$

where

d_b is the bolt or cable diameter [m];
l is the free length of the bolt or cable [m];
T_{bf} is the ultimate load obtained from a pull-out test [MN];
Q is a deformation-load constant for the anchor and head [m/MN];
E_s is Young's Modulus for the bolt or cable [MPa];
s_c is the circumferential bolt spacing [m];
s_l is the longitudinal bolt spacing [m].

Equation (15) assumes that the reaction forces developed by the bolt are concentrated at the ends of the bar therefore the equation should not be applied in the case of grouted bolts —for which the load transfer is distributed throughout the length of the shank.

The circumferential bolt spacing s_c in equations (14) and (15) can be computed as $s_c = 2\pi R /n_b$, where n_b is the total number of equally spaced bolts or cables installed in the cross-section.

Typical values of Young's modulus for the steel are listed in Table 3.

To illustrate how the constants T_{bf} and Q are obtained from testing of bolts, consider the diagram in Figure 7. This shows the results of a pull-out test performed by Stillborg (1994) on a mechanically anchored bolt 16 mm in diameter

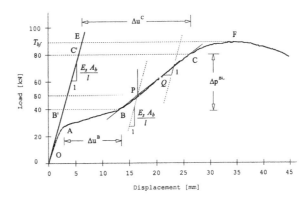

Figure 7. Results of a pull-out test performed on a mechanically anchored bolt (adapted from Stillborg 1994).

and 3 m long installed in a concrete block of compressive strength $\sigma_{cc} = 60$ MPa (details of the test can be found in the original article). The load applied to the bolt is shown on the vertical axis and the bolt deformation on the horizontal axis. The line OE corresponds to the elastic deformation of the shank (see Fig. 6). The curve $OABCF$ shows the total deformation measured at the wall of the concrete block — this includes the deformation of the shank, anchor, plate, washer and nut.

The high rate of displacement over the portion AB of the curve is associated with the initial compliance of the plate, washer and nut assembly. The steeper portion BC is associated with deformation of the bolt shank and the anchor. At point C in the curve, the bolt starts to yield and at point F the bolt fails. The constant T_{bf} in equation (14) is therefore defined by point F in Figure 7.

In practice, the bolt will usually be pre-tensioned during installation, in order to avoid the initial 'flat' segment AB associated to compliance of the plate and its associated components. The level of pre-tension should be sufficient to fully 'seat' the plate, washer and nut. For the results presented in Figure 7, for example, an appropriate level of pre-tensioning would be between 40 and 60 kN. Thus if a pre-tension of 50 kN is applied, the effect is to move the origin of the (subsequent) load-deformation curve from point O to point P.

The constant Q in equation (15) is the rate of deformation for the portion BC in Figure 7—this disregards the elastic deformation of the shank that is already accounted by the first term within brackets in equation (15).

Table 5. Values of the ultimate load T_{bf} and the deformation load constant Q for bolts of different diameters d_b and lengths l (adapted from Hoek and Brown 1980).

d_b [mm]	l [m]	T_{bf} [MN]	Q [m/MN]
16	1.83	0.058	0.241
19	1.83	0.089	0.024
22	3.00	0.196	0.042
25	1.83	0.254	0.143

Values determined for *expansion shell* bolts in field tests. The rock types are: *i)* shale for the 16 mm bolt; *ii)* sandstone for the 22 mm bolt; *iii)* granite for the 25 mm bolt.

204

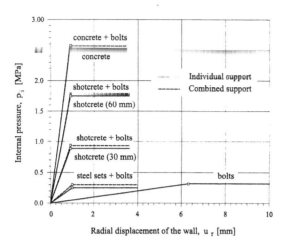

Individual support systems			
Support type	p_s [MPa]	K_s [MPa/m]	u_r^{max} [m]
Shotcrete ($t_c = 30$ mm)	0.89	0.984×10^3	0.90×10^{-3}
Shotcrete ($t_c = 60$ mm)	1.75	2.019×10^3	0.87×10^{-3}
Concrete ($t_c = 75$ mm)	2.53	2.893×10^3	0.87×10^{-3}
Steel sets (127×76)	0.25	0.261×10^3	0.95×10^{-3}
Bolts (19 mm diameter)	0.32	0.050×10^3	6.36×10^{-3}

Combined support systems			
Support type	p_s [MPa]	K_s [MPa/m]	u_r^{max} [m]
Shotcrete (30 mm) + Bolts	0.93	1.034×10^3	0.90×10^{-3}
Shotcrete (60 mm) + Bolts	1.79	2.069×10^3	0.87×10^{-3}
Concrete + Bolts	2.57	2.943×10^3	0.87×10^{-3}
Steel sets + Bolts	0.30	0.311×10^3	0.95×10^{-3}

Mechanical and geometrical properties considered for the supports: *i)* Shotcrete, $\sigma_{cc} =$ 30 MPa, $E_c = 30 \times 10^3$ MPa, $\nu = 0.25$, $t_c = 30$ mm and $t_c = 60$ mm. *ii)* Concrete, $\sigma_{cc} = 35$ MPa, $E_c = 35 \times 10^3$ MPa, $\nu = 0.2$, $t_c = 75$ mm. *iii)* Blocked steel sets, $B = 76$ mm, $D = 127$ mm, $A_s = 1.70 \times 10^{-3}$ m^2, $I_s = 4.76 \times 10^{-6}$ m^4, $E_s = 210 \times 10^3$ MPa, $\sigma_{ys} = 150$ MPa, $S = 1$ m, $\theta = \pi/10$ rad (10 blocks), $t_B = 75$ mm, $E_b = 10 \times 10^3$ MPa. *iii)* Ungrouted bolts, $d_b = 19$ mm, $l = 2$ m, $T_{bf} = 0.1$ MN, $Q = 0.03$ m/MN, $E_s = 210 \times 10^3$ MPa, $s_c = 0.63$ m (10 bolts), $s_l = 0.50$ m.

Figure 8. Support Characteristic Curves for various support systems applied to a tunnel of radius $R = 1$ m.

Considering the magnitudes Δp^{BC}, Δu^B and Δu^C indicated in the diagram, the constant Q can be computed as

$$Q = \frac{\Delta u^C - \Delta u^B}{\Delta p^{BC}} \qquad (16)$$

Ideally the values of T_{bf} and Q should be obtained from pull-out tests performed directly on bolts installed on the rock in situ. The values obtained in this way will depend on the type of rock and the mechanical characteristics of the bolt being tested. Hoek and Brown (1980) list reference values for T_{bf} and Q obtained from tests in different rock types. Some of these values are presented in Table 5 (these are for bolts of diameter 16, 19, 22 and 25 mm and lengths of 1.83 and 3.0 m).

4.4 Combined Effect of Support Systems

If more than one of the support systems described earlier is installed at the same location, their combined effect can be determined by adding the stiffnesses (i.e., the slope of the load vs. deformation curve) for each of the individual supports. This has the effect of increasing the slope of the elastic part of the SCC— the segment KR in Figure 3.

Consider, for example, the case in which two supports —characterized by maximum pressures p_{s1}^{max} and p_{s2}^{max} and elastic stiffnesses K_{s1} and K_{s2}, respectively, are installed in a section of tunnel. The stiffness K_s for the two systems acting together can be computed as $K_s = K_{s1} + K_{s2}$. This value is assumed to remain valid until one of the two supports achieves its maximum possible elastic deformation u_r^{max}—computed as $u_{r1}^{max} = p_{s1}^{max}/K_{s1}$ and $u_{r2}^{max} = p_{s2}^{max}/K_{s2}$, respectively (see equation 9). The combined support system is assumed to fail at that point. The support with the lowest value of u_r^{max} determines the maximum support pressure available for the two supports acting together.

Figure 8 shows the SCC for four types of support: shotcrete (with two different thicknesses), concrete, steel sets and ungrouted bolts for a tunnel of radius $R = 1$ m. The figure also considers the combined effect of two supports acting together: concrete and bolts; shotcrete and bolts; and steel sets and bolts. These SCC's were evaluated using the equations presented in Sections 4.1 through 4.3 (the values considered for the variables are listed in the lower part of the figure.)

For this particular problem, as seen in Figure 8, the SCC for steel sets is considerably below that for shotcrete and concrete linings. This is due to the dominant role played by the (compliant) wood blocking in the system. The lowest values of stiffness K_s —and therefore the maximum deformation u_r^{max}, corresponds to the bolts.

For the combined support systems, the failure is controlled by the stiffer shotcrete, concrete or steel set support types (compare the values of u_r^{max} listed in both tables). There is a slight improvement in the support capacity p_s^{max} and the stiffness K_s when two supports are considered acting together.

5. Construction of the Longitudinal Deformation Profile

The Longitudinal Deformation Profile (LDP) discussed in Section 2 is an important component of the Convergence-Confinement method. It provides insight into how quickly the support begins to interact with the rock-mass behind the face of the tunnel (i.e., it defines the point K in Fig. 3).

When the far-field stresses acting on the rock-mass are assumed to be uniform, the profile of radial displacements along the axis of the tunnel can be computed from numerical models of the problem shown in Figure 9a. The figure represents a longitudinal cross-section of an unlined tunnel of radius R in the vicinity of the face. At a distance x from the face the radial displacement is u_r. When the distance x is large enough, the radial displacement reaches the maximum value u_r^M. For negative values of x (i.e., ahead of the face), the radial displacement decreases and the displacement becomes essentially zero at some finite distance ahead of the face.

From elastic models of the problem represented in Figure 9a, Panet (1995) suggests the following relationship between radial displacements and distance to the face:

$$\frac{u_r}{u_r^M} = 0.25 + 0.75\left[1 - \left(\frac{0.75}{0.75 + x/R}\right)^2\right] \qquad (17)$$

This relationship (17), that applies to positive values of x, is plotted in Figure 9b. The horizontal axis of the diagram represents the ratio x/R and the vertical axis represents the ratio u_r/u_r^M.

Chern et al. (1998) present measured values of convergence in the vicinity of the face for a tunnel in the Mingtam Power Cavern project. The measured data are plotted as dots in Figure 9b. Based on this data, Hoek (1999) suggests the following empirical best-fit relationship between radial displacement of the tunnel and distance to the face:

$$\frac{u_r}{u_r^M} = \left[1 + \exp\left(\frac{-x/R}{1.10}\right)\right]^{-1.7} \qquad (18)$$

The relationship (18) is also plotted in Figure 9b. Analysis of the curves defined by equations (17) and (18) indicates that the maximum radial displacement occurs at approximately 8 tunnel radii behind the face of the tunnel, and that the radial deformation is zero at approximately 4 tunnel radii ahead the face. At the face itself, the radial displacement is approximately 30% of the maximum value. Figure 9b also suggests that the elastic approximation defined by equation (17) overestimates the values of radial displacements when compared with the measured data at the Mingtam Power Cavern project and with the statistical approximation to this data. As seen in Figure 3, this overestimation results in underestimation of the final load transmitted to the support.

Ideally, for tunnels designed according to the Convergence-Confinement method, the LDP should be constructed from measured data such as the one presented by Chern et al. (1998). Where such information is not available, the LDP can be constructed from numerical models considering the same elasto-plastic parameters used in construction of the Ground Reaction Curve. Alternatively, and as a first approximation, the LDP could be evaluated using relationship (18).

6. Example

To illustrate the application of the Convergence-Confinement method in the design of tunnel supports we will re-examine the case of the circular tunnel of radius $R = 1$ m shown in Figure 4a. The uniform far-field stress acting in

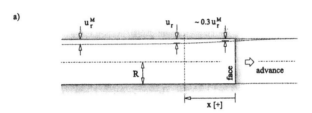

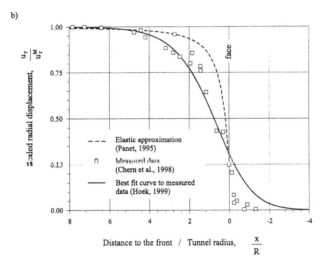

Figure 9. a) Profile of radial displacements u_r for an unsupported tunnel in the vicinity of the tunnel face. b) Deformation profiles derived from elastic models (Panet 1995); measurements in a tunnel (Chern et al. 1998); and best fit to the measurements (Hoek 1999).

206

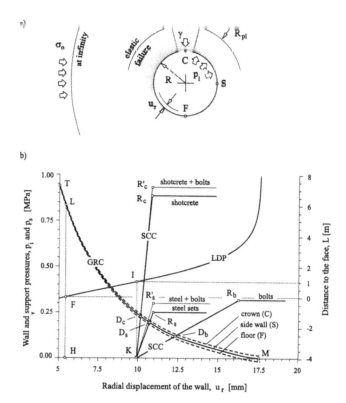

Figure 10. a) Influence of the weight of failed rock in the crown and the floor of a tunnel on the Ground Reaction Curve (points C and F, respectively). b) Rock-support interaction analysis for the tunnel shown in Figure 4 and the support systems shown in Figure 8.

the rock-mass is σ_o =7.5 MPa. A Geological Strength Index equal to 40 is assumed for the rock. The properties of the rock-mass are as indicated in the diagram.

We wish to assess the mechanical behavior of the different support systems considered in Figure 8, assuming that the supports are installed at a distance L=1 m behind the face of the tunnel.

Solution of this problem requires application of the equations presented in Sections 3, 4 and 5, for construction of the GRC, SCC and LDP respectively. Appendix C includes the outline of a spreadsheet for ease in implementation of the equations and construction of the GRC, SCC and LDP. The results of the example are summarized in Figure 10.

For the problem being considered, the GRC is the curve identified as GSI = 40 in Figure 4b. In defining the GRC for a tunnel, it is usual to distinguish between the convergence of side wall, roof and floor of the tunnel (Daemen 1975), since gravitational loading differs for each of these points around the tunnel periphery. The GRC shown in Figure 4b (the curve corresponding to GSI = 40) is considered to be representative of the load-convergence relationship for a point on the side-wall of the tunnel—point S in Figure 10a. For the same internal pressure p_i, the convergence of the

roof can be expected to be larger than that at the side because of the weight of the failed material on top of the tunnel (point C in the figure). The GRC for the roof can then be obtained by adding the same amount $\gamma (R_{pl}$–$R)$ to the internal pressure p_i, where γ is the unit weight of the rock-mass and R_{pl} is the extent of the plastic region —values of γ for different rock types are listed in Table A-1. Similarly, the GRC for the floor of the tunnel can be obtained by subtracting the amount $\gamma (R_{pl}$–$R)$ from the internal pressure p_i (see point F in Figure 10a).

Figure 10b shows an enlargement of the diagram of Figure 4b, representing the GRC curve for GSI = 40. The solid curve that extends from point T to point M represents the radial displacement of point S in Figure 10a. The dotted curves above and below this curve are the GRC's for the crown (point C) and floor (point F) for an assumed rock-mass unit weight γ =24 × 10^{-3} MN/m^3. Note that the pressure transmitted at the crown of the tunnel is slightly higher than the pressure at the floor.

The LDP computed from equation (18) is also shown in Figure 10b. This curve is an enlargement of the solid curve in Figure 9b (note that, in Figure 10b, the distance to the face is shown on the vertical axis on the right side, and the associated radial displacement on the horizontal axis). For the distance $L = 1$ m between the face and the support at installation, the LDP allows to define the point K (horizontal coordinate $u_r \approx 10$ mm), from which the SCC starts (see also point K, Fig. 3). In Figure 10, the face of the tunnel is identified by the vertical line passing through point F, that is defined by the coordinate $L = 0$ m in the vertical axis on the right.

The SCC for the support systems shown in Figure 8 are also represented in Figure 10b. Note that only shotcrete (30 mm thick), steel sets and bolts are considered as support alternatives in this problem. The SCC for the combination of two support systems (shotcrete and bolts *and* steel sets and bolts) are also shown in the figure.

Following the notation in Figure 3, the maximum stress p_s^{max} for each of the different support systems is indicated as points R and R' (for the individual and the combined systems, respectively). For example, the maximum support pressure for steel sets is ≈ 0.25 MPa, for steel sets and bolts ≈ 0.30 MPa, for bolts ≈ 0.35 MPa, for shotcrete ≈ 0.88 MPa, and for shotcrete and bolts ≈ 0.94 MPa.

As mentioned above, the point K is defined by the LDP for a distance to the face of $L = 1$ m. It is evident from this diagram, that the point K cannot lie to the left of point H, which represents the face of the tunnel. Therefore, the maximum possible pressure that the rock-mass can transmit to any given support in this problem will be less than the pressure defined by the vertical coordinate of point L in the figure —this is ≈ 0.84 MPa (see Section 2).

Points D_s, D_c and D_b in Figure 10 represent the final support pressure that the rock-mass will transmit to the different support systems, once the tunnel face has moved well away from the support; for example, the final pressure transmitted by the rock-mass is ≈ 0.12 MPa for bolts (D_b); ≈ 0.20 MPa for steel sets (D_s) and ≈ 0.23 MPa for shotcrete (D_c).

Analysis of the location of points D in the diagram indicates that all the supports considered in this problem are capable of sustaining the final load transmitted by the rock-mass (note that points D lie below the corresponding 'support-capacity' points R). The relative merits of one or other of the support systems will depend on the allowable amount of convergence and/or the value of safety margin against failure desired for the support.

The final convergence for each support system is given by the horizontal coordinate of the various points D; for example, for shotcrete the convergence is ≈ 10.1 mm, for steel sets ≈ 10.6 mm and for bolts ≈ 12.4 mm.

The safety coefficients for the different supports can be obtained as the ratio between the vertical coordinates of points R and D in the figure; for example, for bolts the ratio is 0.30/0.12 [MPa/MPa], indicating a safety coefficient of ≈ 2.5. If minimizing the wall closure is a primary concern, the steel sets and shotcrete are the best alternative (note that the horizontal coordinate of point D_b is significantly larger than the horizontal coordinates of points D_c or D_s). On the other hand, if a large safety coefficient is desired, shotcrete and bolts are superior to steel sets (note that the ratio of vertical coordinates of points R_c and D_c and R_b and D_b). If steel sets are required to be used to line the tunnel anyway, it may be worth considering installation of bolts in combination with the sets. This is justified by the significant improvement of the 'safety' margin —i.e., the ratio of vertical coordinates of points R_s and D_s ($\approx 0.25/0.20 = 1.25$) compared to R'_s and D_s ($\approx 0.30/0.20 = 1.5$).

7. Limits of Application of the Convergence-Confinement Method

The Convergence-Confinement method is based on two important assumptions:

i) the far-field principal stresses normal to the long axis of the tunnel are of constant magnitude σ_o, independent of the radial orientation (such a state of stress is often referred to as *uniform* or *hydrostatic*); and

ii) the tunnel cross-section is circular, of radius R.

This section discusses the validity of the method for cases in which the far-field (principal) stresses are unequal and the tunnel cross-section is non-circular.

The principal stresses that exist at the site prior to excavation of a tunnel depend on the geological history of the site and in general are unequal. Figure 11a, adapted from Hoek and Brown (1980), shows measured values of vertical stresses σ_z as a function of depth z for different regions of the world. The linear function that best fits the measured data is given by the relationship,

$$\sigma_z = 0.027\, z \qquad (19)$$

where σ_z is expressed in MPa and z in meters. If the vertical stress at a depth z is assumed to be associated with the weight of overburden material (i.e., the lithostatic pressure at the depth z), equation (19) suggests that the mean unit weight of the rock where the measurements were made is about 0.027 MN/m³. This value corresponds to the unit weight of silicates, a major component of many rocks (see Table A-1).

The mean horizontal stress σ_x at a depth z is usually expressed in terms of the corresponding vertical stress σ_z. The horizontal-to-vertical stress ratio k (also referred to in soil mechanics as the 'at-rest' coefficient of earth pressure) is defined as

$$k = \frac{\sigma_x}{\sigma_z} \qquad (20)$$

Figure 11b, adapted from Hoek and Brown (1980), shows the values of the coefficient k corresponding to the vertical stresses in Figure 11a. The diagram indicates that the horizontal stresses σ_x are bounded by the two curves shown. The *minimum* value of mean horizontal stress is ≈ 0.5 times the value of vertical stress and the *maximum* value of horizontal stress is ≈ 3.5 times the mean value of vertical stress. Even higher ratios than those indicated in Figure 11b have been recorded. At the Underground Research Laboratory (URL) in Pinawa (Canada), for example, the mean horizontal stress in the Lac du Bonnet granite is 52 MPa, at a depth of 420 m —with a vertical stress of approximately 11 MPa, i.e., $k = 4.6$ (see Martin and Simmons 1993).

The diagrams in Figure 11 suggest that the principal stresses at the site are often unequal. They also indicate that the vertical stress will probably vary with depth according to a lithostatic gradient.

If the dimensions of the cross section of the tunnel are small compared to the depth z of the tunnel, the far-field stresses σ_z and σ_x can be assumed to be constant over the proposed tunnel section (e.g., the vertical in-situ stress at the depth of the crown and invert of the proposed tunnel can be considered to be the same —and similarly for the horizontal in-situ stresses).

Two quantities can be used to characterize a given non-uniform plane stress state (σ_z, σ_x): the dimensionless coefficient k—defined by equation (20), and the mean stress σ_o defined as

$$\sigma_o = \frac{\sigma_x + \sigma_z}{2} \qquad (21)$$

The uniform state of stress assumed by the Convergence-Confinement method can be expressed as $\sigma_o = \sigma_x = \sigma_z$ and $k = \sigma_x / \sigma_z = 1$ (see Fig. 1b).

The elasto-plastic problem of excavating a circular tunnel in a non-uniform stress field has been studied analytically by a few investigators.

208

a)

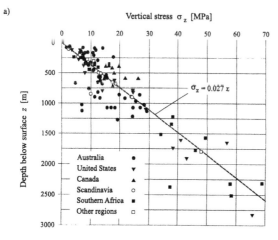

b)

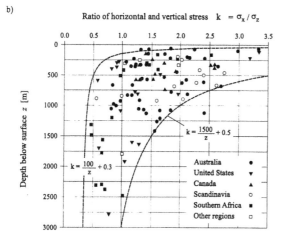

Figure 11. a) Measured values of vertical stress σ_z as a function of depth z in different regions of the world. b) Corresponding values of horizontal-to-vertical stress ratio k (adapted from Hoek and Brown 1980).

Detournay and Fairhurst (1987) considered the case of a circular cavity subject to unequal far-field stresses and excavated in a Mohr-Coulomb material. Figure 12a illustrates the problem considered. A circular cavity of radius R is subject to a uniform internal pressure p_i and horizontal and vertical stresses σ_x and σ_z, respectively; the figure considers the case $\sigma_z > \sigma_x$ (because of the symmetry of the problem, the case $\sigma_x < \sigma_z$ can be obtained by rotating the axes of the cavity through 90°).

In contrast to the Hoek-Brown failure criterion (A-2) discussed in Appendix A, the Mohr-Coulomb failure criterion considered by Detournay and Fairhurst is

$$\sigma_1 = K_p \, \sigma_3 + \sigma_{ci} \qquad (22)$$

where σ_{ci} is the unconfined compressive strength of the intact rock and K_p is the 'passive reaction' coefficient that is

computed from the friction angle ϕ of the intact rock as

$$K_p = \frac{1 + \sin \phi}{1 - \sin \phi} \qquad (23)$$

It was found that, when the stress state represented in Figure 12a is expressed in terms of the parameters σ_o and k defined by equations (21) and (20), respectively, cavities with horizontal-to-vertical stress coefficients k smaller than a limiting value k_{lim} are statically determinate. The authors also made the interesting observation that the mean radius of the plastic region around the tunnel and the mean convergence at the crown and sidewall of the tunnel are the same as the corresponding values for a cavity subject to *hydrostatic* far-field stresses σ_o —i.e., comparing Figures 1b and 12a, $R_{pl} \approx 0.5 (R_{pl1} + R_{pl2})$ and $u_r \approx 0.5 (u_{r1} + u_{r2})$. Cavities with a horizontal-to-vertical stress coefficient k larger than a limiting value k_{lim}, are statically indeterminate and develop a 'butterfly-shaped' failure zone (see Figure 12a); for these cases, the extent of the failure zone and the displacements around the periphery of the tunnel are decidedly non-uniform —and have no apparent relationship to the case of uniform loading σ_o represented in Figure 1b.

The limiting coefficient k_{lim} can be determined from the values of scaled mean pressure σ_o / σ_{ci} and friction angle ϕ in Figure 12b. Note that the diagram in Figure 12b assumes that the cavity is unsupported. Because of the self-similar nature of the problem, the diagram also applies to cases in which there is a uniform internal pressure acting inside the cavity. In such cases, the value of k_{lim} is read from Figure 12b by taking an equivalent ratio of mean pressure and compressive strength $\sigma_o / \sigma_{ci} |_{eq}$, as given by the expression in equation (24), (after Detournay and St. John 1988),

$$\left. \frac{\sigma_o}{\sigma_{ci}} \right|_{eq} = \frac{\frac{\sigma_o}{\sigma_{ci}} \left(1 - \frac{p_i}{\sigma_o} \right)}{\frac{p_i}{\sigma_o} \frac{\sigma_o}{\sigma_{ci}} (K_p - 1) + 1} \qquad (24)$$

Figure 12b can be used to evaluate the applicability of the Convergence Confinement method when the far-field stresses are non-uniform. For tunnels characterized by coefficients k smaller than the value k_{lim} shown in the figure, the Convergence-Confinement method provides a reasonable estimation of the shape of the failure zone and displacements to expect around the cavity. For tunnels characterized by coefficients k larger than the value k_{lim} shown in the figure, the resulting shape and extent of the failure zone and the convergence around the tunnel will be too variable to apply the method. For these cases, numerical analysis should be used for the design.

As mentioned above, Figure 12b applies to Mohr-Coulomb materials characterized by a friction angle ϕ and unconfined compressive strength σ_{ci} (and linear failure envelope). For Hoek-Brown materials considered in this study, an equivalent diagram to that represented in Figure 12b could be constructed by approximating the Hoek-Brown

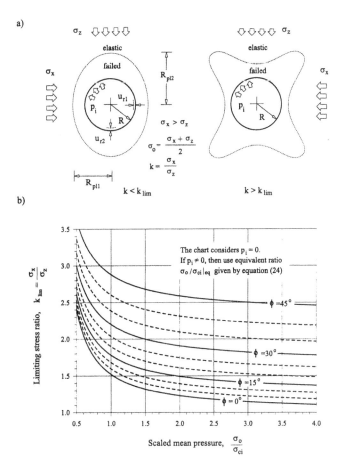

Figure 12. a) Circular cavity in a Mohr-Coulomb material subject to uniform internal pressure and unequal far-field stresses. b) Limiting values of the horizontal-to-vertical stress ratio k_{lim} as a function of the scaled mean stress σ_o/σ_{ci} and friction angle ϕ. For $k < k_{lim}$, the problem is statically determinate, and the mean values of failure extent and wall closure are comparable to those obtained for uniform loading σ_o and $k=1$ (adapted from Detournay and Fairhurst 1987).

parabolas of Figure A-1 as straight lines with 'equivalent' friction angle ϕ and unconfined compressive strength σ_{ci} (examples of this type of approximation can be found in Hoek 1990).

For tunnels driven in Hoek-Brown materials under unequal far-field stresses, an indication of the expected shape of the failure zone —and thus an estimate of the applicability of the Convergence-Confinement method, can be obtained from elastic analysis. Figure 13a presents the results from an analysis of this type. The different curves in the figure represent the extent of the 'over-stressed' regions determined by comparing the maximum and minimum elastic stresses at a point as given by the Kirsch elastic solution, with the Hoek-Brown strength criterion (A-2), i.e.,

$$\sigma_1 \geq \sigma_3 + \sigma_{ci}\left(m_b\,\frac{\sigma_3}{\sigma_{ci}} + s\right)^a \tag{25}$$

If the elastic stresses exceed the strength indicated by equation (25), the rock at that point is *over-stressed*. The diagrams in Figure 13a show cases of unequal far-field stresses characterized by $\sigma_o = 7.5$ MPa and different values of horizontal-to-vertical stress ratio k, together with rock-mass strength parameters $\sigma_{ci} = 20$ MPa, $m_b = 1.8$ and $s = 1.3 \times 10^{-3}$ in the yield criterion (25). The curves on the right side of the diagram correspond to horizontal-to-vertical stress coefficients $k < 1$ and the curves on the left side of the diagram to coefficients $k > 1$. The equations and a spreadsheet needed to construct these curves are shown in Appendix D. It is seen, for example, that for $k < 0.6$, the *over-stressed* region has the 'butterfly' shape discussed in the analysis by Detournay and Fairhurst. The Convergence-

Confinement method should not be used for such cases. Elastic rock-support interaction analyses such as those presented by Einstein and Schwartz (1979) and Matsumoto and Nishioka (1991) are preferable in these cases. Wherever possible, results obtained from these elastic analyses should be verified against results obtained from elasto-plastic numerical analyses. The latter consider stress and displacement changes that occur due to formation of the plastic zone.

However, although approximate, the shape of the plastic zone obtained from the elastic *over-stressed* analysis is generally comparable to the correct one (note that the *extent* of the plastic or failed zone, can be considerably underestimated by this approach). For example, Shen and Barton

(1997) have used the elastically *over-stressed* zone concept to identify the shape of regions where shear failure is likely to occur in heavily jointed rock-masses. They showed that the shape of the regions of 'slipping' joints derived from elastic analysis was comparable to the shape obtained using the discrete element numerical model UDEC (Cundall 1971).

Figure 13b shows a *FLAC*³ᴰ elasto-plastic model of the problem shown in Figure 4; the elasto-plastic model considers the far-field stresses to be $\sigma_z = 10.3$ MPa and $\sigma_x = 4.7$ MPa (i.e., $\sigma_o = 7.5$ MPa and $k = 0.45$). It is seen that the shape of the resulting failure zone in the *FLAC*³ᴰ model is comparable to the shape given by the curve corresponding to $k = 0.45$ in Figure 13a.

The assumption of circular cross-section of the tunnel in combination with hydrostatic far-field stresses guarantees that the displacements around the tunnel will be constant all around the periphery. In practical design of tunnels, the actual cross-section depends mainly on the purpose of the tunnel. For example, Figure 14a shows a 'horseshoe' shape chosen for a motorway tunnel in Germany. The width and height of the cross section depend on the number of lanes and maximum height of vehicles that the tunnel needs to serve.

For cases in which the cross-sectional area of the tunnel is not circular, the Convergence-Confinement method can still be used to provide a first estimate of the extent of the failure zone and the resulting convergence of the walls. Within certain limits, the shape of the tunnel can be regarded as circular with a radius equal to the mean value of the maximum and minimum dimensions of the section. In such cases, the mean extent of the failure zone and the mean convergence at the walls for the non-circular geometries are comparable to the values one would predict for the equivalent circular section.

Figure 14 shows the results of a 3DEC (Itasca Consulting Group 1998a) elasto-plastic analysis of the cross-sectional area shown in Figure 14a for the stress conditions and rock-mass and support properties indicated in the figure. Application of the Convergence-Confinement method assuming the section to be circular of radius $R = 4$ m, leads to a plastic zone $R_{pl} \approx 6$ m in extent. It is seen that this extent is comparable with that obtained from the 3DEC model in which the actual geometry of the cross-section is considered. Results for the loads and the convergence in the lining are also comparable.

8. Discussion

The Hoek-Brown criterion and its scaled form discussed in Appendix A is a convenient and widely used method for describing the strength of rock-masses in situ. When applying the criterion to real cases, the engineer should bear in mind the limitations, pointed out by Hoek and Brown in deriving the criterion. Probably the most important of these is the assumption of elastic

a)

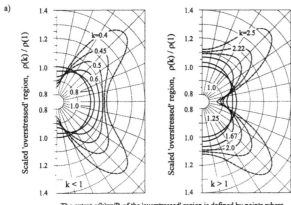

The extent $\rho(k) = r/R$ of the 'overstressed' region is defined by points where,

$$\sigma_1 \geq \sigma_3 + \sigma_{ci} \left(m_b \frac{\sigma_3}{\sigma_{ci}} + s \right)^a$$

b)

R = 1 m
$\sigma_x = 4.65$ MPa
$\sigma_z = 10.34$ MPa
$\sigma_o = 7.5$ MPa
k = 0.45

$\sigma_{ci} = 20$ MPa
$m_i = 15$
GSI = 40
$m_b = 1.8$
$s = 1.3 \times 10^{-3}$

$G_{rm} = 1.0$ GPa
$\nu = 0.25$
$\psi = 30°$

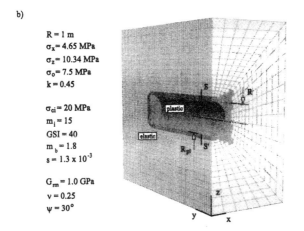

Figure 13. a) Diagrams indicating the extent of the elastically 'over-stressed' regions computed from the elastic analysis of a circular cavity subject to non-hydrostatic loading; the charts are valid for ratios $m_b \sigma_{ci} / \sigma_o = 4.8$ and $s/m_b^2 = 4 \times 10^{-4}$. For example, when $\sigma_o = 7.5$ MPa, $\sigma_{ci} = 20$ MPa, $m_b = 1.8$ and $s = 1.3 \times 10^{-3}$ (see Appendix D). b) FLAC³ᴰ elasto-plastic analysis of the problem represented in Figure 4 for highly unequal far-field principal stresses.

perfectly-plastic isotropic behavior for the material. If the rock-mass contains joints with a preferred orientation, the mechanical behavior can be expected to be anisotropic and the Hoek-Brown criterion may give quite misleading results. In such cases, a treatment with elasto plastic non-isotropic models (such as the ubiquitous joint model implemented in the continuum code *FLAC*) or with discontinuum models (such as *UDEC*) would be more appropriate (Itasca 1996, 1998).

In cases where the rock-mass is reasonably well described as isotropic, such that the Hoek-Brown criterion can be applied with reasonable confidence, the Convergence-Confinement method described in this paper can be used to obtain a useful estimate of the magnitude of loads that the rock-mass will transmit to supports installed behind the face of a tunnel.

As discussed in Section 2, the Convergence-Confinement method relies on several simplifying assumptions, the most important of which are the circular geometry of the tunnel and hydrostatic (or uniform) far-field stresses. In this case, the loads transmitted by the rock-mass to the support are uniform within each section. If the geometry of the tunnel is non-circular or the far-field stresses are non-uniform, the loads will not be uniform and bending moments will be induced in the support.

For situations in which the assumptions mentioned above are violated, the Convergence-Confinement method is still useful in the early stages of the design. As discussed in Section 7, a characteristic tunnel radius and a characteristic uniform stress can be computed from the existing geometry and far-field stresses and the engineer can then make quick comparisons of the mechanical response of different support alternatives—under representative conditions of rock-mass deformability and strength. This can help in deciding on the most convenient support system in the actual case. The final design of the support—which would probably require consideration of the distribution of bending moments and compressive loads induced by the non-uniform far-field stresses— can be made later on the basis of more rigorous numerical analyses of the rock-support interaction problem.

In summary, the Convergence-Confinement method is a useful tool not only for the design of supports in tunnels, but also as a simple illustrative model that allows a better understanding of the complex problem of transference of loads in the vicinity of the tunnel face.

References

AFTES, French Association for Underground Works. (1978). Analysis of tunnel stability by the Convergence-Confinement Method. *Underground Space* **4** (4), 221—223.

Atkinson, J. (1993). *An Introduction to the Mechanics of Soils and Foundations.* New York: McGraw-Hill Book Company.

Bieniawski, Z.T. (1976). Rock mass classification in rock engineering. In Bieniawski (Ed.), *Proc. of the Symp. in Exploration for Rock Engineering, Cape Town*, 97-106. Rotterdam: Balkema.

Brady, B.G.H. and E.T. Brown (1985). *Rock Mechanics for Underground Mining* (Second ed.). New York: Chapman & Hall.

Brown, E.T., J.W. Bray, B. Landayi, and E. Hoek (1983). Ground response curves for rock tunnels. *ASCE J. Geotech. Eng. Div.* **109** (1), 15–39.

Carranza-Torres, C. and C. Fairhurst (1999). The elasto-plastic response of underground excavations in rock masses that satisfy the Hoek-Brown failure criterion. *Int. J. Rock Mech. Min. Sci.* **36** (6), 777–809.

Chern, J.C., F.Y. Shiao, and C.W. Yu (1998). An empirical safety criterion for tunnel construction. In *Proc. Regional Symposium on Sedimentary Rock Engineering, Taipei, Taiwan*, 222–227.

CONSTRADO. 1983. *Steel Designers' Manual*. Granada: Constructional and Steel Research and Development Organisation.

a)

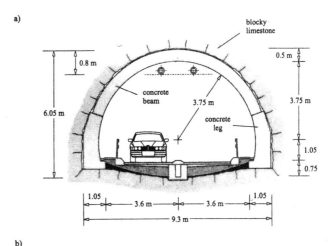

b)

Stress conditions
σ_o = 29 MPa
k = 1.0

Distance to the front
L = 4 m

Rock-Mass
σ_{ci} = 120 MPa
m_i = 9
GSI = 30
m_b = 0.74
s = 4.2 x 10^{-4}
G_{rm} = 1.33 GPa
ν = 0.3

Concrete lining
t = 0.5 m
σ_{cc} = 45 MPa
ν = 0.3
E_c = 36 GPa

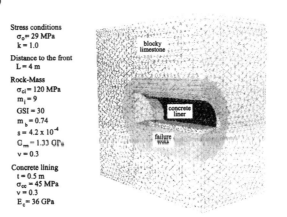

Figure 14. a) Cross-section of a German motorway tunnel in 'hydrostatically' loaded rock. b) 3DEC elasto-plastic model of the rock-support interaction problem. The extent of the failure zone predicted numerically is seen to agree with the value obtained from the Convergence-Confinement method; i.e., in this case, it is ≈ 1.5 times the characteristic radius of the tunnel.

212

Cundall, P.A. (1971). A computer model for simulating progressive large movements in blocky rock systems. *Proceedings of the Symposium of the International Society of Rock Mechanics, Volume 1*, Paper No. II-8.

Daemen, J. J. K. (1975). *Tunnel support loading caused by rock failure*. Ph. D. thesis, University of Minnesota.

Derucher, K. and G. Korfiatis (1988). *Materials for Civil and Highway Engineers*. New York: Prentice Hall.

Detournay, E. and C. Fairhurst (1987). Two-dimensional elasto-plastic analysis of a long, cylindrical cavity under non-hydrostatic loading. *Int. J. Rock Mech. Min. Sci. & Geomech. Abstr.* **24** (4), 197–211.

Detournay, E. and C. St. John (1988). Design charts for a deep circular tunnel under non-uniform loading. *Rock Mechanics and Rock Engineering* **21**, 119–137.

Dodge, M., C.Kinata, and C.Stinson (1997). *Running Microsoft Excel 97*. Redmond. Washington: Microsoft Press.

Einstein, H.H. and C.W. Schwartz (1979). Simplified analysis for tunnel supports. *ASCE J. Geotech. Eng. Div.* **104** (4), 499—518.

Fenner, R. (1938). Untersuchungen zur Erkenntnis des Gebirgsdruckes. *Glückauf* **74**, 681-695 and 705-715.

Franklin, J. A. and E. Hoek (1970). Developments in triaxial testing technique. *Rock Mechanics* **2**, 223–228.

Gieck, K. (1977). *Technische Formelsammlung*. Berlin: Verlag.

Goodman, R. E. (1980). *Introduction to Rock Mechanics*. New York: Wiley and Sons.

Hoek, E. (1983). Strength of jointed rock masses. Rankine Lecture. *Geotechnique* **33** (3), 187–223.

Hoek, E. (1990). Estimating Mohr-Coulomb friction and cohesion values from the Hoek-Brown failure criterion. *Int. J. Rock Mech. Min. Sci. & Geomech. Abstr.* **27** (3), 227–229.

Hoek, E. (1999). Personal communication.

Hoek, E. and E. T. Brown (1980). *Underground Excavations in Rock*. London: The Institute of Mining and Metallurgy.

Hoek, E. and E. T. Brown (1997). Practical estimates of rock mass strength. *Int. J. Rock Mech. Min. Sci. & Geomech. Abstr.* **34** (8), 1165–1186.

Hoek, E., P. K. Kaiser, and W. F. Bawden (1995). *Support of Underground Excavations in Hard Rock*. Rotterdam: Balkema.

Hoek, E., P. Marinos, and M. Benissi (1998). Applicability of the Geological Strength Index (GSI) classification for very weak and sheared rock masses. The case of the Athens Schist Formation. *Bull. Eng. Geol. Env.* **57** (2), 151–160.

Itasca Consulting Group (1996). *UDEC. Universal Distinct Element Code. Version 3.0*. User Manual. Minneapolis: Itasca.

Itasca Consulting Group (1997). *FLAC³ᴰ. Fast Lagrangian Analysis of Continua in 3 Dimensions*. Version 2.0. User Manual. Minneapolis: Itasca.

Itasca Consulting Group (1998a). *3DEC. 3 Dimensional Distinct Element Code. Version 2.0*. User Manual. Minneapolis: Itasca.

Itasca Consulting Group (1998b). *FLAC. Fast Lagrangian Analysis of Continua*. Version 3.4. User Manual. Minneapolis: Itasca.

Jaeger, J.C. and N.G.W. Cook (1979). *Fundamentals of rock mechanics* (Third ed.). New York: John Wiley & Sons.

Labasse, H. (1949). Les pressions de terrains dans les mines de huiles. *Revue Universelle des Mines*, Liège, Belgium, Series 9, Vol. 5, No 3, 78–88.

Lama, R.D. and V.S. Vutukuri (1978). *Handbook of Mechanical Properties of Rocks. Testing Techniques and Results. Volume II*. Trans Tech Publications.

Leonhardt, F. (1973). *Vorlesungen über Massivbau*. Berlin: Springer-Verlag.

Londe, P. (1988). Discussion on the determination of the shear stress failure in rock masses. *ASCE J. Geotech. Eng. Div.* **114**(3), 374–376.

Martin, C.D. and G.R. Simmons (1993). The Atomic Energy of Canada Limited Underground Research Laboratory: An Overview of Geomechanics Characterization. In J.A. Hudson (Ed.), *Comprehensive Rock Engineering*, 915–950. Pergamon: Oxford.

Matsumoto, Y. and T.Nishioka (1991). *Theoretical Tunnel Mechanics*. Tokyo: University of Tokyo Press.

Panet, M. (1995). *Calcul des Tunnels par la Méthode de Convergence-Confinement*. Paris: Press de l'école Nationale des Ponts et Chaussées.

Serafim, J.L. and J.P. Pereira (1983). Consideration of the geomechanical classification of Bieniawski. In *Proc. Int. Symp. on Engineering Geology and Underground Construction. Lisbon, Volume 1 (II)*, 33–44.

Shen, B. and N.Barton (1997). The disturbed zone around tunnels in jointed rock masses. *Int. J. Rock Mech. Min. Sci. & Geomech. Abstr.* **34** (1), 117–125.

Singh, M.M. and S.A. Bortz (1975). Use of special cements in shotcrete. In *Use of Shotcrete for Underground Structural Support*, 200–231. New York: American Society of Civil Engineers.

Stillborg, B. (1994). *Professional Users Handbook for Rock Bolting*. Trans Tech Publications.

Timoshenko, S.P. and J.N. Goodier (1970). *Theory of Elasticity* (Third ed.). New York: McGraw Hill.

Wang, Y. (1996). Ground response of a circular tunnel in poorly consolidated rock. *ASCE J. Geotech. Eng.* **9**, 703–708.

Appendix A. The strength and deformability of rock-masses according to the Hoek-Brown failure criterion.

The Hoek-Brown criterion has found wide practical application as a method of defining the stress conditions under which a rock-mass will deform inelastically and, if not supported adequately, collapse.

The parameters defining the Hoek-Brown criterion can be estimated from a combination of laboratory tests on intact rock cores and an empirical 'adjustment' to account for the reduced strength of the rock-mass due to the presence of weaknesses and jointing.

It must be noted that this criterion assumes continuum-isotropic behavior for the rock-mass and should not be applied to cases in which there is a preferred orientation of jointing, such that the mass would not behave as an isotropic continuum.

Testing of rock specimens under triaxial conditions of loading allows the combination of stresses that lead to failure (or collapse) of the specimen to be determined. According to Hoek and Brown, the failure condition of *intact rock* samples is given by the following parabolic law (Hoek and Brown 1980),

$$\sigma_1 = \sigma_3 + \sigma_{ci} \sqrt{m_i \frac{\sigma_3}{\sigma_{ci}} + 1} \qquad (A\text{-}1)$$

where

σ_3 is the confining stress applied to the sample (e.g., in MPa);

σ_1 is the axial stress that produces failure of the sample (e.g., in MPa);

σ_{ci} is the unconfined compression strength of the intact rock (in MPa);

m_i is a dimensionless parameter, the value of which depends on the type of rock being tested.

In order to characterize the intact rock in terms of equation (A-1), it is necessary to determine the parameters σ_{ci} and m_i. This is done by statistical analysis of strength σ_1 observed for various values of confining stress σ_3 in triaxial tests (Hoek 1983). Appendix B explains this procedure and lists the equations needed to perform the analysis.

To illustrate the application of equation (A-1), let us consider the triaxial test results shown in Figure A-1 obtained by Franklin and Hoek (1970) for samples of different rock types: *i)* granite, *ii)* quartz dolerite and *iii)* marble (details of the tests can be found in the original paper; Appendix B shows a summary only of the results).

The horizontal and vertical axes in the diagram correspond, respectively, to the confining stress σ_3 and the axial stress at failure σ_1 divided by the unconfined compression strength σ_{ci} for each rock type. The dots represent the pairs (σ_3, σ_1) obtained from the triaxial tests (see Appendix B). The solid lines are the corresponding failure envelopes defined by equation (A-1) with the parameters σ_{ci} and m_i computed from equations (B-1) and (B-2) in Appendix B. It can be seen that, although there is some dispersion in the results, the general trend is for the scattered points to align to the parabolas defined by equation (A-1).

Triaxial testing of rock samples is an expensive procedure and, in most cases, results of the extensive tests needed to determine the parameters σ_{ci} and m_i in the relationship (A-1) are not available. In this case, when information on the unconfined compressive strength is available (e.g., from UCS tests or, indirectly, from Point Load Tests), the parameter m_i may be estimated from empirical charts or tables (Hoek et al. 1995).

Table A-1 and Figure A-2, adapted from Lama and Vutukuri (1978), Goodman (1980) and Hoek and Brown (1997), respectively, show typical values of σ_{ci} and m_i for different rock types that could be taken as a reference for use in equation (A-1).

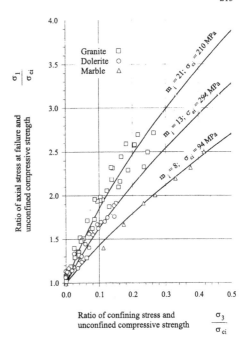

Figure A-1. Failure envelopes obtained from triaxial tests of samples of different rock types (after Franklin and Hoek 1970). The horizontal and vertical axes represent the confining stress σ_3 and the maximum axial stress σ_1 respectively divided by the unconfined compressive strength σ_{ci} of the sample (see Appendix B).

As noted earlier, joints and defects in a rock-mass reduce the strength of the mass below the strength of an intact specimen of the same rock type. By using the so-called *Geological Strength Index* (or *GSI*) as a scaling parameter, the failure criterion defined by equation (A-1) can be adjusted to provide an estimate of the decreased strength of the rock-mass in the field.

According to Hoek and Brown (1997), the GSI is an empirically derived number that varies over a range between 10 and 100 (the GSI is dimensionless), and can be estimated by examination of the quality of the rock-mass in situ—by direct inspection of an outcrop, for example. By definition, GSI values close to 10 correspond to very poor quality rock-masses, while GSI values close to 100 correspond to excellent quality rock-masses.

Figure A-3 (adapted from Hoek and Brown 1997, and Hoek et al. 1998) shows how the GSI can be estimated from the structure and surface conditions of the rock-mass (for example, a rock-mass with *Blocky / Disturbed* structure and *Poor* surface condition will have a GSI close to 30).

The value *GSI* = 25, indicated by a discontinuous line in Figure A-3, is significant in that it defines the limit between rock-masses of very poor quality (*GSI* < 25) and those of good to reasonable quality (*GSI* > 25). For rock-masses of good to reasonable quality (i.e., *GSI* > 25) the Geological Strength Index is equivalent to the *Rock Mass Rating* (*RMR*) introduced by Bieniawski (1976) when the rating for *Groundwater* is assessed as 'dry' and the rating for *Joint Orientation* as 'favorable'.

Table A-1. Reference values for the unconfined compressive strength σ_{ci}, Poisson's ratio n, Young's Modulus E, Shear Modulus G and unit weight γ for intact rock (adapted from Lama and Vutukuri 1978 and Goodman 1980).

Rock type	σ_{ci} [MPa]	ν	E [MPa]	G [MPa]	γ [MN/m^3]
1) Andesite	130.6	0.16	44.3×10^3	19.1×10^3	25.2×10^{-3}
2) Basalt	148.0	0.32	33.9×10^3	12.8×10^3	27.8×10^{-3}
3) Conglomerate	30.3	0.12	1.3×10^3	0.6×10^3	24.2×10^{-3}
4) Diabase	321.3	0.28	95.8×10^3	37.4×10^3	28.8×10^{-3}
5) Dolomite	46.9	0.29	29.0×10^3	11.2×10^3	24.5×10^{-3}
6) Gneiss	165.0	0.27	76.3×10^3	30.0×10^3	26.8×10^{-3}
7) Granite	141.1	0.22	73.8×10^3	30.3×10^3	26.4×10^{-3}
8) Limestone	51.0	0.29	28.5×10^3	11.1×10^3	23.3×10^{-3}
9) Quartzite	320.1	0.11	88.4×10^3	39.8×10^3	25.7×10^{-3}
10) Sandstone	73.8	0.38	18.3×10^3	6.6×10^3	21.4×10^{-3}
11) Siltstone	122.7	0.22	26.2×10^3	10.7×10^3	25.4×10^{-3}
12) Tuff	11.3	0.19	3.7×10^3	1.5×10^3	23.5×10^{-3}

Origin of the samples: *1)* Palisades Dam, Idaho, USA; *2)* Nevada Test Site, USA; *3)* Mc Dowell Dam, Arizona, USA; *4)* New York, USA; *5)* Minneapolis, Minnesota, USA; *6)* Graminha Dam, Brazil; *7)* Nevada Test Site, USA; *8)* Bedford, Indiana, USA; *9)* Baraboo, Wisconsin, USA; *10)* Amherst, Ohio, USA; *11)* Hackensack, N.Y., USA; *12)* Nevada Test Site, USA.

When the scaling factor GSI is introduced, the Hoek-Brown failure criterion for the *rock-mass* is given by the following relationship (Hoek and Brown 1997):

$$\sigma_1 = \sigma_3 + \sigma_{ci} \left(m_b \frac{\sigma_3}{\sigma_{ci}} + s \right)^a \qquad \text{(A-2)}$$

The parameter m_b in equation (A-2) depends on both the intact rock parameter m_i, of equation (A-1), and the value of GSI, as defined by the equation

$$m_b = m_i \exp\left(\frac{GSI - 100}{28}\right) \qquad \text{(A-3)}$$

The parameters s and a also depend empirically on the value of *GSI* as follows, for $GSI \geq 25$,

$$s = \exp\left(\frac{GSI - 100}{9}\right) \qquad \text{(A-4)}$$

$$a = 0.5$$

and for $GSI < 25$

$$s = 0 \qquad \text{(A-5)}$$

$$a = 0.65 - \left(\frac{GSI}{200}\right)$$

Table A-2 lists the values of m_b, s and a obtained from equations (A-3), (A-4) and (A-5) for different values of GSI. It can be seen that when $GSI = 100$ (the hypothetical case in which the rock-mass has the same strength as the intact rock sample), the parameters are $m_b = m_i$, $s = 1$ and $a = 0.5$. With these values, the yield condition for the rock-mass, equation (A-2), and for the intact rock, equation (A-1), are the same.

Londe (1988) showed that the Hoek-Brown failure criterion defined by equation (A-2) can be transformed into a 'general' failure envelope that is independent of the parameters σ_{ci}, m_b and s.

The transformation suggested by Londe applies to the particular case $a = 0.5$ and involves dividing the stress magnitudes by $m_b \sigma_{ci}$ and adding the term

s/m_b^2. Considering the parameters introduced in equation (A-2), the scaled stresses S_1 and S_3 can be defined as,

$$S_1 = \frac{\sigma_1}{m_b \, \sigma_{ci}} + \frac{s}{m_b^2} \qquad \text{(A-6)}$$

$$S_3 = \frac{\sigma_3}{m_b \, \sigma_{ci}} + \frac{s}{m_b^2} \qquad \text{(A-7)}$$

Rock type	Class	Group	Texture		
			Coarse	Medium	Fine
SEDIMENTARY	Clastic		Conglomerate (22)	Sandstone (19)	Siltstone (9)
	Non-clastic	Organic	.... Coal .... (8~21)		
		Carbonate	Breccia (20)	.... Limestone .... (8~10)	
METAMORPHIC	Non-foliated		Marble (9)	Hornfels (19)	Quartzite (24)
	Slightly foliated		Migmatite (30)	Amphibolite (25~31)	Mylonites (6)
	Foliated		Gneiss (33)	Schists (4~8)	Phyllites (10)
IGNEOUS	Light		Granite (33)		Rhyolite (16)
			Granodiorite (30)		Dacite (17)
	Dark		Gabbro (27)	Dolerite (19)	Basalt (17)
	Extrusive pyroclastic		Agglomerate (20)	Breccia (18)	Tuff (15)

Figure A-2. Reference values for the coefficient m for different rock types (adapted from Hoek and Brown 1997). The value of m_i is shown in parentheses below the name of the rock.

With the stresses σ_3 and σ_1 replaced by the scaled stresses S_3 and S_1 from equations (A-6) and (A-7), the failure criterion for the rock-mass, equation (A-2), can be written in the form (Londe 1988)

$$S_1 = S_3 + \sqrt{S_3} \qquad (A\text{-}8)$$

Note that in this 're-scaled' form of the failure criterion the parameters σ_{ci}, m_b and s are 'hidden' within the scaled stresses S_1 and S_3, the relationship applies then to any type of rock that is assumed to obey the Hoek-Brown criterion[b].

To illustrate the use of Londe's transformation, we will re-examine the triaxial test results for the samples of granite, quartz dolerite and marble presented in Figure A-1. Note that the results for intact rock samples can be equally approximated by equation (A-2), taking $GSI = 100$, $s = 1$, $m_b = m_i$ and $a = 0.5$. Figure A-4 represents the scattered pairs (σ_3, σ_1) of Figure A-1 together with the Hoek-Brown failure criterion—equation (A-2) or (A-1)—plotted in terms of scaled principal stresses (i.e., with the axes representing the transformed stresses S_1 and S_3 defined by equations A-6 and A-7). It is seen that the stresses at failure for all three types of rocks align now to the 'general' form of the Hoek-Brown criterion defined by equation (A-8).

The use of equation (A-8) rather than equation (A-2) can lead to important simplifications in mechanical analyses involving the Hoek-Brown criterion. Carranza-Torres and Fairhurst (1999) have applied the transformation (A-8) in solving the problem of excavating cylindrical and spherical openings in rock-masses that satisfy the Hoek-Brown failure criterion. This solution is the basis for construction of Ground Reaction Curves in the Convergence-Confinement method, discussed in Section 3.

Just as the strength of the rock-mass is usually lower than the strength of the intact rock, the (elastic) deformation modulus of the rock-mass is also usually lower than that of the intact rock. Serafim and Pereira (1983) have proposed an empirical relationship to compute the deforma-

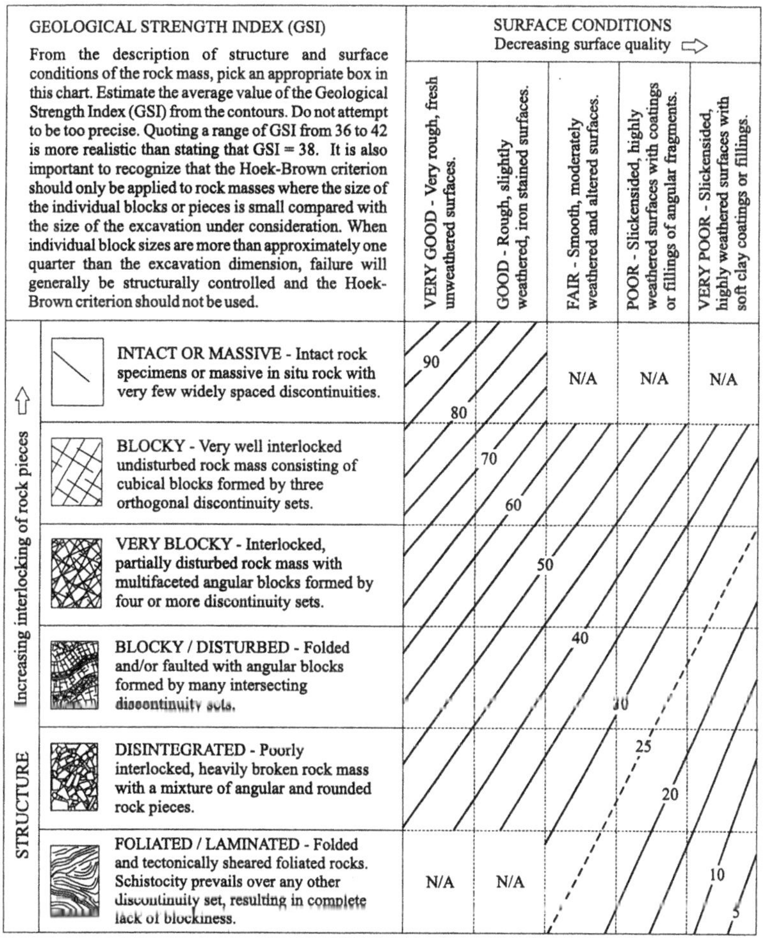

Figure A-3. Empirical chart for the estimation of the Geological Strength Index (GSI) based on the characteristics of the rock-mass (adapted from Hoek and Brown 1997 and Hoek et al. 1998).

Table A-2. Values of coefficients m_b, s and a as a function of the Geological Strength Index (GSI), computed from equations (A-3), (A-4) and (A-5), respectively. (Note that the second column represents ratio m_b/m_i; values of m_i for different rock types are given in Figure A-2).

GSI	m_b/m_i	s	a
100	1.00	1.00	0.5
75	40.95×10^{-2}	621.77×10^{-4}	0.50
50	16.77×10^{-2}	38.66×10^{-4}	0.50
25^+	6.87×10^{-2}	2.40×10^{-4}	0.50
25^-	6.87×10^{-2}	0.00	0.53
10	4.02×10^{-2}	0.00	0.60

tion modulus of the rock-mass from the unconfined compressive strength of the intact rock sample and the value of the Rock Mass Rating (RMR) by Bienawski (1976). Based on the original equation by Serafim and Pereira, Hoek and Brown (1997) propose the following relationship between the rock-mass modulus E_{rm} and the Geological Strength Index GSI:

$$E_{rm} = 1000\, C\,(\sigma_{ci})\, 10^{\frac{GSI-10}{40}} \qquad \text{(A-9)}$$

where

$$C\,(\sigma_{ci}) = 1 \qquad \text{if } \sigma_{ci} \geq 100 \text{ MPa}$$

$$= \sqrt{\frac{\sigma_{ci}}{100}} \qquad \text{if } \sigma_{ci} \leq 100 \text{ MPa}$$

In equation (A-9), both the unconfined compressive strength σ_{ci} and the rock-mass modulus E_{rm} are expressed in MPa.

In elasto-plastic analyses of deformations—such as the one presented later in Section 3, the rock-mass shear modulus G_{rm} is used rather than the deformation modulus E_{rm} given by equation (A-9). The shear modulus of the rock-mass can be estimated from the deformation modulus using the classic relationship from isotropic elasticity,

$$G_{rm} = \frac{E_{rm}}{2\,(1+\upsilon)} \qquad \text{(A-10)}$$

In equation (A-10), υ is Poisson's ratio for the rock-mass, and is usually considered to vary between 0.1 and 0.3 (Hoek and Brown 1980).

To illustrate the application of equations (A-9) and (A-10), let us consider the properties of the granite sample listed in Table A-1. The unconfined compressive strength of the intact rock is approximately $\sigma_{ci} = 141$ MPa; if the Geological Strength Index of the rock-mass is $GSI = 50$, then the deformation modulus of the rock-mass is, from equation (A-9), $E_{rm} = 1187$ MPa. If Poisson's ratio for the intact rock and rock-mass are both assumed to be equal to 0.22 then, from equation (A10), the shear modulus of the rock-mass is

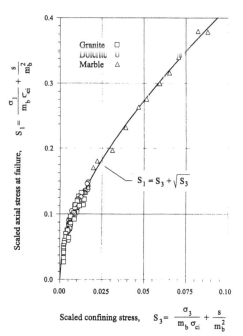

Figure A-4. Results from triaxial tests shown in Figure A-1 with the principal stress axes σ_1 and σ_3 normalized according to transformations (A-6) and (A-7). Note that in this reference system results for all three rock types fall on the 'general' failure envelope given by equation (A-8) (Londe 1988).

$G_{rm} = 486$ MPa. The elastic constants E_{rm} and G_{rm} for the rock-mass are seen to be significantly lower than the corresponding constants E and G for the intact rock sample listed in Table A-1.

Notes for Appendix A

[a] The observed dispersion appears to be proportional to the number of specimens tested, with the highest dispersion for the 48 samples of granite and the lowest for the 14 samples of marble (see Appendix B).

[b] It should be emphasized though that the equation (A-8) is strictly valid only when the parameter a in equation (A-2) is equal to 0.5. According to equations (A-4) and (A-5), $a = 0.5$ for the broad range of situations in which $GSI = 25$.

Appendix B. Determination of the parameters σ_{ci} and m_i from results of triaxial tests on intact rock samples.

The parameters σ_{ci} and m_i in equation (A-1) can be obtained from statistical analysis of the (σ_3, σ_1) results obtained from triaxial tests on intact rock specimens —σ_3 is the constant confining stress applied to the sample and σ_1 is the axial stress that produces collapse of the sample. Hoek (1983) describes the basis of the statistical analysis and the needed equations. A brief summary of the equations is presented below.

If n rock cores have been tested and n pairs (σ_3, σ_1) have been obtained from the tests, the parameters σ_{ci} and m_i can be found from the following expressions (Hoek 1983):

$$\sigma_{ci}^2 = \frac{\Sigma(\sigma_1 - \sigma_3)^2}{n} - \frac{\Sigma\sigma_3}{n}\left[\frac{n\,\Sigma\sigma_3(\sigma_1-\sigma_3)^2 - \Sigma\sigma_3\Sigma(\sigma_1-\sigma_3)^2}{n\,\Sigma\sigma_3{}^2 - (\Sigma\sigma_3)^2}\right]$$

(B-1)

$$m_i = \frac{1}{\sigma_{ci}}\left[\frac{n\,\Sigma\sigma_3(\sigma_1-\sigma_3)^2 - \Sigma\sigma_3\Sigma(\sigma_1-\sigma_3)^2}{n\,\Sigma\sigma_3{}^2 - (\Sigma\sigma_3)^2}\right]$$

(B-2)

A regression coefficient r can be evaluated to estimate the 'goodness of fit' of the parabolic approximation. The coefficient is computed as,

$$r^2 = \frac{\left[n\,\Sigma\sigma_3(\sigma_1-\sigma_3)^2 - \Sigma\sigma_3\Sigma(\sigma_1-\sigma_3)^2\right]^2}{\left[n\,\Sigma\sigma_3^2 - (\Sigma\sigma_3)^2\right]\left[n\,\Sigma(\sigma_1-\sigma_3)^4 - (\Sigma(\sigma_1-\sigma_3)^2)^2\right]}$$

(B-3)

For example, Table B-1 lists the pairs (σ_3, σ_1) obtained by Franklin and Hoek (1970) from triaxial tests on samples of granite, quartz dolerite and marble —these results have been discussed in Appendix A (see Fig. A-1).

The parameters σ_{ci}, m_i and r^2 computed using the equations B-1, B-2 and B-3 are indicated below the values of σ_3 and σ_1 for each rock type.

Table B-1. Stress pairs (σ_3, σ_1) obtained from triaxial tests on different rock samples (after Franklin and Hoek 1970). The parameters σ_{ci}, m_i and r^2, obtrained with equations (B-1) through (B-3) are indicated below the tabulated (σ_3, σ_1) values.

Granite (48 samples from Blackingstone quarry, Devon, UK)

σ_3 [MPa]	σ_1 [MPa]	σ_3 [MPa]	σ_1 [MPa]	σ_3 [MPa]	σ_1 [MPa]
9.3	309.3	43.5	541.0	29.4	485.8
43.2	539.8	0.0	179.3	12.8	269.9
55.8	569.5	0.0	111.3	49.3	488.4
2.8	249.0	1.6	234.3	18.0	410.7
0.0	197.3	19.7	406.5	0.0	195.9
21.4	407.6	0.0	201.5	28.8	458.0
10.7	316.0	25.8	431.1	0.0	196.1
34.8	512.7	0.0	193.4	5.3	270.5
0.0	171.5	38.1	543.7	21.7	415.9
6.5	273.3	22.9	409.9	33.0	440.1
0.0	135.9	27.1	392.8	16.4	340.4
5.0	293.1	13.2	359.0	8.7	318.5
7.8	276.8	28.5	486.8	9.9	312.1
33.9	453.9	7.8	283.0	45.5	566.0
0.0	213.6	39.2	480.8	7.4	284.4
51.7	523.9	15.1	330.1	17.7	362.3

$$\sigma_{ci} = 210 \text{ MPa}; m_i = 21; r^2 = 0.91$$

Quartz Dolerite (38 samples from Northumberland, UK)

σ_3 [MPa]	σ_1 [MPa]	σ_3 [MPa]	σ_1 [MPa]	σ_3 [MPa]	σ_1 [MPa]
5.0	333.8	44.1	561.2	0.0	331.8
0.0	315.0	34.7	498.9	7.4	344.6
23.9	453.6	20.2	410.9	3.5	284.7
0.0	311.4	0.0	305.1	0.0	267.5
0.0	314.4	10.3	341.9	0.0	299.7
6.9	390.5	34.5	497.8	27.6	489.0
20.7	457.1	0.0	272.9	2.4	341.0
1.3	328.3	42.9	514.8	37.0	512.7
0.0	315.7	13.7	380.7	0.0	214.4
28.3	474.6	0.0	210.7	13.9	364.0
31.0	496.4	0.0	275.8	0.0	273.7
17.2	422.6	0.0	312.0	0.0	278.9
42.1	552.2	21.7	461.2		

$$\sigma_{ci} = 294 \text{ MPa}; m_i = 13; r^2 = 0.92$$

Carrara Marble (14 samples from Italy)

σ_3 [MPa]	σ_1 [MPa]	σ_3 [MPa]	σ_1 [MPa]	σ_3 [MPa]	σ_1 [MPa]
30.9	205.9	0.0	90.3	35.2	217.2
3.9	119.1	39.1	234.4	51.7	262.2
0.0	93.1	21.8	179.2	25.2	188.1
16.2	156.4	10.5	131.1	0.0	93.8
2.2	111.6	47.5	263.1		

$$\sigma_{ci} = 94 \text{ MPa}; m_i = 8; r^2 = 0.99$$

Appendix C. Spreadsheet implementation of the Convergence-Confinement method.

Figures C-1 through C-3 present a computer spreadsheet for implementation of equations in Sections 3, 4 and 5, in order to construct the GRC, SCC and LDP curves, respectively.

The spreadsheet is divided in two main parts: *a)* Input of data and *b)* Output of results.

In part *a)* (Fig. C-1), the geometry, rock-mass properties, loading conditions, distance to the face and properties of the three support systems discussed in Section 4 are entered.

In part *b)* (Figs. C-2 and C-3), the elastic and plastic parts of the GRC, the LDP (from equation 18) and the SCC for individual and combined supports are defined.

Linking of the cells in the spreadsheet is accomplished by giving a name to each cell and range of cells, and expressing formulae within the spreadsheet in terms of the named cells (see for example, Dodge et al. 1997).

The names given to *individual* cells are indicated within parentheses at the side of the cells in Figures C-1 through C-3.

For *ranges* of cells, defined by a box outlined by dashed lines, the name is indicated at the bottom of the box.

The values in the shaded cells in the spreadsheet are computed with formulas.

Tables C-1 and C-2 define the formulas for Figures C-2 and C-3—expressed in terms of cell/range names—that need to be entered in these cells.

Note that the values shown in the spreadsheet are those for the practical example discussed in Section 6. To construct the different curves shown in the diagram of Figure 10, the following ranges must be plotted:

- for the GRC, the ranges ure_grc and urp_grc in the horizontal axis and the ranges pip_grc, pip_grc, pip_r_grc and pip_f_grc in the vertical axis; the last two ranges correspond to the internal pressure at the crown and floor of the tunnel—see points C and F in Figure 10a.
- for the LDP, the range ur_ldp in the horizontal axis and the range lf in the vertical axis.
- for the SCC of individual supports, the ranges ur_sc, ur_ss and ur_sb in the horizontal axis and the ranges pi_sc, pi_ss and pi_sb in the vertical axis.
- for the SCC of combined supports, the ranges ur_sc_b, and ur_ss_b in the horizontal axis and the ranges pi_sc_b and pi_ss_b in the vertical axis.

(a) INPUT					
(a1) Geometry			**(a2) Rock-Mass**		
R [m]	1.0	(R_t)	σ_{ci} [MPa]	20.0	(sig_ci)
			m_i	15	(m_i)
(a3) Loading			ν	0.25	(nu)
σ_0 [MPa]	7.5	(sig_0)	ψ [deg]	30	(psi)
			GSI	10	(GSI)
			γ [MN/m^3]	2.40E-02	(gamma)
(a4) Face Effect					
L [m]	1.0	(L_f)			
(a5) Support					
(a5-1) Blocked Steel			**(a5-2) Shotcrete or Concrete**		
B [mm]	76	(B_s)	σ_{cc} [MPa]	30.0	(sig_cc)
D [mm]	127	(D_s)	E_c [GPa]	30.0	(E_c)
A_s [m^2]	1.70E-03	(A_s)	ν_c	0.25	(nu_c)
I_s [m^4]	4.76E-06	(I_s)	t_c [mm]	30	(t_c)
E_s [GPa]	210.0	(E_s)			
σ_{ys} [MPa]	150.0	(sig_ys)	**(a5-3) Bolts**		
S [m]	1.0	(S_s)	d_b [mm]	19	(d_b)
n_B	10	(n_B)	l [m]	2.0	(l_b)
t_B [mm]	90	(t_B)	T_{bf} [MN]	0.10	(T_bf)
E_B [GPa]	10.0	(E_B)	Q [m/MN]	0.03	(Q_b)
			E_b [GPa]	210.0	(E_bolt)
			n_{bolt}	10	(n_bolt)
			s_l [m]	0.5	(s_lb)

Figure C-1. First page of the spreadsheet. Data input.

(b) OUTPUT

(b1) Ground Reaction Curve

m_b	1.76	(m_b)
s	1.27E-03	(s_coeff)
E_{rm} [GPa]	2.51	(E_rm)
G_{rm} [GPa]	1.01	(G_rm)
K_ψ	3.00	(K_psi)
S_0	2.14E-01	(S_0)
P_i^{cr}	7.58E-02	(Pi_cr_s)
p_i^{cr}	2.65	(pi_cr)

Elastic part

point	p_i [MPa]	u_r^{el} [mm]
1	7.50	0.0
2	2.65	2.4
	(pie_grc)	(ure_grc)

Plastic part

point	p_i [MPa]	p_i^r [MPa]	p_i^f [MPa]	P_i	R_{pl}/R	u_r^{pl} [mm]
1	2.65	2.65	2.65	7.58E-02	1.00	2.41
2	2.41	2.41	2.41	6.90E-02	1.03	2.55
3	2.17	2.17	2.17	6.21E-02	1.05	2.73
4	1.93	1.93	1.93	5.53E-02	1.08	2.96
5	1.69	1.69	1.68	4.84E-02	1.12	3.27
6	1.45	1.45	1.44	4.15E-02	1.15	3.67
7	1.21	1.21	1.20	3.47E-02	1.20	4.20
8	0.97	0.97	0.96	2.78E-02	1.24	4.91
9	0.72	0.73	0.72	2.10E-02	1.30	5.91
10	0.48	0.49	0.47	1.41E-02	1.37	7.40
11	0.24	0.25	0.23	7.27E-03	1.46	9.96
12	0.00	0.02	-0.02	4.11E-04	1.57	17.65
(pt_grc)	(pip_grc)	(pip_r_grc)	(pip_f_grc)	(pipa_grc)	(xl_grc)	(ur_max) (urp_grc)

(b2) Longitudinal Deformation Profile

point	L_f/R	L_f [m]	u_r [mm]
1	-4.00	-4.00	0.03
2	-2.91	-2.91	0.18
3	-1.82	-1.82	0.79
4	-0.73	-0.73	2.83
5	0.36	0.36	7.04
6	1.45	1.45	11.82
7	2.55	2.55	15.05
8	3.64	3.64	16.61
9	4.73	4.73	17.28
10	5.82	5.82	17.51
11	6.91	6.91	17.61
12	8.00	8.00	17.64
(pt_lpd)	(lf_r)	(lf)	(ur_ldp)

For a distance [m]= 1 (Lf_2) -> ur [mm]= 9.93 (ur_0)

Figure C-2 (at left). Second page of the spreadsheet. Computation of the GRC and LDP.

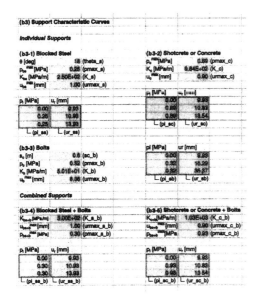

Figure C-3. Third page of the spreadsheet. Computation of the SCC.

Table C-1. Formulas to be entered in the shaded cells in Figure 2.

m_b =m_i*EXP((GSI-100)/28)
s_coeff =IF(GSI>=25,EXP((GSI-100)/9),0)
E_rm =SQRT(sig_ci/100)*10^((GSI-10)/40)
G_rm =E_rm/2/(1+nu)
K_psi =(1+SIN(psi*PI()/180))/(1-SIN(psi*PI()/180))
S_0 =sig_0/(m_b*sig_ci)+s_coeff/m_b^2
Pi_cr_s =1/16*(1-SQRT(1+16*S_0))^2
pi_cr =(Pi_cr_s-s_coeff/(m_b^2))*m_b*sig_ci
pie_grc =sig_0 (first row), =pi_cr (second row)
ure_grc =(sig_0-pie_grc)/2/(G_rm*1000)*R_t*1000
pip_grc =pi_cr*(12-pt_grc)/11
pip_r_grc =pip_grc+gamma*(xi_grc*R_t-R_t)
pip_f_grc =pip_grc-gamma*(xi_grc*R_t-R_t)
pips_grc =pip_grc/(m_b*sig_ci)+s_coeff/m_b^2
xi_grc =EXP(2*(SQRT(Pi_cr_s)-SQRT(pips_grc)))
urp_grc =((K_psi-1)/(K_psi+1)+2/(K_psi+1)*xi_grc
　　　^(K_psi+1)+(1-2*nu)/4/(S_0-Pi_cr_s)*LN(xi_grc)^2
　　　-((1-2*nu)/(K_psi+1)*(SQRT(Pi_cr_s))/(S_0-Pi_cr_s)
　　　+(1-nu)/2*(K_psi-1)/(K_psi+1)^2/(S_0-Pi_cr_s)
　　　*((K_psi+1)*LN(xi_grc)-(xi_grc)^(K_psi+1)+1))
　　　R_t(sig_0-pi_cr)/2/(G_rm*1000)*1000
If_r =-4+(pt_ldp-1)*12/11
If =If_r*R_t
ur_ldp =ur_max*(1+EXP(-If/1.1/R_t))^-1.7
Lf_2 =L_f
ur_0 =ur_max*(1+EXP(-L_f/1.1/R_t))^-1.7

Table C-2. Formulas to be entered in the shaded cells in Figure C-3.

theta_s =180/n_B
pmax_s =(3*A_s*I_s*sig_ys)/(2*S_s*R_t*theta_s*PI()/180
　　　*(3*I_s+D_s/1000*A_s*(R_t-t_B/1000-0.5*D_s/1000)
　　　*(1-COS(theta_s*PI()/180))))
K_s =1/(((S_s*R_t^2)/(E_s*1000*A_s))+(S_s*R_t^4/(E_s*1000*I_s)
　　　*((theta_s*PI()/180*(theta_s*PI()/180+SIN(theta_s*PI()
　　　/180)*COS(theta_s*PI()/180)))/(2*(SIN(theta_s*PI()/180)
　　　^2))-1)+(2*S_s*theta_s*PI()/180*t_B/1000*R_t)/(E_B*1000
　　　*(B_s/1000)^2))
urmax_s =pmax_s/K_s*1000
pi_ss =0 (first row), =pmax_s (second and third rows)
ur_ss =ur_0 (first row), =ur_0+urmax_s (second row)
　　　=ur_0+urmax_s*4 (third row)
pmax_c =sig_cc/2*(1-(R_t-t_c/1000)^2/R_t^2)
K_c =E_c*1000/(1+nu_c)*(R_t^2-(R_t-t_c/1000)^2)/((1-2*nu_c)*
　　　R_t^2+(R_t-(t_c/1000))^2)/R_t
urmax_c =pmax_c/K_c*1000
pi_sc =0 (first row), =pmax_c (second and third rows)
ur_sc =ur_0 (first row), =ur_0+urmax_c (second row)
　　　=ur_0+urmax_c*4 (third row)
sc_b =2*PI()*R_t/n_bolt
pmax_b =T_bf/sc_b/s_lb
K_b =1/(sc_b*s_lb)*(PI()*(d_b/1000)^2*E_bolt*1000)/(4*l_b+
　　　Q_b*PI()*(d_b/1000)^2*E_bolt*1000)
urmax_b =pmax_b/K_b*1000
pi_sb =0 (first row), =pmax_b (second and third rows)
ur_sb =ur_0 (first row), =ur_0+urmax_b (second row)
　　　=ur_0+urmax_b*4 (third row)
K_s_b =K_s+K_b
urmax_s_b =IF(urmax_s<urmax_b,urmax_s,urmax_b)
pmax_s_b =K_s_b*urmax_s_b/1000
pi_ss_b =0 (first row), =pmax_s_b (second and third rows)
ur_ss_b =ur_0 (first row), =ur_0+urmax_s_b (second row)
　　　=ur_0+urmax_s_b*4 (third row)
K_c_b =K_c+K_b
urmax_c_b =IF(urmax_c<urmax_b,urmax_c,urmax_b)
pmax_c_b =K_c_b*urmax_c_b/1000
pi_sc_b =0 (first row), =pmax_c_b (second and third rows)
ur_sc_b =ur_0 (first row), =ur_0+urmax_c_b (second row)
　　　=ur_0+urmax_c_b*4 (third row)

Appendix D. Elastic analysis of circular tunnels in non-uniform stress conditions.

The elastic solution of a circular cavity of radius R subject to non-uniform horizontal and vertical far-field stresses σ_x and σ_z, respectively, and internal pressure p_i is given by the classical Kirsch solution (see, for example, Timoshenko and Goodier 1970).

Considering the mean stress σ_o and the horizontal-to-vertical stress coefficient k defined by equations (21) and (20), respectively, the scaled radial stress σ_r/σ_o at a distance r from the center of the cavity is given by,

$$\frac{\sigma_r(r)}{\sigma_o} = 1 - \left(\frac{R}{r}\right)^2 + \frac{k-1}{k+1}\left[1 - 4\left(\frac{R}{r}\right)^2 + 3\left(\frac{R}{r}\right)^4\right]\cos(2\theta) + \frac{p_i}{\sigma_o}\left(\frac{R}{r}\right)^2 \quad \text{(D-1)}$$

The scaled 'hoop' stress σ_θ/σ_o is

$$\frac{\sigma_\theta(r)}{\sigma_o} = 1 + \left(\frac{R}{r}\right)^2 - \frac{k-1}{k+1}\left[1 + 3\left(\frac{R}{r}\right)^4\right]\cos(2\theta) - \frac{p_i}{\sigma_o}\left(\frac{R}{r}\right)^2 \quad \text{(D-2)}$$

and the scaled shear stress $\sigma_{r\theta}/\sigma_o$ is

$$\frac{\sigma_{r\theta}(r)}{\sigma_o} = -\frac{k-1}{k+1}\left[1 + 2\left(\frac{R}{r}\right)^2 - 3\left(\frac{R}{r}\right)^4\right]\sin(2\theta) \quad \text{(D-3)}$$

The maximum and minimum principal stresses —σ_1 and σ_3, respectively, corresponding to the stresses σ_r, σ_θ and $\sigma_{r\theta}$ defined above—are computed using the classic relationships (see, for example, Jaeger and Cook 1979):

$$\sigma_3^1(r)/\sigma_o = \frac{\sigma_\theta/\sigma_o + \sigma_r/\sigma_o}{2} \pm \sqrt{\left(\frac{\sigma_\theta/\sigma_o - \sigma_r/\sigma_o}{2}\right)^2 - (\sigma_{r\theta}/\sigma_o)^2} \quad \text{(D-4)}$$

The extent r of the 'over-stressed' region discussed in Section 7 can be found from the condition that the strength of the material —defined by the Hoek-Brown failure criterion (A-2)— is exceeded. Considering the parameter a to have the value 0.5 in equation (A-2), the condition is written as,

$$\sigma_1(r) \geq \sigma_3(r) + \sigma_{ci}\left(m_b\,\frac{\sigma_3(r)}{\sigma_{ci}} + s\right)^a \quad \text{(D-5)}$$

The maximum value of r, can be obtained from the condition that the left and right sides of expression (D-5) are equal. Thus, dividing both sides by $m_b\,\sigma_{ci}$ and multiplying and dividing each term by σ_o, the inequality (D-5) transforms into

$$\frac{\sigma_1(r)/\sigma_o}{m_b\,\sigma_{ci}/\sigma_o} = \frac{\sigma_3(r)/\sigma_o}{m_b\,\sigma_{ci}/\sigma_o} + \sqrt{\frac{\sigma_3(r)/\sigma_o}{m_b\,\sigma_{ci}/\sigma_o} + \frac{s}{m_b^2}} \quad \text{(D-6)}$$

The non-linear equation (D-6) must be solved for the unknown r. Because of the method of scaling used in equation (D-6), the values obtained for the extent r are valid for constant ratios of $m_b\,\sigma_{ci}/\sigma_o$ and s/m_b^2 (see example in Fig. 13).

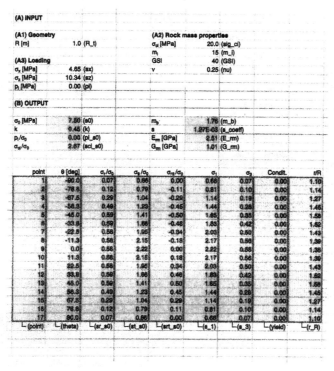

Figure D-1. Spreadsheet for evaluation of the extent of the 'over-stressed' region around a circular cavity subject to uniform internal pressure and unequal far-field principal stresses.

The solution of the non-linear equation (D-6) can be found with a spreadsheet such as shown in Figure D-1. Linkage between cells in the spreadsheet is accomplished using the formulae listed in Table D-1. Note that, as in the spreadsheet presented in Appendix C, cell and range-of-cell names are indicated within parentheses in Figure D-1.

The equation solver implemented in most spreadsheet packages has to be used to evaluate the cells in the range r_R. In the commercial spreadsheet program *Excel* (see for example, Dodge et al. 1997) the 'Goal seek' tool has to be applied specifying '*Set cell:*' to point to cells in the range yield, '*To value:*' to be equal to 0.0 and '*By changing cell:*' to point to the cells in the range r_R. Note that in this spreadsheet program, the 'Goal seek' tool applies to individual cells, so that a macro should be recorded in order to find the solution in all cells of the range r_R simultaneously.

Diagrams such as those presented in Figure 13a can then be constructed by plotting the ranges theta and r_R in radial coordinates.

Table D-1. Formulae to be entered in the shaded cells of the spreadsheet shown in Figure D-1.

s0 =0.5*(sx+sz)
k =sx/sz
pi_s0 =pi/s0
sci_s0 =sig_ci/s0
m_b =m_i*EXP((GSI-100)/28)
s_coeff =IF(GSI>=25,EXP((GSI-100)/9),0)
E_rm =SQRT(sig_ci/100)*10^((GSI-10)/40)
G_rm =E_rm/2/(1+nu)

theta =-90+180*(point-1)/16
sr_s0 =1-1/r_R^2+(k-1)/(k+1)*(1-4*1/r_R^2+3*1/r_R^4)
    *COS(2*theta*PI()/180)+pi_s0*1/r_R^2
st_s0 =1+1/r_R^2-(k-1)/(k+1)*(1+3*1/r_R^4)
    *COS(2*theta*PI()/180)-pi_s0*1/r_R^2
srt_s0 =-(k-1)/(k+1)*(1+2*1/r_R^2-3*1/r_R^4)
    *SIN(2*theta*PI()/180)
s_1 =(sr_s0+st_s0)/2+SQRT((sr_s0-st_s0)^2/4+srt_s0^2)
s_3 =(sr_s0+st_s0)/2-SQRT((sr_s0-st_s0)^2/4+srt_s0^2)
yield =s_1/(m_b*sci_s0)-s_3/(m_b*sci_s0)
    -SQRT(s_3/(m_b*sci_s0)+s_coeff/m_b^2)
r_R (see main text)

Papers from

Geotextiles and Geomembranes

Data base of field incidents used to establish HDPE geomembrane stress crack resistance specifications[☆]

Y. Grace Hsuan[*]

GRI, Drexel University, 475 Kedron Ave., Folsom, PA 19033, USA

Received 15 January 1999; received in revised form 13 May 1999; accepted 15 May 1999

Abstract

Geomembranes made from high-density polyethylene (HDPE) have a relatively high percent crystallinity, and are therefore of concern with regard to stress cracking. The stress crack resistance (SCR) of HDPE geomembranes must be properly evaluated and specified accordingly. A research project was initiated in 1990 to investigate the stress cracking behavior of HDPE geomembranes in the field. A total of 16 field sites were located, in which HDPE geomembranes showed some degree of stress cracking. The SCR of retrieved field samples was evaluated by both the bent strip and notched constant tensile load (NCTL) tests. The test data indicates that the bent strip test could not adequately predict the stress cracking field performance of HDPE geomembranes. In contrast, the NCTL test could distinguish the SCR of the retrieved field geomembranes. Based on the NCTL test data obtained from the field geomembranes, a 100 hour transition time is recommended to be the minimum value for an acceptable HDPE geomembrane specification. In addition, a single point notched constant tensile load (SP-NCTL) test was also established for quality control and quality assurance purposes. The recommendation for the SP-NCTL test is set at 200 hours failure time under an applied load of 30% room temperature yield stress. © 1999 Elsevier Science Ltd. All rights reserved.

Keywords: Geomembrane; High density polyethylene; Stress crack resistance; NCTL test

---

[☆]This paper was originally presented at the 12th Geosynthetics Research Institute Conference. It was received and subsequently revised.

[*] Tel.: (610) 522-8440; fax: (610) 522-8441.

E-mail address: ghsuan@coe.drexel.edu (Y. Grace Hsuan)

1. Introduction

High-density polyethylene (HDPE) is a widely used polymer for manufacturing of geomembranes used in liquid and waste containment facilities. The relatively high crystallinity (40 to 50%) of the material provides both high chemical resistance and low diffusion rates, which are required in most containment facilities. However, along with this high crystallinity is an increase in the tendency for the polymer to stress crack. Stress cracking refers to failure of the geomembrane under stress in a brittle manner exhibiting little or no elongation adjacent to the failure surface. Such cracking phenomenon has been observed in some field HDPE geomembranes. As a result, a research project was initiated to evaluate stress crack resistance (SCR) by laboratory testing and field investigation of stress cracking occurrences involving HDPE geomembranes.

At the time the project was started, the standard test method used to evaluate SCR of HDPE geomembrane was ASTM D1693, the bent strip test. In this test, notched specimens are confined under constant strain during the entire testing period. The configuration also allows for stress relaxation to take place. An alternative test is the notched constant tensile load (NCTL) test. In this test, notched specimens are subjected to a constant tensile load during the testing period. This paper briefly describes the development of NCTL and SP-NCTL tests, and counterpoints them to the bent strip test.

Regarding the field investigation part of the project, the detail of each field case history is reported, e.g., geographical location, application, geomembrane properties, and service duration. The basic material properties and SCR were evaluated on 16 retrieved field samples, and 18 as-manufactured geomembranes. An evaluation of the NCTL test data leads to a recommendation for an acceptance criterion for HDPE geomembrane insofar as SCR is concerned. Also, establishing the minimum value for the SP-NCTL test is discussed. Finally, the paper summarizes the current stress cracking specification for HDPE geomembranes by incorporating these recommended values.

2. Stress cracking phenomena

When semi-crystalline polymers deform at room temperature, they tend to be relatively tough and undergo considerable plastic deformation before rupture. For example, high-density polyethylene (HDPE) exhibits a break elongation of at least 500%. However, this type of polymer can also fail in a brittle manner, referred to as stress cracking (Williams, 1987). Stress cracking is defined in ASTM D 883 as "an external or internal rupture in a plastic caused by tensile stress less than its short-term mechanical strength". The transition from ductile-to-brittle failure requires knowledge of stress level, stress concentration factor, temperature and surrounding environment. However, the fundamental governing factor is the polymer characteristics, among which crystallinity and molecular weight are the most important.

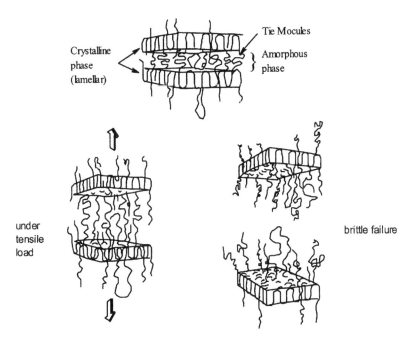

Fig. 1. Schematic diagram to illustrate the brittle failure at the molecular level.

There are two different stress crack mechanisms that may occur in semi-crystalline polymers. They have been classified as rapid crack propagation (RCP) and slow crack growth (SCG). RCP is associated with cracking that occurs at extremely high velocities, generally over 300 m/sec, and subzero temperatures (Plastic Pipe Line, 1982). The extent of the cracking can be over hundreds of meters in length. In contrast, SCG occurs at velocities less than 0.1 m/sec (Kinloch and Young, 1983). Crack lengths can be initially quite short but they can grow with time. Eventually the extent of the cracking can lead to excessive leakage in the system, defeating the design function of the geomembrane. The focus of this paper is on the SCR mechanism, although in some field cases RCP was also observed.

SCG of semi-crystalline polymers takes place by breaking the interlamellar tie molecules in the amorphous phase (Lustiger, 1983; Keith et al., 1971). Lustiger and Rosenberg (1988) illustrate such a mechanism using a conceptual model, as shown in Fig. 1. The number of tie molecules increases with molecular weight, but it decreases as the crystallinity increases. Due to the simplistic linear molecular structure of the polyethylene, a relatively high crystallinity can be achieved in comparison to other polymers. (The density of the polymer is directly proportional to the crystallinity). In order to produce polyethylene with different density ranges, co-monomers are added to lower the density. It has found that the type and amount of co-monomer can also influence the SCR of material (Huang and Brown, 1990; Lu and Brown, 1990,1991). For HDPE geomembranes, the density of opaque resin ranges between 0.929 to

0.938 g/ml. Regarding the molecular weight, it is estimated by an indirect method, ASTM D 1238 melt index (MI) test. Polyethylene with a low MI value, which indicates a high molecular weight, generally exhibits a greater SCR than that with a high MI value. The HDPE geomembrane has a maximum MI value of 1.0 g/10 min. Although SCR is closely related to the density and MI of the polymer, these two material parameters could not predict the SCR of the polymer (Hsuan et al., 1993). It is essential to evaluate the SCR using a specific test.

3. Literature review

Although there is extensive literature available on stress cracking of plastics, limited information is published on HDPE geomembranes. Interestingly, three state-of-the-art reviews on stress cracking in HDPE geomembranes, in three different countries (USA, Canada and Germany), all appeared in the mid-1980 when the material started to be used in environmental application. Under EPA sponsorship, Telles et al. (1984) provided an annotated bibliography on the interrelated topics of HDPE geomembrane stress cracking and various seaming methods. Not only was the bent strip test referenced, but also the Lander's test was discussed.

Fruch et al. (1986) prepared a report for Alberta Environmental Research Trust, which was a very complete treatise on various test methods for PE geomembranes. Included in the stress crack portion of the report was a description of the molecular structure of HDPE, its tie molecule arrangements, the mechanisms of cracking and crazing, and related theories. They found that the current test methods were "in a state of disarray" and that additional research and development was warranted.

Lastly, the Hoechst group in Germany under the leadership of Koch (1988), developed lifetime prediction methods for HDPE geomembranes. Their extrapolation method was based on accelerated testing of ductile-to-brittle behavior of HDPE material. The predicted site temperature data was then associated with a field strain value and stress relaxation data of the material to deduce a long-term design stress value.

A few technical papers on the subject of HDPE geomembrane stress cracking have subsequently appeared in the literature. Fisher (1989) reported on several field failures noting the importance of manufacturing, installation and seaming in preventing the phenomenon from occurring. Peggs (1988,1989) reviewed several field failures along with the actual remedial measures taken at the various sites. Finally, a survey made by GeoSyntec, Inc. (1989) resulted in the information provided in Table 1. Seventeen (17) stress crack incidents were discovered, all of which were on exposed HDPE geomembranes. The conclusion of the GeoServices study, as well as Telles et al. (1984) and Fruch et al. (1986) was that a more appropriate testing method was needed to evaluate SCR of the HDPE geomembrane than the existing bent strip test. The recommendations for preventing the phenomenon were suggested to focus in four distinct categories: materials and seams, design, construction and regulations.

Table 1
Summary of field stress cracking, after GeoSyntec Inc. (parentheses indicate number of sites with stress cracks)

Mfgr.	Type of installation			Environment			Exposed liner	Covered liner	Years in service
	Solid waste	Liquid waste	Total	Hot	Cold	Both			
1	2 (0)	7 (3)	9 (3)	1 (0)	3 (3)	5 (0)	9 (3)	0 (0)	7–10
2	17 (2)	6 (2)	23 (4)	16 (3)	4 (1)	3 (0)	19 (4)	4 (0)	2–6
3	4 (0)	14 (10)	18 (10)	11 (9)	0 (0)	7 (1)	18 (10)	0 (0)	1–8

4. Tests used to evaluate stress cracking in HDPE geomembranes

4.1. Previous test method

In the past, the test that was used to evaluate HDPE geomembranes was the ASTM D 1693 "Bent Strip" test. The test is one in which a surface notched rectangular test specimen is bent in a 180° arc and placed within the flanges of a small metal channel, as shown in Fig. 2. The entire assembly is immersed in a 10% Igepal/90% water solution, which is maintained at 50°C. Ten test specimens are evaluated in each test; five from the machine direction and five from the cross machine direction. The usual criterion was that no cracking should occur in less than 1500 hours of testing.

Although the bent strip test is very simple to perform, it has two major problems associated with it. The first is stress relaxation during the progress of the test. The rate

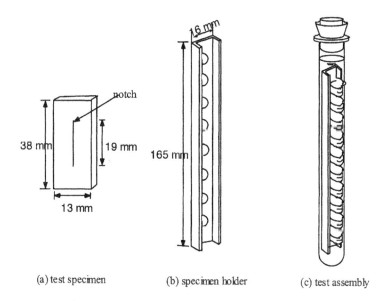

(a) test specimen   (b) specimen holder   (c) test assembly

Fig. 2. Test set up for the ASTM D 1693 bent strip test.

of stress relaxation is variable between different materials and thicknesses, and is completely unknown as to its magnitude and behavior. To merely require the test to be run for a long time period does not necessarily make the test more challenging to the material. Due to the stress relaxation issue, a comparison of different HDPE geomembranes is questionable, and should be viewed with considerable caution. The other problem has to do with nature of the test procedure. The test is monitored from time-to-time by laboratory personnel. If specimens fail before 1500 hours, their failure times are not accurate known. The SCR of different materials cannot be meaningfully distinguished; thus, a test that can provide quantitative data is desired.

4.2. Current test method

An alternative test was developed in the form of constant tensile load test, which was initially proposed by Lander (1960). The test used unnotched dumbbell shaped specimens placed under various tensile stresses. The tensioned specimens were immersed in a bath containing 10% Igepal/90% water solution at 50°C to accelerate the crack growth. When a specimen failed, its failure time was recorded by a timer. Unfortunately, the full test took a long time to complete (generally over 10,000 hours). Furthermore, the test results indicated considerable scatter due to variations of defects in each test specimen. A statistical method was required to analyze the data and a large number of tests were necessary for each material being evaluated. Therefore, the test was not well suited for quality control purposes, and was essentially abandoned.

As a result of the EPA project, the above described test was modified with a defined initiator, i.e, a notch, being introduced on each test specimen. Thus, the initiation of the crack no longer relies on the material defects, which vary from specimen to specimen. Hence, the test is called a notched constant tensile load (NCTL) test. In conducting NCTL tests, dumbbell shaped specimens are taken parallel to the cross machine direction of the geomembrane. This will also be the direction of general stress application, and is also the most sensitive orientation of an extruded geomembrane with respect to stress cracking. The notch is introduced at the central constant width section on the face of the specimen. The depth of the notch is such that the ligament thickness is equal to 80% of the nominal sheet thickness, as shown in Fig. 3. A set of notched specimens (usually 20) is then immersed in 10% Igepal/90% water solution at a constant temperature of 50°C. The applied stresses typically range from 20% to 50% of the room temperature yield stress in increments of 5%. Three replicate specimens are tested at each stress level, and the failure time of each individual specimen is recorded to an accuracy of 0.1 hour. The test data is presented by plotting the applied stress versus average failure time on a log-log scale.

The typical response curve is given in Fig. 4. There are two distinct regions having different slopes. Specimens in the high stress region with short failure times fail mainly in a ductile mode, whereas those in the low stress region with long failure time fail mainly in a brittle mode. The region between the ductile and brittle behavioral trends is called the "transition" region. The transition region can be either be very abrupt, have a "nose" shape, or have a "step" shape. Each has precedent in the polymer literature (Hsuan et al., 1993).

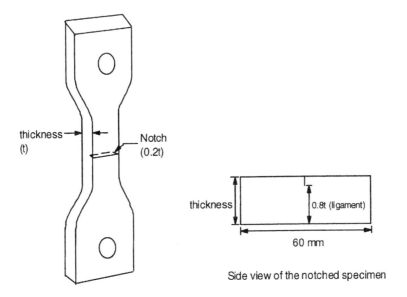

thickness (t)

Notch (0.2t)

thickness

0.8t (ligament)

60 mm

Side view of the notched specimen

Isometric view of the notched specimen

Fig. 3. NCTL test specimen configuration.

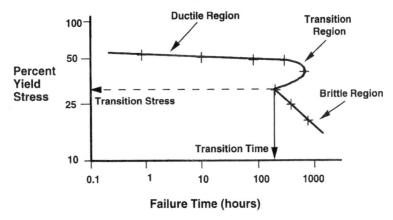

Fig. 4. Ductile-to-brittle transition curve resulting from the NCTL test.

There are four quantitative values that can be obtained from the NCTL test response curves, as shown in Fig. 4. They are the following:

- slope of ductile portion of the curve;
- slope of brittle portion of the curve;
- transition time; and
- transition stress level.

The two slope values can be utilized in describing a failure time as Eq. (1).

$$t_f = \sigma^{1/s_i} \tag{1}$$

where t_f is the failure time (hour), σ is the applied stress (percent yield stress or actual stress value); and s is the slope (either in the ductile or brittle region).

Of significantly greater importance than the slopes, however, is the transition point, which is defined as the beginning of the brittle region of the curve. Shown on the behavioral curve of Fig. 4 are the specific targeted points for transition time (T_t) and transition stress (T_σ), respectively. Materials with low transition stress and long transition time particularly the latter indicate a high SCR. Thus, the specification will be based on the transition time of the material. The recommended numeric value of transition time will be discussed in the later section of the paper.

5. Field investigation of stress cracking of HDPE geomembranes

One of the tasks in the EPA funded research project was to investigate stress cracking of HDPE geomembranes in the field at available sites. Material from sixteen sites were investigated. However, not every site was visited. In some cases, samples were sent to our laboratory for evaluation by the site personnel. In some cases, photos were provided by site personnel without any field samples.

The general information regarding the investigated field sites is presented in Table 2(a). The table is summarized as follows:

- All geographic regions of the USA (and two in Canada and one in Europe) are included. Six sites were in the Southwest, four in the Northwest and Canada, and the other five were spread throughout the Central, Southeast and Northeast regions. The site in Europe is located in Italy.
- The majority of the stress cracked samples were found in exposed geomembranes. They included the exposed liner of liquid impoundments above the liquid level, during construction and before backfilling, and during repair after liquid was removed. No opportunities were available to examine buried HDPE geomembranes such as those used for landfill liners.
- The geomembrane thicknesses vary from 1.0 to 2.5 mm.
- Geomembranes from five HDPE geomembrane manufacturers were involved.
- In addition to direct installation by the five manufacturers, several cases occurred where independent installation firms had been involved.
- In general, the specification for the type of resin was difficult to identify. Where it is known, its identity is numerically listed. There appear to be five resin suppliers involved.
- All four major HDPE seaming methods were included, i.e., flat extrusion, fillet extrusion, hot wedge and hot air. However, the majority of failures were associated with the two extrusion types of seams (flat and fillet). In addition, five failures involved double seaming, i.e., a fillet extrusion seam being placed over a flat extrusion or hot wedge seam.

- The year that the geomembranes were manufactured and installed varied from 1981 (which coincides with the earliest HDPE installations) to 1991.
- The length of service time varied from three months to seven years.

The majority of the retrieved field samples were evaluated for their physical material properties, tensile properties and SCR. For the SCR evaluation, both the bent strip and NCTL tests were performed. The test results are presented in Table 2(b), and are summarized as follows:

- Density of the carbon black compounded geomembranes ranged from 0.948 to 0.962 g/ml.
- Carbon black content ranged from 2.0 to 2.5%.
- Melt index values, under condition 190/2.16, ranged from 0.16 to 0.57 g/10 min.
- Flow rate ratio, condition P/condition 190/5.0, ranged from 2.9 to 6.4.
- Yield stress ranged from 16.9 to 24.2 MPa.
- Geomembranes from nine of the twelve sites failed to pass the 1500 hr bent strip test specification.
- Geomembranes from ten sites were evaluated by the NCTL test. Their transition times ranged from 4 to 97 hours. Furthermore, the transition stresses ranged from 20% to 45% of the yield stress of the material at room temperature.

Information related to the cracking of the geomembranes in the field including the location of the cracks and the likely cause of the cracking. This information is presented in Table 2(c). Summary comments taken from the table are as follows:

- The morphology of failure surfaces (as viewed in scanning electron micrographs) were either fibril or flake structures, as shown in Figs. 5 and 6, respectively. The fibril morphology indicates a slow crack growth (SCG) mechanism, and flake morphology reflects a rapid crack propagation (RCP) mechanism, (Halse et al., 1990).
- Both slow and rapid cracking were involved. Cracking in six sites was dominated by RCP, although in two sites, cracking was initiated by SCG, and then transformed to RCP. Cracking in nine other sites was caused by SCG.
- The estimated severity of cracking is listed with values ranging from minor (four sites) to major (ten sites).
- The location of the cracks when they occurred at seams (which was the general situation for SCG) was commonly in the lower sheet. The crack initiated at the junction between extrudate and the lower sheet, and eventually propagated through the entire thickness of the lower sheet, as shown in Fig. 7. In three sites, cracking took place at the extrudate of a fillet extrusion seam, as shown in Fig. 8.
- The location of the cracks within the facility itself was always in the exposed runout length or along the side slopes above the liquid level for liquid impoundments. In landfill cover applications, cracking was observed in the exposed section When cracking was observed in the bottom of a facility, it was either during initial construction or remediation, i.e., when the geomembrane was exposed.
- The cause of the tensile stresses was almost always thermally induced, which resulted in contraction of the geomembrane. However, there are a few anomalies, e.g., hydraulically induced or by a heavy and rapid loading.

234

Table 2

(a) General description of the field sites evaluated in this study

Site No.	Region of USA	Purpose	Thickness (mm)	Mfgr.	Install.	Resin Mfgr.	Seam Type	Const. Date	Service (years)
1	SW	SI (Sludge)	2.0	#1	#1	#4	#1 & #2	Fall 1981	4 to 7
2	SW	SI (Raffinate)	2.5	#2	#4	-	#2	June 1987	1
3	NE	SI (Leachate)	1.5	#4	#8	-	#4	1986	2
4	NW	SI (Gypsum sludge)	1.5	#2	#2	#1	#2	Summer 1986	3 months (Nov.–Jan.)
5	NW	SI (Leachate)	2.0	#3	#3	#2	#3 & #2	Spring 1983	4 months (Nov.–Feb.)
6	SW	SI (Brine)	2.5	#1	#1	#3	#1 & #2	Spring 1983	2 to 6
7	SW	SI (Brine)	2.5	#1	#1	#3	#1 & #2	Spring 1982	2 to 6
8	SW	SI (Evaporation pond)	2.0	#1	#1	#5	#1	Fall 1987	1
9	SW	SI	1.5	#2	#6	-	#2	-	-
10	SE	SI (Water)	1.5	#2	#6	-	#2	1987	1
11	MW	SI (Sewage)	1.0	#5	#5	-	#4	1985	3
12	SE	Cover liner	2.0	#3	#3	-	#3	1989	4
13	NW (Canada)	SI	1.5	#1	#7	-	#2 & #3	-	-
14	NW (Canada)	SI (Black liquor)	2.0	#1	#7	-	#2 & #3	Summer	2
15	SW	-	1.5	#3	#3	#3	#2	1991	< 1
16	Italy	Cover liner	2.0	#1	#9	-	#1	1989	4

(b) Physical and mechanical properties of exhumed HDPE geomembranes

Site No.	Density (g/ml)	C.B (%)	M.I. (190/2.16) (g/10 min)	FRR (2.16/5.0)	Yield stress (MPa)	Bent strip (hour)	NCTL Ts (%sy)	NCTL Tt (hr)
1	0.952	2.5	0.23	5	24.2	600 (7/10 failed)	–	–
2	–	–	–	–	–	–	–	–
3	0.956	2.5	0.17	4.7	19.7	1000 (no failure)	35	97
4	0.962	2.4	0.27	5	19.8	400 (3/5 failed)	–	–
5	0.949	2.4	0.57	4.2	20.3	300 (2/5 failed)	43	7
6	0.954	2.2	0.19	6.4	23	700 (9/10 failed)	41	4
7	0.956	2	0.22	4.7	22.4	72 (10/10 failed)	41	4
8	0.955	2.4	0.28	4.2	21.4	1000 (1/10 failed)	35	27
9	–	–	–	–	–	–	–	–
10	0.958	2.2	0.18	4.7	22.5	48 (10/10 failed)	–	–
11	–	–	–	–	–	–	–	–
12	0.948	2.6	0.49	4.33	19.3	170 (10/10)	40	8
13	0.954	2.3	0.16	5	24.1	2000 (0/20)	30	55
14	0.948	2.3	0.56 to 0.48	2.9 to 3.4	18.8	2000 (7/10)	42	17
15	0.949	2.4	0.57	4.3	16.9	500 (10/10)	42.5	11
16	–	–	–	–	18.4	–	27.5	5.7

Table 2 (continued)

(c) Description type and likely cause of stress cracking at field sites

Site No.	Morph.	Crack type	Crack severity	Crack Loc. at seam*	Crack Loc. at site**	Cause of stress	Cause of failure
1	Fibrous	Slow	Major (hundreds of cracks short and long)	LS & Ext.	2,3	Thermal	Heavy grinding
2	-	-	Moderate	-	2,3	Thermal	High temperature during seaming
3	Flake	Rapid	Minor (one 600 mm crack)	LS	3	Hydraulic	Probably due to an impact at the seam
4	Fibrous	Slow	Major (tens of cracks, short and long)	LS	2,3,4	Thermal	Heavy grinding and high temperature curing seaming
5	Fibrous and Flake	Slow then Rapid	Major (entire section cracked spreading throughout sheets and seams)	LS	3,4	Thermal	Extreme cold ($-15°C$) initiated at seam spread throughout. Also high temperature seaming
6	Fibrous	Slow	Major (tens of cracks, short and long)	LS & Ext.	2,3	Thermal	Heavy grinding and high temperature seaming
7	Fibrous	Slow	Moderate (10 cracks - 20 to 500 mm in length	LS	3,4	Thermal	Heavy grinding and high temperature seaming
8	Fibrous	Slow	Major (many cracks, short and long)	US	6	Thermal and Bending	Anchored in position, heavy grinding
9	Fibrous	Slow	Minor	-	3	-	-

No.	Morph.		Crack Description	Crack loc. at seam	Crack loc. at Site		Cause
10	–	–	Minor (two 450 mm cracks)	LS	2,3	–	–
11	–	Slow	Minor	LS	2	Thermal	Overheating and poor workmanship
12	Flake	Rapid	Major (entire section cracked spreading throughout sheets and seams)	LS	all area of the cover	Thermal	large temperature change
13	Damaged face	Rapid	Major (large area cracked spreading throughout sheets and seams)		4,5,6	Thermal	Extreme cold ($-40°C$)
14	Fibrous	Slow	Major (large area cracked spreading throughout sheets and seams)	LS	4,5,6	Thermal	Intermittent hot liquid at 65°C created folds in the liner. Cracks formed at apex of folds
15	Flake	Rapid	Major (single crack at 48 m long)	LS and/or US	6	Heavy loading	Dynamic loading during heap leach filling
16	Fibrous Flake	Slow and Rapid	Major (small cracks at seams and long cracks in sheet)	Ext and Sheet	exposed area	Thermal	Probably due to cool temperature and poor seaming

Notes: Mfgr. = Manufacturer, Const. = Construction, SI = Liquid impoundment, Install. = Installer, Service = Length of service.

Seam Type: #1 = fillet extrusion, #2 = flat extrusion, #3 = hot air, #4 = hot wedge

C.B. = Carbon black content, M.I. = Melt index, Condition 190/2.16, FRR = Flow rate ratio (Condition 190/5.0/Condition 190/2.16), * = tested at 50°C in 10% Igepal solution

Morph. = Morphology

* Crack location at seam: LS = lower sheet at seam, US = upper sheet at seam, Ext. = through the seam extrudate

** Crack location at Site: 1 = trenched, 2 = horizontal run out, 3 = exposed slope, 4 = liquid fluctuation zone, 5 = below liquid zone, 5 = bottom of the impoundment

238

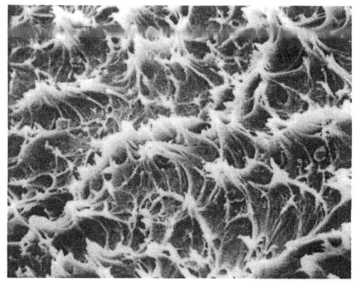

Magnification = 400X

Fig. 5. A fibril morphology resulted from a slow crack growth mechanism.

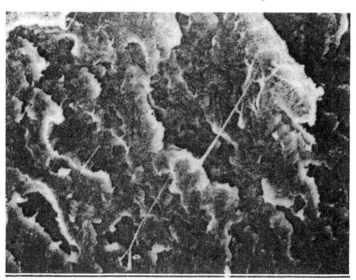

Magnification = 400X

Fig. 6. A flake morphology resulted from rapid crack propogation.

- The cause of the initiation of the failure varied, but heavy grinding (for extrusion seams) and/or high thermal energy (from double seaming) were generally involved.

In summarizing the findings presented in Table 2(a), (b) and (c), it is seen that majority of the field failures investigated were associated with exposed geomembranes.

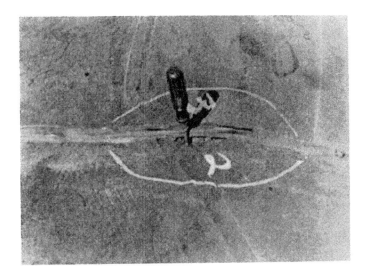

Fig. 7. Cracking took place at the edge of fillet extrusion seam.

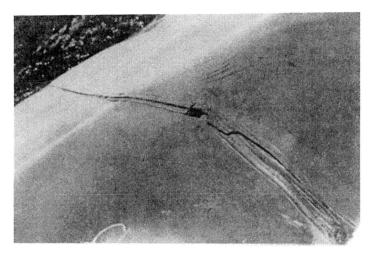

Fig. 8. Cracking within and at the edge of a fillet extrusion seam.

Temperature induced stresses played a significant role in that they induced the tensile stresses without which cracking may not occur in HDPE geomembranes. Furthermore, cracking generally proceeded via a SCG mechanism. RCP was only observed when the geomembrane was subjected to large change in ambient temperature. Cracking mostly occurred at the discontinuity formed by overlapping seams and was associated with heavy grinding and/or overheating. Figs. 9 and 10 show a series of cracks that were initiated at the grinding marks. Proper construction quality control and quality assurance is an important factor in minimizing features that stress cracking may initiate.

240

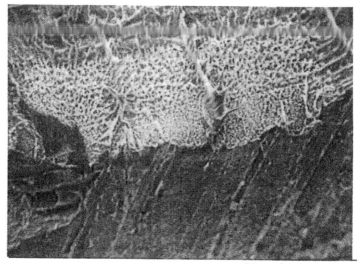

Magnification = 40X

Fig. 9. Cracking initiated from the grindng marks of a seam.

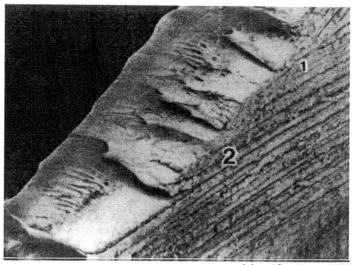

Magnification = 50X

Fig. 10. Cracking initiated from the grindng marks of a seam.

Regarding the original acceptance of these failed geomembranes, it is not known if bent strip testing via ASTM D 1693 was performed on the original manufactured sheet, and passed according to the common criterion. However, the majority of the retrieved field samples failed to pass the specification of 1500 hours. Geomembranes from Sites 3, 8 and 13, passed the bent strip test, but cracking still occurred in the field.

This indicates that the bent strip test was unable to adequately assess the SCR of the geomembrane. For the NCTL test, all ten evaluated field geomembranes showed transition times less than 100 hours, whereas the transition stress was between 27.5 to 43% of the yield stress.

6. Stress crack resistance of as-manufactured HDPE geomembranes

In addition to the evaluation of SCR of geomembranes that exhibited cracking in the field, the EPA study also involved an assessment of as-manufactured HDPE geomembranes available at the time. Eighteen as-manufactured geomembranes were tested for their basic material properties and SCR. The test results are given in Table 3. The density of the geomembrane reflects the percent crystallinity, since these two parameters are proportional to each other. The melt index values were in the range of 0.12 to 0.58 g/10 min., which was very similar to the retrieved field geomembranes. In general, the yield stress of the commercially available new geomembranes was slightly lower than the retrieved field samples; the values ranging from 15.2 to 23.2 MPa.

Regarding the SCR tests, only one out the 18 geomembranes failed the bent strip test. In contrast, a large variation was observed in the transition time of the 18 geomembranes resulting from the NCTL test. The transition times ranged from 10 to

Table 3
Material properties and SCR test data of 18 commercially new geomembranes

Commercially sample	Crystallinity (%)	M.I. (190/2.16) (g/10 min)	Yield stress (MPa)	Bent strip* no. failed/no. of tested specimens	NCTL T_σ (%σy)	T_t (hr)
1	43	0.17	18.3	0/10	35	420
2	58	0.12	20	0/10	35	70
3	55	0.09	23.2	0/10	30	115
4	51	0.23	17.2	0/10	38.5	30
5	49	0.55	17.2	0/10	35	50
6	48	0.5	15.2	0/10	40	70
7	44	0.58	20.3	10/10	35	10
8	51	0.15	19.4	0/10	30	5000
9	53	0.52	16	0/10	40	60
10	48	0.15	16.9	0/10	32.5	300
11	49	0.5	19.2	0/10	28	500
12	46	0.32	16.6	0/10	40	70
13	55	0.38	17.2	0/10	35	600
14	56	0.16	17.2	0/10	35	600
15	52	0.17	18.6	0/10	35	100
16	44	0.58	17.9	0/10	37.4	110
17	50	0.23	19.9	0/10	35	17
18	47	0.32	18.6	0/10	35	200

* = Test duration is 1,500 hours.

242

5000 hours, and the transition stresses were between 28% to 40% of the respective yield stresses.

7. Recommendation of transition time for the NCTL test

As shown in Fig. 11, the results of the NCTL test reveal a wide range of transition times in both field and as-manufactured geomembranes. In comparison, the transition stresses fall within a narrow range of the yield stress of the material. Thus, the transition time seems to be the property of choice for comparing the SCR of different HDPE geomembranes. A geomembrane with a long transition time signifies better SCR than one with a short transition time. In making a recommendation for minimum transition time, Hsuan et al. (1993) have studied seven of the ten failure sites, and recommended the minimum time to be 100 hours. This is because the longest failure time of the seven field geomembranes was 97 hours; hence, the minimum recommended time must be greater than that value. The nearest round number that is higher than 97 hours was selected to be 100 hours. (Note that the second highest field failure time was 55 hours, and 100 hours would be a factor of safety of two.)

Also to be noted is that approximately half of the new geomembranes have transition times less than 100 hours; thus, they fail to comply with the above criterion and should not be used. Clearly, stress cracking of HDPE geomembrane should continue to be monitored; thus, the minimum transition time could be modified to reflect the future performance of the HDPE geomembranes in the field.

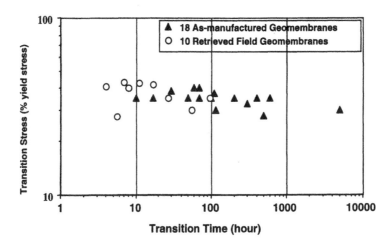

Fig. 11. Transition points of retrieved field geomembranes and as-manufactured geomembranes.

8. Single point notched constant load test

In order to obtain the transition time value from the NCTL test, minimum ten different applied stresses should be evaluated. The total testing time can be more than three weeks and many test stations are required. This is not practical for use as a quality control (QC) test. Subsequently, an abbreviated version of the test was developed. It is called the single point notched constant tensile load (SP-NCTL) test. The test was adopted as an appendix to ASTM D 5397. The focus of the test is on the brittle region of the curve, as indicated in Fig. 4. The concept is to select a stress level near, but slightly less than the initial portion of the brittle region and specifying the minimum failure time at that stress is to be sustained, as described by Hsuan and Koerner (1995).

Fig. 12 shows an idealized NCTL curve with transition point located at the recommended minimum transition time value of 100 hours, and at a transition stress of 35% of the yield stress. The latter value is the transition stress of the majority of the qualified laboratory samples. Using the brittle slope of all qualified samples, a zone of acceptability can be established, as illustrated by the shaded area in Fig. 12. This brittle zone is now used to determine the minimum failure time for the SP-NCTL test. The highest stress level that can assure brittle failure is 30% yield stress. A straight line drawn at 30% yield stress will intersect the minimum and maximum boundaries of the brittle zone at 120 and 200 hours, respectively. Any failure time in excess of the maximum value (i.e., 200 hours) indicates that the test material has a better SCR than those, which minimally qualify via the full NCTL test.

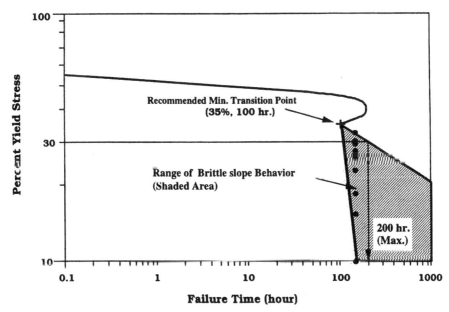

Fig. 12. Illustration of logic behind the single point notched constant tensile load (SP-NCTL) test (Note: 200 hours is the minimum recommended time at 30% yield stress).

Table 4
Various SP-NCTL test criteria pass/fail decisions

Test cycle	Yield stress (based on ASTM D 638)	Number of test specimens	Passing criteria	If noncompliance occurs
A	Manufacturer's mean value via MQC testing	5	4 out of 5 with $R_t > 200$ hr (noncomplying specimen with $R_t > 100$ hr)	Retest using cycle B
B	Manufacturer's mean value via MQC testing	5	4 out of 5 with $R_t > 200$ hr (noncomplying specimen with $R_t > 100$ hr)	Reject railcar or perform full test using cycle C
C	Manufacturer's mean value via MQC testing	30 or more	Onset of brittle portion of curve $R_t > 100$ hr	Reject railcar

For the SP-NCTL test, the recommended applied stress is 30% of the yield stress of the geomembrane at room temperature. The minimum failure time at that stress must be 200 hours, or longer. Furthermore, to gain a statistical validity, five specimens should be tested.

9. Stress crack specfication

In 1996, the Geosynthetic Research Institute (GRI) established a specification for the SCR of HDPE geomembrane smooth sheet by incorporating both NCTL and SP-NCTL tests, i.e., GRI GM-10. The specification addresses the test conditions and resulting pass/fail criteria for manufacturing quality control (MQC). The specification criteria are summarized in Table 4 for a sample taken every 90,000 kg of resin.

10. Summary

The stress crack resistance of HDPE geomembranes was evaluated on 16 retrieved field samples and 18 as-manufactured geomembranes, using both the bent strip and NCTL tests. The 16 field samples had a wide geographic distribution, and included a variety of resins and sheet manufacturers. The majority of the samples were taken from the exposed regions of the liner. The extent of the cracking varied from a single crack to extensive cracking. Slow crack growth was the dominant cracking mechanism. Significantly, a number of these sites still had acceptable bent strip test results, yet clearly showed stress cracking in the field. The deficiencies of the bent strip test for qualifying HDPE geomembrane was fully demonstrated.

The NCTL test overcomes the limitations of the bent strip test. The transition time obtained from the test can be used to distinguish the SCR of various types of geomembranes. The transition time of ten tested field samples is within 100 hours. For the 18 as-manufactured geomembranes, the transition time ranged from a few hours to thousands of hours. Based on the results of field samples, the acceptable criterion for an HDPE geomembrane when tested in a NCTL test is a transition time equal or greater than 100 hours.

However, the NCTL test is not suitable for QC/QA purposes due to the long required testing duration. Thus, the SP-NCTL test was developed. A single applied stress of 30% yield stress is utilized. The failure time at that stress level is required to be 200 hours based on data obtained from samples that met the full NCTL test requirement.

The above two criteria are adopted in the GRI GM-10 specification for the SCR of HDPE geomembranes.

11. Conclusions

The advantage of the SP-NCTL test over the bent strip test has been clearly demonstrated in this paper. It is essential for HDPE geomembranes to possess adequate SCR so that premature cracking does not occur during the service lifetime of the material.

Acknowledgements

This project was funded by the EPA under Cooperative Agreement No. CR-815692-01. Our sincere appreciation is extended to the agency. In addition, sincere thanks are extended to GRI members, since considerable internal funding was also provided to support the project. Particularly the author wishes to thank the HDPE resin suppliers and HDPE geomembrane manufacturers for their technical support. Their interaction was essential in bring these new specifications into existence.

References

Fruch, B., Dixon, P., Hunt, L.W., 1986. Review and development of physical test methods for polyethylene geomembranes. Alberta Environmental Research Trust, Final Report by Hanson Materials Engineering, Edmonton, Alberta, Canada, 150pp.

GeoSyntec, Inc., 1989. Preliminary assessment of stress cracking of polyethylene liners. Draft Report to US EPA, Task 6, Work Assignment 68-3, for NUS Corp., 27pp.

Halse, Y.H., Koerner, R.M., Lord Jr., A.E., 1990. Stress cracking morphology of polyethylene (PE) geomembrane seams. In: Peggs, I.D. (Ed.), Geosynthetics: Microstructure and Performance ASTM STP 1076. ASTM, Philadelphia, pp. 78-99

Hsuan, Y.G., Koerner, R.M., Lord Jr., A.E., 1993. Stress cracking resistance of high density polyethylene geomembranes. Journal of Geotechnical Engineering, ASCE 119 (11), 1840-1855.

Hsuan, Y.G., Koerner, R.M., 1995. The single point-notched constant tensile load test: A quality control test for assessing stress crack resistance. Geosynthetics International 2 (5), 831-843.

Huang, Y.-L., Brown, N., 1990. The dependence of Butyl branch density on slow crack growth in polyethylene: Kinetics. Journal of Polymer Science: Part B: Polymer Physics 28, 2007-2021.

Keith, H.D., Padden, F.J., Vadimsky, R.G., 1971. Intercrystalline links: Critical evaluations. Journal of Applied Physics 42, 4585-4592.

Kinloch, A.J., Young, R.J., 1983. Fracture Behaviour of Polymers, Elsevier Applied Science Publishers, 496pp.

Koch, R., Gaube, E., Hessel, J., Gondro, Ph.C., Heil, H., 1988. Long term resistance of dump sealing sheets of polyethylene. Report TR-89-0055, Translated by Sitran, Santa Barbara, CA, 19pp.

Lander, L.L., 1960. Environmental stress rupture of polyethylene. SPE Journal, pp. 1329-1332

Lu, X., Brown, N., 1990. The transition from ductile to slow crack growth failure in a copolymer of polyethylene. Journal of Materials Science 25, 411-416.

Lu, X., Brown, N., 1991. Unification of ductile failure and slow crack growth in an ethylene-octene copolymer. Journal of Materials Science 26, 612-620.

Lustiger, A., 1983. The molecular mechanism of slow crack growth in polyethylene. In: Eighth Plastic Fuel Gas Pipe Symp. American Gas Association, Arlington, VA, pp. 54-56.

Lustiger, A., Rosenberg, J., 1988. Predicting the service life of polyethylene in engineering applications. In: Koerner, R.M. (Ed.), Durability and Aging of Geosynthetics. Elsevier Applied Science, pp. 212-229.

Peggs, I.D., 1988. Failure and remediation of a geomembrane lining system. Geotechnical Fabrics Report 6 (6), 13-16.

Peggs, I.D., Carlson, D.S., 1989. Stress cracking of polyethylene geomembrane seams: Field experience. In: Koerner, R.M. (Ed.), Durability and Aging of Geosynthetics. Elsevier Applied Science, pp. 195-211.

Plastic Pipe Line, 1982. Published by Gas Research Institute, Vol. 3, No. 2, December, 3pp.

Telles, R.W., Lubowitz, H.R., Unger, S.L., 1984. Assessment of environmental stress corrosion of polyethylene liners in landfills and impoundments. Report on Contract 68-03-3218, US EPA, Cincinnati, OH, 89pp.

Williams, J.G., 1987. Fracture Mechanics of Polymers, J. Wiley and Son, New York, 302pp.

Evaluation and suggested improvements to highway edge drains incorporating geotextiles

G.P. Raymond[a],*, R.J. Bathurst[b], J. Hajek[c]

[a]*Department of Civil Engineering, Queen's University, Kingston, Ont., K7L 3N6, Canada*
[b]*Department of Civil Engineering, Royal Military College of Canada, Kingston, Ont., K7K 7B4, Canada*
[c]*Ministry of Transportation, Toronto, Ont., M3M 1J8, Canada*

Received 3 November 1998; received in revised form 24 February 1999; accepted 26 February 1999

Abstract

Design recommendations for retrofitted highway edge drains are presented. The recommendations are based on results from excavations made to evaluate the performance of various types of geosynthetic edge drains at selected locations on Ontario's highways. The evaluations include observations from excavations at highways built on both clay and sand subgrades. The geosynthetics excavated include geocomposite edge drains, geotextile-wrapped pipe edge drains and geotextile-wrapped aggregate edge drains. The edge drains were installed using various techniques that included: ploughed-in-place, trench excavation, and mechanical trencher and boot. The excavated edge drains were installed either as part of pavement construction or retrofitted several years after the original construction. All excavated edge drains used the pre-installed (existing) excavated or displaced shoulder Granular 'A' as backfill to the installed drain. Two main observations are: (a) drains that were installed adjacent to and in contact with the pavement edge (in North American terminology the pavement surface refers to the driving surface) soon became separated from the pavement edge by eroded/pumped fine soil particles seriously compromising the performance of the pavement subdrain systems, particularly where an open-graded drainage layer (OGDL) was used; (b) the granular backfill was considerably (up to 1000 times) less permeable than the geotextiles used for the edge drains. These and other problems form the background for the recommendations for the design and installation of retrofitted edge drains to ensure satisfactory field performance. © 1999 Elsevier Science Ltd. All rights reserved.

Keywords: Highway; Geotextile; Geocomposite; Edge drain; Drainage

---

* Corresponding author. Tel.: + 1 613 533 2131; fax: + 1 613 533 2128.
E-mail address: ray2@civil.queensu.ca (G.P. Raymond)

Reprinted from *Geotextiles and Geomembranes* **18 (1)**, 23-45 (2000)

1. Introduction

Fig. 1 illustrates the typical known water infiltration locations for a pavement structure (in North American terminology the pavement surface refers to the driving surface). Effective highway drainage is vital to minimise the effects of water damage. This includes surface drainage, groundwater lowering and internal drainage. Initially the Ministry of Transportation of Ontario (MTO) adopted the use of open grade drainage layers (OGDL) below the pavement surface that extended under the highway shoulder surface so as to daylight on the ditch slope surface. Hajek et al. (1992) reported that this technique for removing internal water was unsatisfactory, and that free water was retained at the pavement edge shoulder. To allow for the gravity drainage of this free water MTO modified their internal drainage practice by providing retrofitted edge drains at the pavement edge. This work presents investigations from six sites, each involving numerous excavations, dealing with one aspect of internal drainage, namely edge drainage, and then describes the new MTO practice adopted for retrofitted edge drain installations. The work updates/finalizes the preliminary work presented earlier (Raymond et al., 1995a, 1995b, 1996a, 1996b). Internal drainage is the collection and discharge of water that may enter the pavement structure through the surface course, surface cracks, granular shoulders or from the subgrade. Internal drainage is helped by subdrains, French drains, geocomposite drains and/or open-graded drainage layers (OGDL) and by sloping the subgrade. In Ontario the subgrade is sloped at 3% for most types of highway sections. In flat landscape areas the edge drains are required to be installed with a minimum gradient of 0.1% (1000H:1V). To assess the effectiveness of the Ontario practice, a research contract was established ("Performance of Geotextile-Based Hydraulic Pavement Systems") between Queen's University and MTO. The representatives of the University

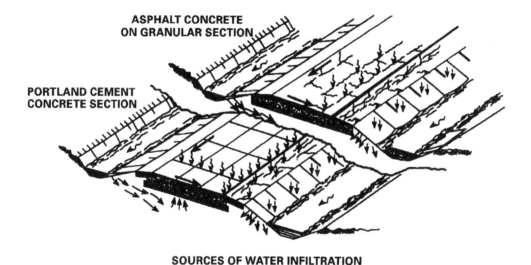

SOURCES OF WATER INFILTRATION

Fig. 1. Typical highway water infiltration sources.

included members of the Royal Military College of Canada. The project involved field investigation of several sites (six report herein), and included exhumation of edge drains and the surrounding granular materials and their testing in the laboratory.

2. Properties of geotextile found at investigated sites

Three types of geotextiles were excavated from the sites visited. The sock geotextile used to wrap the 100 mm internal diameter pipe drains were made from 4.5 tex polyester yarn knitted to give a Filtration Opening Size (FOS), (CGSB 148.1, 1993) $\leqslant 600$ µm; grab strength (ASTM D-1682) $\geqslant 225$ N with elongation at break $\geqslant 75\%$; Mullen burst (ASTM D-3786) $\geqslant 690$ kPa. The aggregate geotextile-wrap used for French drain type drainage and for open graded drainage layer unbound aggregate surround were nonwoven needlepunched geotextiles with the properties of FOS $\leqslant 100$ µm; mass/unit area $\geqslant 230$ gm^2; grab strength $\geqslant 550$ N with elongation at break $\geqslant 75\%$. The installations where a geocomposite was used consisted of a plastic core 300 mm wide (placed vertically) with 25 mm high cusps all wrapped in a non-woven geotextile. This resulted in an installed vertical height of 300 mm and an horizontal width of 25 mm. The geotextile had a measured FOS $\leqslant 200$ µm; mass/unit area $\geqslant 135$ g/m^2; grab strength $\geqslant 400$ N with elongation at break $\geqslant 50\%$.

3. Site 1. Geotextile sock-wrapped pipe edge drain and expanded polystyrene protected clay subgrade

This example involves the performance of a section of highway reconstructed in 1981 because of severe frost damage. The site has an average Freezing Index of about 1400°C days (The Freezing Index is the difference between the maximum and minimum values on a plot of cumulative degree-days of below-freezing temperature for one freezing season. The value of a degree-day is the difference between the average daily air temperature and freezing temperature, which in SI units is 0°C). This site recorded the largest Freezing Index of all sites investigated. All other sites had average Freezing Index values less than about 600°C days. Fig. 2 shows the condition of the reconstructed pavement section. As seen the section, although intended to be frost damage resistant, exhibited severe frost damage. The highway has two asphaltic concrete paved lanes (the pavement). Each lane was measured to be about 4.1 m wide (total pavement width 8.2 m) and each had an outer 3 m wide partially paved shoulder. The partially paved portion was about 0.6 m wide. The reconstruction drawings showed the asphaltic concrete to be 100 mm deep. At the locations investigated however, the asphaltic concrete was about 300 mm thick because of the subsequent frost damage repairs. Typically the damage resulted from ice lenses causing frost heave and then loss of bearing support on thawing. Below the asphaltic concrete was a 150 mm thickness of compacted Granular 'A'. Fig. 3 shows the grading limits of the MTO Granular 'A' specification which is similar to that of ASTM D-2940 (1974)). Found below the Granular 'A' was a variable depth of compacted clean well

Fig. 2. Photo of typical pavement conditions at Site 1.

graded sand (SW classification according to ASTM D-2487 (1983)). A 40 mm thick expanded polystyrene sheet was embedded 300 mm above the subgrade in order to prevent/reduce frost penetration. The expanded polystyrene extended horizontally 2 m beyond the edge of the driving lanes. A lean clay (CL) formed the subgrade. The minimum measured depth of sand was about 750 mm. Typical properties of the clay measured during this investigation were Liquid Limit, LL = 34; Plastic Limit, PL = 20; Plasticity Index, PI = 14; and 100% passing the 75 μm sieve. A geotextile-wrapped perforated pipe subdrain was installed just beyond the edge of the expanded polystyrene. According to the construction drawings the pipe should have been installed with its mid-cross section at the subgrade surface elevation.

Three excavations were done at this highway location. At two of the locations the subdrain pipe invert was below the subgrade surface. The subgrade surface was free of water and its shear strength, as measured by a Torvane, was about 100 kPa. At the other excavation the pipe was above the subgrade surface and free water existed below the pipe. The subgrade surface was soft and the shear strength of the clay about 20 mm below the subgrade surface was about 30 kPa. The excavation was extended along the pipe through relatively impermeable natural soil to the discharge ditch. The pipe invert maintained an elevation above the subgrade surface. There was no outlet for the water trapped on top of the subgrade in the clean sand below the pipe invert. Fig. 3 (Site 1) shows the results of washed sieve analyses (ASTM C-117, 1984) on the

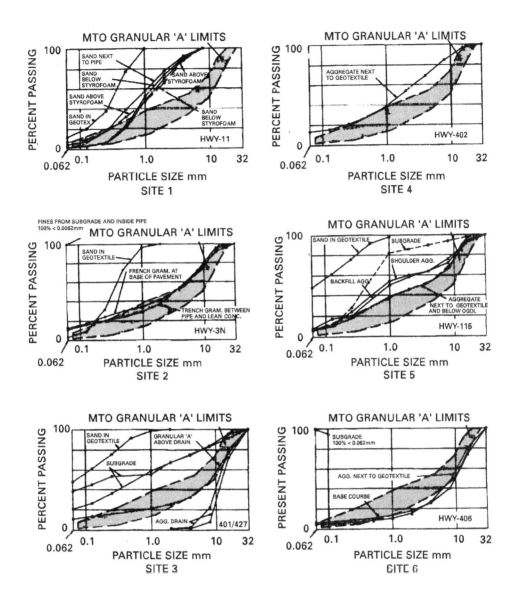

Fig. 3. Typical grading curves of soil sampler from sites investigated.

sand samples recovered from the excavations. Contrary to the construction draw-
ings/specifications some samples contained up to 10% of particles passing the 62 μm
sieve. The sand samples that contained a high fines content are unlikely to meet
Casagrande's (1932) non-frost susceptible criterion of less than 3% particles passing
the 20 μm size. Examination of the geotextile sock showed considerable damage in the
form of small holes up to 10 mm in diameter, probably caused by compaction

penetration of coarse aggregate during installation. Ultrasonic cleaning in water of the recovered geotextile indicated a grading of trapped particles ranging down from a maximum size of 1 mm. Fig. 3 (Site 1) includes the grading of the trapped fines. No measurements were taken of the edge drain pipe gradient which appeared to have been installed at a constant depth below the flat pavement surface. This would have resulted in zero gradient.

3.1. Summary of Site 1 findings

The findings from Site 1 indicate that:
(a) the trench backfill used for the geotextile-wrapped perforated drainage pipe should be clean sand (< 5% passing 75 μm sieve). Table 1 illustrates a recommended grading for this material;
(b) the base elevation of the drainage pipe invert should be located well below the sub-surface the pipe is intended to drain and the pipe's trench base must be able to ensure full trench drainage; and
(c) below flat ground surfaces subdrains should be installed with a gradient dropping towards the outlet locations.

4. Site 2. Geotextile sock-wrapped pipe edge drain and clay subgrade

This example involves the performance of a concrete pavement surface on a clay subgrade. A 15.5 km length of experimental pavement was built in 1982. It incorporated four sections with different Portland cement concrete pavement designs. All were provided with a 3 m wide shoulder of which the first 0.6 m was paved. The partial shoulder paving included either an extended Portland cement concrete pavement slab (cast at and with the pavement surface) or an added asphaltic concrete surface. Edge drainage was installed next to the edge of the Portland cement concrete slab just after paving operations. Placement was either below the partially paved asphaltic concrete shoulder or next to the Portland cement concrete shoulder. The edge drainage consisted of a longitudinal geotextile sock-wrapped perforated polyethylene pipe. The pipe was installed by the ploughed-in-place method at a constant depth below the pavement surface. Shoulder pavement junction distress was noted at a number of the sites visited and is discussed further in Chapter 9. The internal invert of the pipe was specified to be installed a minimum of 50 mm below the lowest subgrade surface elevation. Outlets were provided at a maximum 100 m intervals and at all low spots. The shoulder was constructed of Granular 'A' material and was considered a permeable media for pavement drainage to the pipe. After several years the outlet pipes either discharged muddy water or were non-functional during wet weather. In November 1991, nine years after the installation, excavations were made to assess the performance of the pavement drainage system.

The observations at excavations made through the asphaltic concrete shoulder were quite similar to each other. Longitudinal cracks ran along the asphalt shoulder/concrete junction. Subgrade clay fines could be observed in the cracks. The partially paved asphaltic concrete shoulders were distressed. The surface had dropped from between 40 mm to 55 mm in relation to the concrete pavement surface. As seen from the edge of the excavation the pavement consisted of either: 200 mm of Portland cement concrete over 100 mm of open-graded drainage layer (OGDL) placed directly on the clay subgrade or; 300 mm of Portland cement concrete cast directly on the clay subgrade. The OGDL had a bituminous binder and was contaminated with clay fines. The OGDL asphalt cement treatment, visible from all excavations, was stripped. Dirty water (with clay fines) was observed to be pumping laterally from the area where there was an OGDL and from the concrete slab/subgrade interface at excavations where there was no OGDL present. The pipe was located longitudinally about 300 mm from the edge of the concrete slab. Its internal invert elevation varied in each excavation from 75 mm below to 50 mm above the lowest subgrade surface elevation. All pipes had sediment and all geotextile sock-wraps were damaged. The pipes above the subgrade were generally underlain by a thin mixture of granular material and lumps of lean clay. The pipes below subgrade elevation were founded in the clay subgrade. Pieces of clay could be found above and around the pipe in all excavations. Permeability tests showed the backfill to be about 1000 times less permeable than the same material tested after washing on a 75 μm sieve. Fig. 3 (Site 2) includes typical grading curves for some of the recovered samples and for the soil trapped in the geotextile. The grading of the geotextile trapped soil is similar to that for Site 1.

All excavations made next to the partially paved Portland cement concrete shoulder showed similar performances. The Portland cement concrete shoulders and pavements were generally in good condition, highlighting the superior performance of integral constructed partially paved shoulders. Some transverse joints were cracked and working. The Portland cement concrete thickness was 180 to 200 mm. Below the Portland cement concrete slab was 125 mm thick lean concrete. The lean concrete was built directly on the subgrade. Free water discharged from between the Portland cement concrete slab and the lean concrete support. Fig. 4 shows that a knife blade could easily be inserted between the two layers. Any lean concrete that had extended beyond the concrete shoulder edge had been displaced during pipe installation. The subgrade was a stiff overconsolidated lean clay (CL). Typical properties were: LL = 40; PL = 20; PI = 20; and 100% passing the 75 μm sieve. The subgrade 'Torvane' shear strength was greater than 100 kPa (the limit of the Torvane). Conditions relating to the pipe trench backfills, their permeability and the pipe elevations were similar to those found below the asphaltic concrete shoulder excavations.

4.1. Summary of Site 2 findings

The findings from Site 2 indicate that:

(a) all excavations showed pumping dirty water with clay fines indicating that the edge drains did not stop subgrade erosion by pumping water;

Fig. 4. Photo showing gap between concrete slab and lean concrete base (Site 2).

(b) both trench bases and open graded drainage layers must be protected from the intrusion of fine grained soils and fine sands by an appropriate graded granular filter base and vertical backfill layer respectively;

(c) the base elevation of the invert of the excavation for the drainage pipe should be located below that of the base elevations of the layers they are intended to drain along with drainage of the pipe trench base;

(d) the ploughed-in-place method of pipe installation caused mixing of subgrade clay lumps with the edge drain backfill such that no clean backfill was observed;

(e) all recovered geotextile sock material were severely damaged with numerous holes 5 to 10 mm in size;

(f) all edge drain pipes were partly filled or filled with sediments;

(g) partially paved shoulders should, where practical, be made as an integral extension of the pavement layer/slab.

5. Site 3. Geotextile-wrapped open-graded aggregate edge drain

This example involves the performance of a pavement on a compacted clayey gravel (GC) subgrade (sometimes called till, or boulder clay). Full details for this site are presented by Raymond et al. (1998). The highway was built as a freeway in 1970. A geotextile-wrapped aggregate drainage system was retrofitted in 1982. It was located below the shoulder. In June 1992 excavations were made to evaluate the drainage performance. At the excavations the subgrade comprised a high fill of recompacted clayey gravel (GC). Typical soil properties were: LL = 25; PL = 16; PI = 9; and approximately 25–35% passing the 75 μm sieve. The Portland cement concrete surface slab was 230 mm thick and was built on a 150 mm cement treated base. The base was directly on the subgrade. Next to the pavement edge was a fully paved 3 m wide asphaltic concrete shoulder with curb and gutter. At the sites investigated the highway's longitudinal grade was 2%; the cross-pavement slope was 2.5%; and the cross-shoulder slope was 6%. The drainage system was installed in a 300 mm wide trench below the shoulder. The trench was located next to the edge of the cement treated base resulting in its centre line being 760 mm from the edge of the concrete pavement. A geotextile-wrapped open-graded aggregate containing a nominal 100 mm diameter perforated corrugated plastic pipe formed the drainage system. The pipe was placed 50 mm above the trench base (and base portion of the geotextile). The top of the wrapped stone was level with, or slightly above, the top of the cement treated base. The wrapped stone layer depth varied between 300 to 325 mm. Its depth was 150 mm below subgrade elevation.

The shoulder surface at the site had settled at least 25 mm below the pavement surface. The area above the trench had settled more. The observed shoulder/pavement junction distress was noted at a number of the sites visited and is discussed further at the end of Chapter 9. The cement treated base was of poor quality either due to poor initial mix or deterioration. The aggregate inside the geotextile-wrap that was beside and above the pipe was clean. Fig. 3 (Site 3) includes typical grading curves from recovered samples including particles trapped in the base portion of the geotextile. The grading of trapped particles is similar to the grading of trapped fines measured at Sites 1 and 2. Fig. 5 illustrates how the trapped aggregate on the trench bottom was cemented to the geotextile by migrated subgrade fines. The fines had been transported by seepage water through the clear stone and deposited on the geotextile. The sides of the exhumed geotextile were clean compared with the bottom portion which was also wet. After removal of the geotextile a very thin (< 1 mm) depth of ponded water was evident on small areas of the trench base. The geotextile was undamaged and no punctures or worn through areas were evident. One excavation was at a transverse joint which revealed erosion of a portion of the Portland cement concrete slab at the joint. In the eroded portion was a black-coloured granular material deposit that was probably lines from the asphaltic concrete shoulder.

One of the excavations was made where an outlet pipe discharged to a catch basin. The drainage pipe was turned down slightly at the outlet. The outlet was connected through a T-junction to a non-perforated outlet pipe that sloped steeply at right angles to the pavement edge. The pipe allowed water to discharge into a catch basin.

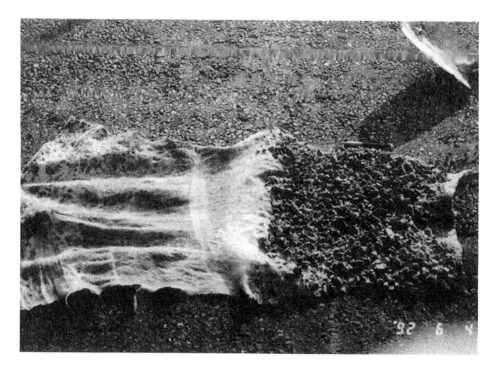

Fig. 5. Photo of typical condition of trench geotextile-wrap (Site 3).

The T-junction was wrapped with a geotextile. It was bedded on about 150 mm of Granular 'A' containing about 12% passing the 75 μm sieve. The 150 mm thick Granular 'A' layer had a very low relative permeability compared with the open-graded trench aggregate. Fig. 6 illustrates the outlet detail found. Shown are the turn down details of the T-junction edge drain pipe. Without depressing the longitudinal subdrain pipe at the junction, water would be trapped in the open-grade aggregate by the less permeable Granular 'A'.

5.1. Summary of Site 3 findings

The findings from Site 3 indicate that:
(a) no evidence was obtained that the geotextile or the pipe had sustained any damage;
(b) the geotextile at the sides and top of the aggregate drain were clean enough to be permeable;
(c) geotextile fouling had occurred at and above the base of the edge drain trench;
(d) the clear stone bedding below the pipe was performing as intended to act as a trap for soil fines;
(e) the edge drains did not stop subgrade erosion by pumping water through the lean concrete base course.

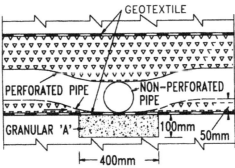

TURN DOWN DETAIL FOR EDGE DRAIN.

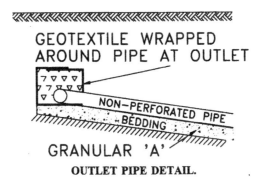

OUTLET PIPE DETAIL.

Fig. 6. Turn down details at outlet connection.

6. Site 4. Geocomposite edge drain and clay subgrade

This example involves the performance of a pavement on a clay subgrade that incorporates a geocomposite edge drain. Full details for this site are presented by Raymond et al. (1994). The highway was built as a freeway in 1981. In 1990, a 15 km length was repaired/rehabilitated. The 1981 construction drawings show a wearing surface of 80 mm of asphaltic concrete overlay on a 175 mm thick Portland cement concrete slab. A base of 125 mm thick lean concrete was built on a stiff clay subgrade (c_u = 50 to 100 kPa). Typical subgrade properties were: LL = 40; PL = 20; PI = 20; and 100% passing the 75 µm sieve. A 3 m wide shoulder was provided with an initial 0.6 m wide partially paved portion having a thickness of 80 mm of asphaltic concrete. The repair work of 1990 involved 15 km of highway built in two phases. Phase 1 involved many short sections totalling 1.5 km. Each short section entailed the complete removal of pavement materials to the subgrade surface below the driving lane only. The removed materials were replaced starting with a 100 mm thick geotextile-wrapped unbound open-graded drainage layer (OGDL). The OGDL was overlain with 300 mm of Portland cement concrete covered by 80 mm of asphaltic concrete. In Phase 2 a geocomposite edge drain was placed along the entire 15 km length of pavement edge (i.e. Portland cement concrete slab). It was placed at a constant depth from the pavement surface. The installation involved a trench excavated to 25 mm below the lowest subgrade elevation.

The geocomposite edge drain consisted of a plastic core 300 mm wide (placed vertically) with 25 mm high cusps all wrapped in a nonwoven geotextile. Outlets to the ditch were provided at all low spots or every 100 m. After compaction of the trench backfill, the partially paved shoulder was replaced. During the first winter after repair (1990/91), shoulder/pavement differential frost heave of 50 mm was observed. The winter heave had dropped the paved shoulder area transversely towards the outer edge. The observed shoulder/pavement junction distress was noted at a number of the sites visited and is discussed further at the end of Chapter 9. Excavations were made in the summer of 1991. The repaired pavement exhibited intermittent centre line cracks; intermittent severe to moderate mid-lane cracks; secondary transverse cracks; and transverse 'rout and seal' cracks that were starting to develop into 'alligator' cracks. Shoulder edge cracks ran parallel to the Portland cement concrete pavement edge.

On excavation, all the recovered geocomposite edge drains exhibited crushed bases. Fig. 7 illustrates the condition of a typical damaged edge drain. They all contained some clay sediment. Some were completely plugged with clay. The geotextile-wrap of some sections was observed to be cut by the plastic core along the line of the geocomposite invert. Water pumped clay fines, several millimetres thick, were frequently observed deposited between the pavement structure and the geocomposite edge drains rendering the drainage system non-functional. Thick deposits of clay fines were particularly evident near the base between the OGDL and geocomposite. It is not known if the thicker clay deposits at the geocomposite base were due to forced particle migration between the two surfaces (geocomposite and OGDL geotextile) or due to installation of the geocomposite with a large void space between the two surfaces. The coefficient of permeability of the recovered backfill was about 0.5 μm/s. After washing on a 75 μm sieve this value increased to about 100 μm/s. For comparison, the uncontaminated geotextile permeability is about 400 μm/s. Fig. 3 (Site 4) includes typical grading curves of the backfill particles. On flat areas of the highway, the measured elevations showed both the highway surface and edge drain invert to have zero longitudinal slope. This is consistent with the installation specification requiring a uniform trench depth below the pavement edge. Clearly this results in poor hydraulics for drainage.

Excavations made at Phase 1 locations (driving lane rebuilt) generally found the subgrade elevation to be below the invert elevation of the crushed geocomposite edge drain. The crushing occurred during frost heaving or during the installation. There was always a build up of clay between the OGDL and geocomposite edge drain as described above. The OGDL was often partially filled with water. Fig. 8 shows the conditions found at one section where the cusps of the geocomposite edge drain core were visible through the geotextile. The geotextile was believed to have worked against the slab face during frost heave.

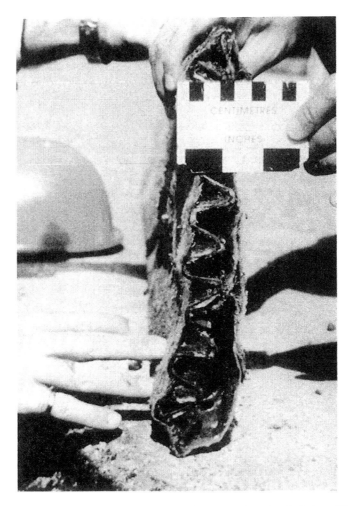

Fig. 7. Photo of typical crushed condition of geocomposite edge drain base (Site 4).

6.1. Summary of Site 4 findings

The findings from Site 4 indicate that:

(a) differences exist between the design intentions and actual performance since the geotextile-wraps of the edge drains and open graded drainage layers were unable to filter subgrade fines from pumping water;

(b) the permeability of the backfill was about 800 times smaller than the edge drain geotextile which is contrary to conventional practice that requires the soil permeability be less than the permeable of the geotextile by a factor of about 10 to 25;

(c) in all the excavations the lower cusps of the geocomposite edge drain core were crushed and some geotextiles were cut by the core along the line of the geocomposite invert;

Fig. 8. Photo of holes in geotextile of geocomposite edge drain at Site 4.

(d) the specification relating to the geocomposite core should include a clause related to the vertical compression and buckling resistance of the core and should have a rounded shape along the line of its invert;

(e) neither drainage nor a cement stabilized base nor a geotextile could prevent erosion of subgrade fines by pumping water;

7. Site 5. Geocomposite edge drain with sand subgrade

This example involves the performance of a highway on a sand subgrade that incorporates a geocomposite edge drain. Full details for this site are presented by Raymond et al. (1996a). The example involves a highway built as a freeway in 1990.

The investigation was undertaken in June 1992, two years after construction. The pavement surface was a 200 mm thick Portland cement concrete. This was built on a 100 mm thick asphalt cement bound open-graded drainage layer (OGDL). Below the OGDL was a 100 mm thick layer of Granular 'A' constructed on a minimum 500 mm of sand, compacted over the natural subgrade. The compacted sand is considered the highway subgrade at this site. Next to the pavement was a 3 m wide shoulder. A width of 0.6 m of the shoulder was Portland cement concrete slab built integral with the pavement. A geocomposite edge drain was installed using a mechanical trencher (vermeer cutter) and boot next to the edge of the Portland cement concrete slab (partially paved shoulder). The geocomposite edge drain was the same as that described at Site 4.

Several excavations were made. Two were made 2 m apart and were located up-slope and close to an outlet. Both were initially made without disturbing the geocomposite edge drain. Water was then poured into the up-slope excavation furthest from the outlet until the water level was at the base of the OGDL elevation. This water level was maintained for about thirty minutes. Little water had to be added. No water appeared at the second excavation or at the outlet pipe during the entire procedure. This suggested that the geotextile was fouled. It also confirmed the low permeability of both the backfill and sand subgrade.

The geocomposite edge drain was then removed from the down-slope excavation, and was observed to be clean and dry. Fig. 9 shows the condition, as found, of the cut

Fig. 9. Photo of typical fouled condition of pavement/edge drain trench wall (Site 6).

face of the OGDL and Granular 'A' underneath. As first observed the cut face of the OGDL could not be seen. It was plugged with fines from the granular materials surrounding the geocomposite edge drain. Also noted was the fact that the partially-sawed contraction joint visible on the concrete slab face extended fully through the slab which, if not properly sealed, would allow entry of water to the OGDL. The lowest elevation of the geotextile at the first excavation (up-slope) was then purposely punctured. Within seconds water discharged into and through the second excavation. Water appeared at the ditch. The geocomposite edge drain at the up-slope excavation was now removed. It was clean inside. After cleaning the cut faces of the OGDL (composed of particles up to 13.5 mm in size) and Granular 'A' (composed of particles up to 26.5 mm in size) the faces were seen to be very rough and there was a visible crack extending from the pre-sawed joint through the full depth of the slab. Fines had penetrated into the open pores of the OGDL and could not be mechanically cleaned. Should any water have entered into and then discharged from the very permeable OGDL it probably would have been resisted by the less permeable backfill and geotextile materials allowing possible water turbulence. Even with the geocomposite edge drain installed tight against the edge of the OGDL during original construction there would be a continuous vertical plane of void spaces between the rough OGDL face and smooth geotextile surface. In addition, the base of the mechanically cut trench is unlikely to have a square corner due to debris preventing the geocomposite from being tight against the base of the excavation wall, again leaving a void between the lower portions of the excavation wall and geotextile. These voids would aggravate any water turbulence that might occur from any large flow of water out of the highly permeable OGDL meeting the considerable less permeable geotextile. What occurred is difficult to establish since sampling of any deposited material in these voids was difficult. What is clear is that fines from the granular materials were pumped and/or deposited between the OGDL and the geotextile and also forced into the OGDL pores. The accumulated fines fouled the irregular vertical surface between the OGDL and the geotextile and impeded drainage. Fig. 3 (Site 5) includes typical grading curves for recovered samples. Permeability tests on the soil next to the geocomposite edge drain gave coefficients of permeability about 70 times less than the same soil after washing on a 75 μm sieve, and 150 times less than the geotextile.

7.1. Summary of Site 5 findings

The findings from Site 5 indicate that:
(a) the initially cut wall of the trench that contains the geocomposite was very rough and the interface between open graded draining layers and geotextiles was found to be badly fouled;
(b) the open graded drainage layer/geocomposite trench interface must be protected from the intrusion of fine grained soils and sands. A better installation practice would be to place the geocomposite edge drain on the shoulder side of the trench and fill the trench with a high permeability graded granular filter backfill (e.g., Table 1).

Table 1
Grading curve for filter sand to prevent pumping of silt or clay

Sieve size micrometre (μm)	ASTM E-11 (1981) sieve size	Percent passing by weight
9500	3/8″	100
4750	No. 4	95–100
2360	No. 8	80–100
1180	No. 16	50–85
600	No. 30	25–60
300	No. 50	10–30
150	No. 100	2–10
75	No. 200	0–2

8. Site 6. Geocomposite edge drain with clay subgrade and low fines granular base

This example involves the performance of a highway on a clay subgrade that incorporates a geocomposite edge drain and a manufactured (crushed) coarse graded Granular 'A'. The example involves a two lane highway built to standards that would allow later conversion to a freeway. The investigation was undertaken in July 1992, one year after the completion of construction. At the sites selected, the highway was built over a very stiff (c_u = 100 to 200 kPa) clay subgrade. Subgrade properties were: LL = 35; PL = 19; PI = 16; and 100% passing the 75 μm sieve. The subgrade was covered with 300 mm of manufactured coarse graded Granular 'A'. This was followed by 100 mm of an asphalt bound OGDL. The pavement surface was 225 mm of Portland cement concrete. The shoulder was 3.0 m wide of which 500 mm was an extension of the pavement slab (partially-paved shoulder). A geocomposite edge drain was installed along the partially paved shoulder edges. Two different geocomposite edge drain products were used. One was the same as that described for Site 4. The other was similar except the core was in the form of a hollow oblong shape. Both were installed using a vermeer cutter and boot. The surface was then finished using the aggregate material displaced during the installation.

All three excavations showed similar conditions. The geocomposite edge drains and their geotextiles were in good condition. The longitudinal slope of the highway was 0.65%. The Granular 'A' was at the coarse limit of the grading specification and was very permeable. Fig. 3 (Site 6) includes typical grading curves for the samples recovered. All the granular material samples exhibited little material passing the 75 μm sieve. Only two of the granular materials sampled exhibited more than 6% passing the 125 μm sieve (the maximum was 8.1%). All the samples were quarried materials (i.e. 100% crushed). A water permeability test was performed in much the same way as at Site 5. A modification involved the use of a plastic sheet placed in the excavation in an attempt to prevent water escaping through the Granular 'A' base course. The Granular 'A' was too permeable to allow retention of water. Water quickly seeped to the subgrade surface. Since the surface was well sloped the water then ran quickly alone the subgrade surface to the ditch. The low quantity of fine material combined with 100% crushed particles resulted in the granular base material having a high coefficient of permeability. Permeability tests on the soil next to the geocomposite edge drain gave coefficients of permeability 9 times less than the same soil after washing on a 75 μm sieve and 22 times less than the geotextile.

The visual observations and tests done during this site investigation showed that one year after construction there had been no fine material deposited between the OGDL and the geocomposite edge drain geotextile. The results showed that when the base course is highly permeable, the pavement water permeates quickly to the subgrade and by-passes the drainage system. Unfortunately the Granular 'A' at this site does not meet all the grading requirements of the American Society for Testing and Materials base course specifications (ASTM D-1241 (1968) or ASTM D-2940 (1974)). These require the quantity of particles by weight passing the 425 μm or 600 μm sieve, respectively, to be more than 1.5 times the quantity passing the 75 μm sieve. Since these rules generally involve a factor of safety it is hoped that future pumping, subgrade particle migration, or intrusion of granular particles into the subgrade, will not occur. No evidence was noted of pumping beginning to occur, however all the excavations were at the pavement edge and had only been subject to one winter of freeze and thaw. It is the writers' recommendation that all OGDL should be protected from fouling by pumping fines of clay and silt subgrades. This protection is best achieved by a granular material meeting known granular filter and base course criteria even if these may be conservative.

8.1. Summary of Site 6 findings

(a) Although no damage to the geotextile was observed at this site the maximum crushed particle size in the trench backfill should be controlled to minimise the possibility of damage to the geotextile (e.g. ASSHTO M228 (1996) recommendations).
(b) The trench backfill should be manufactured to a sand grading similar to that given in Table 1.

9. General comments

It is evident that the impermeable granular trench backfills found at all but one site prevented effective internal drainage of the area below the pavement. The results of this poor drainage were seen at the excavations next to transverse joints and cracks. At these locations there was: (a) pumping water; (b) erosion of the subgrade; (c) settlement (heave) of the shoulder and; (d) deterioration of the quality of the base/OGDL material. In future retrofit installations the drainage system should be connected directly to the OGDL or pavement granular support using permeable granular material. As illustrated at Site 6, where a coarse granular base was used, the backfill should be a well graded non-cemented clean granular material (less than 5% passing the 75 μm sieve) continuously graded down to the 75 μm sieve. A suggested grading is a concrete sand meeting ASTM C-33 specification plus a limit of 0 to 2% passing at the 75 μm sieve (Table 1).

A geotextile used with a geocomposite edge drain or pipe will generally have a much smaller permeability than the OGDL. This is contrary to the theoretical idea that the permeability of a geotextile should be at least 10 times greater than the permeability of the soil. On the other hand, it is necessary to prevent internal clogging of the edge drain conduit (i.e. area inside the geotextile). Clearly the idea of the permeability of the geotextile being greater than the permeability of the OGDL (soil) was not realised. Fines that can clog will further reduce the permeability of the geotextile as noted at all sites in this study but one. Clearly it is essential to develop a drainage system that prevents fouling of the geotextile.

Some resulting questions may be stated: Can a better design be devised? Can a greater, faster water discharge into the drainage conduit be achieved? Can an alternative design result in a more reliable installation technique? If so, would better protection from water-related damage result for the pavement structure? And, as always, what about the cost/benefit?

As an alternate drainage system, consider allowing water from the granular pavement support (sub-base, base or OGDL) to discharge into a high permeability backfilled trench containing the conduit. Here the water from the pavement support would discharge into the high permeability trench backfill. The surface area of the conduit geotextile receiving water would be considerably greater than that of the typical present installation geotextile surface area. In the new arrangement there is less chance of a void developing between the granular support (base or OGDL) and the trench backfill material. A further concern is the method of installation where the conduit or trench is at a constant depth below the pavement surface. This results in a zero longitudinal slope of the conduit on flat highway alignments. A 100 mm sewer pipe to be self cleaning typically requires a slope of about 2.5%. A 100 mm diameter pipe protected by a geotextile sock-wrap carrying only water should not require such a large slope. It is recommended that a minimum slope of 1% be specified where practical. Installation should be in an excavated open trench or by mechanical excavator and boot with placement of a geocomposite, if used, on the shoulder side of the excavation. The installation should be undertaken with great care. Excavated soil should be removed from the trench opening to prevent excavated soil from falling back into the trench as debris. Trenches should be closed as soon as practical, and no later than the same day as opened. The backfill should consist of an extra clean sand, preferably manufacture graded as per Table 1 or similar. The trench base should comprise a bedding layer where practical. This is to protect the conduit from infiltration of subgrade fines. Similar conclusions have been arrived at by Koerner et al. (1994) and Koerner and Koerner (1995). Koerner et al. (1994) describe one method of installation in an open trench that has been in use by the Kentucky Department of Transportation since 1992 (Allen and Fleckenstein, 1993).

A further argument for placing the drainage trench adjacent to the pavement edge, and draining the OGDL, base or subbase into a permeable clean sand backfill rather than directly into the geocomposite edge drain was demonstrated by the observed shoulder/pavement junction distress at Sites 2, 3 and 4 along with the observed winter shoulder/pavement heave observed at Site 4. Fig. 10 illustrates the effect of frost heave between unloaded shoulder and loaded pavement. The upper left diagram shows the

266

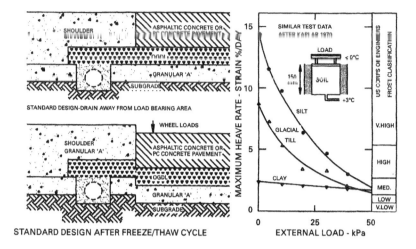

Fig. 10. Illustrated effect of frost heave between unloaded shoulder and loaded pavement. Upper left: initial construction; Lower left: winter heave; Right: laboratory experiments (Kaplar, 1970).

initial construction, while the lower left diagram shows the effect of winter heave and the diagram on the right shows laboratory experimental observations by Kaplar (1970) of the reduced frost heave resulting from surface loading. Snow ploughs will remove portions of the granular shoulder during pavement snow clearance if there is no partial paving of the shoulder or if the partial paved shoulder is built integral with the pavement. During spring thaw this removed shoulder granular is generally replaced during spring maintenance. In addition, freezing/thaw results in particle loosening and during thaw lateral displacement occurs during subsequent compaction.

10. General recommendations for edge drain installations

Based on the general comments made above it is recommended that all pavement edge drainage conduits should be placed in an open trench. Fig. 11 illustrates how either a geocomposite edge drain or a pipe edge drain is now installed by MTO. The drainage conduit, as shown in the figure, is installed either on the shoulder side (geocomposite) or centre (pipe) of the excavation and backfilled with extra clean sand. It is protected from subgrade infiltration by a 50 mm minimum bedding of the backfill material. MTO still maintain a minimum longitudinal grade of 0.1% (1000H : 1V) on flat landscape areas. A greater gradient should be used where possible. An acceptable backfill would be a graded material in conformity with concrete sand meeting ASTM C-33 (1984) plus 0–2% at the 75 μm sieve (Table 1). Outlet pipes should be turned down in a similar manner to that shown in Fig. 6.

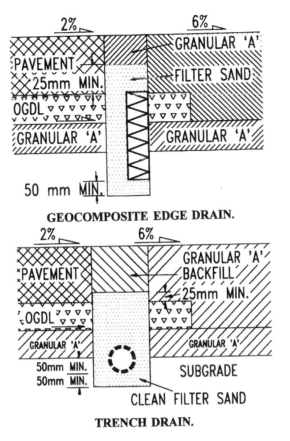

GEOCOMPOSITE EDGE DRAIN.

TRENCH DRAIN.

Fig. 11. Recommended typical retrofit edge drain details.

11. General conclusions and recommendations

This study has shown that the installation of a drainage system, in itself, does not prevent pumping. Also shown was the fact that neither a lean concrete nor a cement treated base nor a geotextile will prevent the migration of clay or silt sized subgrade fines. Lean concrete and cement treated base layers should only be used for structural support. Prevention of migration of clay or silt sized subgrade fines requires suitable graded granular materials. Both the bases and edges of OGDL must be protected from the intrusion of subgrade fines. Typically below an OGDL this would be a Granular 'A' material similar in specification to ASTM D-2940. In cold climates, soil classification of this material would conform to a clean well graded gravel (GW). Where frost is of no importance, classification of the soil could be GW–GM. Below and around edge drainage conduits a concrete sand conforming to ASTM C-33 plus 0–2% at the 75 μm sieve is recommended (Table 1).

All recovered geotextile sock-wrapped pipe installed using the ploughed-in-place method were severely damaged with many holes of 10 mm size. Some holes were larger. All pipes installed in this manner were filled to various depths with sediment. Clearly the damage was more than desirable for installations that ranged from 1 to 11 years old. Nevertheless, the plastic pipe was observed to be in excellent condition at all excavations.

The edge drain invert elevations of drains installed by the ploughed-in-place method were often found to be above the lowest elevations of the layers they were intended to drain. It is therefore recommended that a retrofit drainage system be installed in an open trench. The invert elevation of the trench should be a minimum of 100 mm below that of the lowest elevations of the layers it is intended to drain. The trench invert should have a minimum 1% gradient to outlets. It should be covered with a minimum 50 mm depth of filter soil before placement of the conduit.

At one site all the geocomposite edge drains exhibited crushed or bent cores. The geotextile was often cut along the geocomposite invert line. These conditions have been reported by others (referenced earlier), whether installed in open trenches or by mechanical trenchers and boot. They result, in part, from the inability of mechanical equipment to cut square trench corners during any type of installation. It is recommended that any geocomposite edge drain specification include a clause related to the vertical compression, buckling resistance, and invert roundness of the core.

Outlet pipe trenches must have slopes and inverts low enough to discharge all edge drain trench water. Conduits must be turned down at the outlet or the outlet trench must be filled with a permeable backfill.

Acknowledgements

Gratefully acknowledged is the considerable cooperation and help received by Ministry of Transportation of Ontario personnel who supervised and provided the excavation equipment and manpower. Particular thanks are extended to Ms. P. Marks of the MTO for her significant contributions to this paper. The views expressed in this paper are those of the authors and do not necessarily reflect those of the MTO.

References

Allen, D.L., Fleckenstein, L.J., 1993. Field and laboratory comparison of pavement edge drains in Kentucky. Transportation Research Record No. 1425, National Research Council of USA, pp. 1–10.

ASTM C-33, 1984. Specification for concrete aggregates. American Society for Testing Materials, Philadelphia, PA, USA.

ASTM C-117, 1984. Test method for material finer than 75 μm (No. 200) sieve in mineral aggregates by washing. American Society for Testing Materials, Philadelphia, PA, USA.

ASTM D-1241, 1968. Specification for materials for soil-aggregate subbase, base, and surface courses. American Society for Testing Materials, Philadelphia, PA, USA.

ASTM D-2487, 1983. Classification of soils for engineering purposes. American Society for Testing Materials, Philadelphia, PA, USA.

ASTM D-2940, 1974. Specification for graded aggregate material for bases or subbases for highways or airports. American Society for Testing Materials, Philadelphia, PA, USA.

ASTM E-11, 1981. Specification for wire-cloth sieves for testing purposes, American Society for Testing Materials, Philadelphia, PA, USA.

ASSHTO M228, 1996. Standard Specification for Geotextiles, Designation M228, America Association of State Transportation and Highway Officials, Washington, D.C.

Casagrande, A., 1932. A New Theory of Frost Heave. Discussion, Proceedings, Highway Research Board 11 (1), 168–172.

CGSB 148.1. 1993. Method of Testing Geosynthetics. Geotextiles Filtration Opening Size. Canadian General Standards Board, Ottawa, Canada.

Hajek, J.J., Kazmierowski, T.J., Sturm, H., Bathurst, R.J., Raymond, G.P., 1992. Field Performance of open-graded drainage layers. Transportation Research Record 1354, 55–64.

Kaplar, C.W., 1970. Phenomenon and mechanism of frost heaving. Highway Transportation Research Board, National Research Council, Washington, D.C., pp. 1–13.

Koerner, R.M., Koerner, G.R., Fahim, A.K., Wilson-Fahmy, R.F., 1994. Long term performance of geosynthetics in drainage applications. National Cooperative Highway Research Program Report 367, Transportation Research Board, National Research Council of USA, 54p.

Koerner, R.M., Koerner, G.R., 1995. Lessons learned from geocomposite edge drain clogging due to improper installation. Geosynthetics: Lessons Learned from Failures, Industrial Fabrics Association International, Preprint Volume, pp. 135-141.

Raymond, G. P., Bathurst, R. J., Hajek, J., 1994. Geocomposite edge drain evaluation on Ontario Highway 402 at Strathroy. Proceedings 5th. International Geotextile and Related Products Conference, 2, pp. 743–746.

Raymond, G.P., Bathurst, R.J., Hajek, J., 1995a. Evaluation of pre-1994 Ontario Highway edge drain practice. Recent Developments in Geotextile Filters and Prefabricated Drainage Geocomposites, ASTM Special Technical Publication 1281, (Bhatia and Suits, Eds.), American Society for Testing Materials, Philadelphia, PA 1996, 20 June 1995, Denver Colorado, pp. 222–233.

Raymond, G.P., Bathurst, R.J., Hajek, J., 1995b. Evaluation of highway edge drains incorporating geosynthetics. International Symposium on Unbound Aggregates in Roads UNBAR4, Nottingham (UK), 17–19 July 1995, pp. 75–85.

Raymond, G.P., Bathurst, R.J., Hajek, J., 1996a. Evaluation of geocomposite edge drain on sand subgrade. Geofilters'96, 2nd International Conference on Filtration and Drainage in Geotechnical Engineering, Montreal, Quebec, 29–31 May 1996, pp. 369–378.

Raymond, G.P., Bathurst, R.J., Hajek, J., 1996b. Geosynthetics use in cold climate pavement drainage. 49th Canadian Geotechnical Conference, St. Johns, Newfoundland, 23–25 September 1996, pp. 369–378.

Raymond, G.P., Bathurst, R.J., Hajek, J., 1998. A geotextile-wrapped aggregate highway drain evaluation, 6th International Conference on Geosynthetics, Atlanta, GA, March 1998, Vol. 2, pp. 929–938.

Diffusion of sodium and chloride through geosynthctic clay liners

C.B. Lake, R.K. Rowe*

Department of Civil and Environmental Engineering, University of Western Ontario, London, Ontario, Canada N6A 5B9

Received 4 February 1999; received in revised form 19 July 1999; accepted 21 August 1999

Abstract

The diffusion coefficients deduced from GCL diffusion tests performed with 3 g/L to 5 g/L (0.05 M to 0.08 M) NaCl solutions decrease linearly with decreasing final bulk GCL void ratio. However, this diffusion coefficient is shown to be dependent on the source solution and, when the NaCl concentration is increased significantly, the diffusion coefficient deduced is also shown to increase. For the range of products examined, it is shown that the method of GCL manufacture did not significantly affect the diffusion coefficients. Different methods of performing the test are examined and it is shown that the type of diffusion test has little effect on the diffusion coefficient deduced at a given void ratio for the range of NaCl concentrations examined (3 g/L to 5 g/L). However, as the NaCl concentration increases to 0.6 M or 2.0 M, a constant stress applied to the sample is shown to mitigate increases in diffusion coefficients compared to the case of the sample being tested at a constant void ratio. © 2000 Elsevier Science Ltd. All rights reserved.

Keywords: GCL; Landfill; Diffusion; Contaminant transport; Chloride; Sodium; Leachate

1. Introduction

The advective migration of contaminants through geosynthetic clay liners (GCLs) and GCL composite liner systems has been examined by numerous investigators (see Rowe, 1998 for a review). For typical design situations the advective flow rates through a GCL liner system are very low and chemical (molecular) diffusion may be

* Corresponding author. Tel.: 001-519-661-2126; fax 001-519-661-3942.
E-mail address: r.k.rowe@uwo.ca (R.K. Rowe)

Reprinted from *Geotextiles and Geomembranes* **18 (2-4),** 103-131 (2000) the dominant transport mechanism through the GCL liner system. Unfortunately there is a paucity of research relating to diffusion of contaminants through GCLs (see Lo, 1992; Rowe et al., 2000; Lake and Rowe, 2000).

When GCLs are placed as part of municipal solid waste landfill bottom (base) liners, the GCL will be subjected to a variety of different stresses ranging from low stresses before addition of the waste ($\sim$ 5–20 kPa) to very high stresses after the landfill has been filled with waste and capped ($\sim$ 100–400 kPa). The GCL may also be subjected to a variety of hydration conditions such as being hydrated under relatively low stresses by the moisture from underlying soil and being subsequently consolidated by the weight of the overlying waste. Further complicating matters is the fact that the GCL may be subjected to leachates with different concentrations and compositions. The different possible combinations of stress conditions, hydration conditions and permeating fluids will result in a wide range of final bulk GCL void ratios (see Appendix A for a definition) and one may anticipate that this may have an effect on the diffusive behavior of contaminants migrating through the GCLs.

This paper will examine the effect of different stress conditions and corresponding different GCL final bulk void ratios, different contaminant source solutions (NaCl and a synthetic municipal solid waste leachate), two different diffusion test methods, and the type of GCL manufacturing process on GCL diffusion coefficients. Based on these results, factors to consider when assessing diffusion coefficients for GCL liner systems are discussed.

2. Materials tested

The three GCLs examined in this study are summarized in Table 1 together with the generic symbols used to identify the products in the remainder of the paper. They

Table 1
Properties of thermally treated, needlepunched GCLs examined[a]

Symbol used in this paper	GCL type	Middle bentonite layer	Carrier geotextile	Cover geotextile	Minimum mass/area tested, M_{GCL} (g/m^2)	Mean, standard deviation of M_{GCL} (g/m^2)
NWNWT	NW	Granular sodium bentonite	PP nonwoven[b]	PP nonwoven	5270	5898, 312
WNWT	NS	Granular sodium bentonite	PP woven	PP nonwoven	5665	5795, 113
WNWBT	BFG5000[c]	Powder sodium bentonite	PP woven	PP nonwoven	5481	5578, 85

[a]*Notes*: PP: Polypropylene
[b]Scrim reinforced with slit film woven geotextile.
[c]Cover geotextile filled with 500 g/m^2 of powder bentonite.

included Bentofix NW and NS GCLs (distributed by Terrafix Geosynthetics Inc., based in Rexdale, Ontario, Canada and Naue Fasertechnik in Lubbecke, Germany under slightly different product identifiers, Bentofix B4000 and NSP 4900-3) and BFG5000 GCLs (distributed by Naue Fasertechnik in Germany). The generic symbols used to describe these GCLs throughout the rest of the paper are such that the type of manufacturing of each GCL can be ascertained from the symbols (see Table 1). The first 'NW' or 'W' in the generic name refers to whether the bottom geotextile (sometimes called the "carrier" geotextile) is nonwoven or woven. The second 'NW' describes the top ("cover") geotextile which was nonwoven for all GCLs examined herein. The 'B' in the generic name WNWBT refers to if bentonite is impregnated in the top geotextile and the 'T', if present, refers to the needlepunched fibres being thermally treated.

The sodium bentonite used in these GCLs is predominately smectite ($\sim 90\%$) with initial bentonite pore water chemistry showing Na^+ (0.26 g/L or 0.011 M) and SO_4^{-2} (0.46 g/L or 0.005 M) to be the predominant ions in the pore fluid when the soil was wetted to its saturation moisture content and squeezed at a pressure of 10 MPa (Rowe et al., 2000).

3. Procedures and tests performed

Two types of diffusion tests were performed to examine the influence of contaminant source solution, applied stress and the type of GCL on the diffusion coefficients for sodium and chloride. Specified volume diffusion (SVD) tests allowed comparison of diffusion results by controlling the final bulk GCL void ratios while constant stress diffusion (CSD) tests allowed comparison of diffusion coefficients at similar specified applied stresses. The test procedures and methods of data analysis for both types of diffusion tests have been described in detail by Rowe et al. (2000) and are summarized below for clarity.

The specified volume diffusion (SVD) testing apparatus shown in Fig. 1 consists of a lower ring (receptor reservoir), a middle ring (GCL sample holder) and an upper ring (source reservoir), all made of acrylic. Rigid porous steel plates, the same diameter as the outside edge of the middle ring, are placed above and below the middle ring to prevent the GCL from swelling beyond the thickness of the middle ring. When the diffusion cell is assembled with the threaded rods (see Fig. 1), the GCL is hydrated and begins to swell with the spacer rods placed in the upper and lower reservoirs assisting in constraining the movement of the porous steel plates. By manufacturing different ring heights (5.6 mm, 7.1 mm, 9.1 mm, and 11.0 mm), control over the bulk GCL void ratio can be maintained during testing.

Similar to the SVD apparatus, the constant stress diffusion (CSD) test apparatus shown in Fig. 2 consists of an acrylic lower ring (receptor reservoir) and an acrylic upper ring (source reservoir). However, instead of placing the sample in a middle ring, the sample is supported by a porous steel plate, which rests on a ridge in the lower ring and the lower spacer rod. Unlike the SVD test apparatus, the top porous steel plate is the same diameter as the upper reservoir and is free to move in the upper ring when

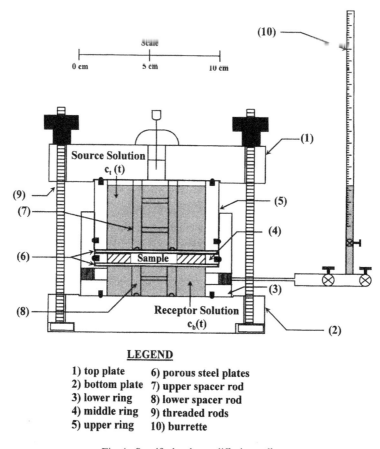

Fig. 1. Specified volume diffusion cell.

the sample hydrates (or contracts due to changes in pore fluid composition). The purpose of the upper spacer rod, which sits on the upper porous steel plate, is to support a plunger that extends through the top of the cell. This plunger supports a load frame that allows application of a stress to the sample during hydration and diffusion testing of the GCL.

Prior to diffusion testing, GCL samples were cut and placed in the diffusion cell as described in Rowe et al. (2000). The GCL samples were hydrated with de-ionized, de-aired water (DDW) from the bottom of the sample and the amount of water uptake was recorded using a burette attached to the bottom reservoir of the GCL diffusion testing apparatus. When water uptake ceased, 1 cm of DDW was added to the source compartment and the sample was left until chemical equilibrium was reached in the source and receptor compartments. A diffusion test was then initiated by replacing the 1 cm of DDW in the upper chamber by the contaminant source solution of interest such that there was no head difference across the GCL. Samples of both the source and receptor (3 ml) were taken daily and replaced with equal volumes of de-ionized,

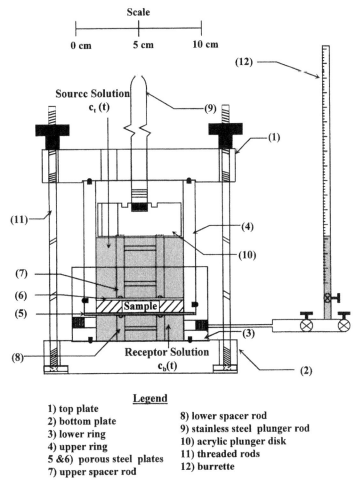

Fig. 2. Constant stress diffusion cell.

distilled water. Cation analyses of samples were performed using an atomic absorption spectrometer and anion analyses were performed using ion chromatography.

Two different source solutions were used for diffusion testing: (a) an NaCl solution at a range of concentrations to be discussed, and (b) a synthetic leachate solution similar to the composition of that observed at the Keele Valley Landfill in Toronto, Ontario, Canada. The nominal concentrations of the major constituents of the synthetic leachate are given in Table 2.

The diffusion tests are summarized in Table 3. In the "test number" column, the first part of the test number symbol (e.g. NWNWT, WNWBT) corresponds to the generic GCL symbols used in Table 1. This is followed by a test number (1,2, …, 6; A,B, … etc.) followed by a designation –SV or –CS which indicates the type of test performed with –SV indicating that it was a specified volume diffusion (SVD) test

Table 2
Composition of synthetic leachate used for GCL diffusion testing[a]

Constituent	Concentration/value (mg/L unless otherwise noted)
Acetic acid	6000
Propionic acid	4000
Butyric acid	600
Na^+	4705
Ca^{+2}	1040
Mg^{+2}	405
NH_4^+	555
K^+	320
Cl^-	3800
SO_4^{-2}	125
CO_3^{-2}	330
NO_3^-	35
HCO_3^-	4260
HPO_4^{-2}	15
pH	6 [−]
Eh	− 120 [mV]

[a]*Notes*: Trace metal solution added to leachate containing Fe^{+2}, Zn^{+2}, Cu^{+2}, Mn^{+2}, Al^{+3}, Co^{+2}. Concentration of Na^+ varies due to addition of NaS_2 reducing agent to obtain desired Eh and NaOH to obtain desired pH.

or −CS indicating a constant stress diffusion test (CSD). In the column "Test type", the number in parentheses denotes either the height to which the GCL was allowed to swell (SVD) or the stress at which the diffusion test was performed (CSD). All samples were hydrated with de-aired, de-ionized water and diffusion tests were conducted with a 4.6 g/L NaCl source solution unless otherwise noted.

To obtain experimental diffusion coefficients, it is necessary to fit a theoretical curve to the experimental data. All modeling was performed with the latest version of the finite-layer contaminant transport program POLLUTE (Rowe and Booker, 1999). The source and receptor reservoirs were specified as finite mass boundary conditions with provisions to account for the mass of contaminant removed by sampling. The initial concentration profile in the sample was the concentration of the receptor solution at the end of the hydration period. The details of the modeling procedure are given in Rowe et al. (2000).

Figs. 3 and 4 show examples of chloride and sodium concentration profiles plotted in terms of the normalized concentration (i.e. source and receptor values are divided by the initial source concentration, c_o) for test NWNWTA-CS. The data points (solid circles) in each figure represent the experimentally measured concentrations for chloride (Fig. 3) and sodium (Fig. 4). The bottom data and curve in each figure are the observed and calculated concentration increase in the receptor. The net increase in mass of contaminant in the receptor is the difference between the increase in mass due to diffusive flux from the sample minus the decrease in mass due to sampling and

Table 3
GCL Diffusion Tests (all performed with 4.6 g/L NaCl solution unless otherwise noted)[a]

Test number	Test type	Chloride diffusion coefficient, D_t (m²/s)	Sodium diffusion coefficient[f], D_t (m²/s)	Final bulk GCL void ratio, e_B(−)	Bentonite total porosity, n_t(−)
NWNWT1-SV	SVD (7.1 mm)[b]	1.5×10^{-10}	2.5×10^{-10}	2.1	0.73
NWNWT2-SV	SVD (11.1 mm)	3.0×10^{-10}	5.0×10^{-10}	3.2	0.80
NWNWT3-SV	SVD (5.6 mm)[b]	3.5×10^{-11}	6.0×10^{-11}	1.1	0.56
NWNWT4-SV	SVD (9.1 mm)	2.8×10^{-10}	4.4×10^{-10}	2.7	0.78
NWNWT5-SV	SVD (7.1 mm)	1.4×10^{-10}	2.2×10^{-10}	1.7	0.67
NWNWT6-SV[e]	SVD (7.1 mm)[c]	2.2×10^{-10}	N/A	1.9	0.70
WNWT1-SV	SVD (7.1 mm)[b]	1.6×10^{-10}	2.7×10^{-10}	2.0	0.69
WNWT2-SV	SVD (9.1 mm)	2.7×10^{-10}	4.0×10^{-10}	2.9	0.77
WNWT3-SV	SVD (5.6 mm)	7.2×10^{-11}	1.1×10^{-10}	1.3	0.59
WNWBT1-SV	SVD (7.1 mm)[b]	1.3×10^{-10}	2.5×10^{-10}	2.1	0.74
WNWBT2-SV	SVD (11.1 mm)	2.9×10^{-10}	4.3×10^{-10}	3.6	0.83
WNWBT3-SV	SVD (9.1 mm)	2.8×10^{-10}	4.4×10^{-10}	2.8	0.79
NWNWTA-CS	CSD (22 kPa)	2.9×10^{-10}	4.8×10^{-10}	2.8	0.78
NWNWTB-CS	CSD (145 kPa)	1.3×10^{-10}	2.1×10^{-10}	1.8	0.69
NWNWTC-CS	CSD (145 kPa)[d]	2.3×10^{-10}	4.0×10^{-10}	2.6	0.77
NWNWTD-CS	CSD (145 kPa)	See Table 5			
NWNWTE-CS[e]	CSD (22 kPa)[c]	3.7×10^{-10}	N/A	2.6	0.77
NWNWTF-CS[e,g]	CSD (145 kPa)[c]	2.2×10^{-10}	N/A	1.8	0.69
		2.0×10^{-10}		1.6	0.67

[a] SVD: Specified volume diffusion test; CSD: Constant stress diffusion test. All samples hydrated with deionized, distilled water.
[b] Source was 3.3 g/L NaCl.
[c] Source was synthetic leachate.
[d] Hydrated at 3 kPa and then consolidated to 145 kPa before diffusion test.
[e] Test is not plotted in Fig. 3.
[f] Based on $K_d = 0$ ml/g.
[g] Diffusion coefficients for sample thickness at beginning and end of diffusion test (see text).

replacement of the extracted fluid with distilled water (which is considered in the modelling). Referring to Fig. 4, it is noted that after approximately ten days, both the observed and calculated concentration of sodium in the receptor begins to decrease slightly due to dilution from sampling and replacement of fluid. The different initial concentrations of sodium and chloride in the receptor (Figs. 3 and 4) give rise to the different shapes of chloride and sodium receptor curves due to different amounts of mass being removed during the sampling process. The top data and curve shows the observed and calculated decreases in concentration in the source solution due to diffusive flux into the GCL and sampling. A good fit to both the source and receptor experimental data were obtained using the parameters given in Figs. 3 and 4.

As discussed by Rowe et al. (2000) batch tests performed with NaCl solutions and granular bentonite in these tests found an upper bound value of K_d of 0.12 ml/g for

278

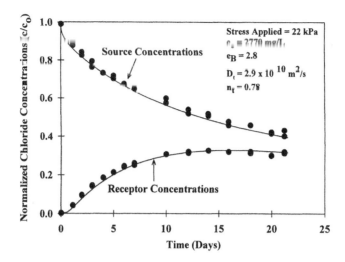

Fig. 3. NWNWTA-CS chloride diffusion.

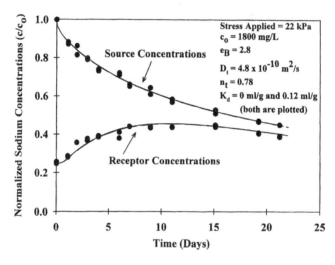

Fig. 4. NWNWTA-CS sodium diffusion.

sodium. Fig. 4 includes the theoretical curves for both K_d of 0 ml/g and K_d of 0.12 ml/g ($D_t = 4.8 \times 10^{-10}$ m²/s for both). The curves are usually indistinguishable for these two K_d values which suggests that sorption does not have a significant effect on deduced Na$^+$ and Cl$^-$ diffusion coefficients and corresponding fluxes for the cases considered. Rowe et al. (2000) noted that some osmotic flow is observed in diffusion tests such as this but also showed that this osmosis had no significant effect on the diffusion of sodium chloride for these tests. Rowe et al. (2000) also discussed the process of anion exclusion and how it could affect GCL diffusion test results.

The tests summarized in Table 3 utilized finite mass boundary conditions for the source and receptor which essentially means that the mass of contaminant in the source (or receptor) at any time, t, is equal to the initial mass in the source (or receptor) minus the mass flux into the sample (plus mass flux out of the sample for receptor) minus the mass lost due to sampling. With no advective transport though the sample, this flux, f, at any point in the GCL can be expressed in terms of the porous media diffusion coefficient $D_p = n_e D_e$,

$$f = -n_e D_e \frac{dc}{dz} = -D_p \frac{dc}{dz} \tag{1}$$

In principle, the diffusion through a porous media is a function of both the effective porosity n_e, and the effective diffusion coefficient, D_e. However, it was found that except at very small times (of no practical consequence), the precise values of the effective porosity, n_e, and effective diffusion coefficient, D_e, are not critical to predicting transport through a single GCL provided the range of values of n_e and D_e correspond to the same product $D_p = n_e D_e$. There was very little latitude in the selection of a value of D_p that would fit the experimental data (Rowe et al., 2000). This approach of using D_p is convenient since generally only the total porosity, n_t, is known for a GCL and the effective porosity, n_e, may vary from one contaminant to another or with other factors such as the void ratio. Therefore, the diffusion coefficients reported in this paper, D_t, are the values deduced from the experimental data for $n = n_t$ (i.e. $D_p = n_t D_t$) and are not effective diffusion coefficients, D_e.

To demonstrate that the combination of n and D giving rise to D_p is not critical for a relatively thin sample such as a GCL, two samples of different lengths (sample A = 10 mm (GCL), sample B = 100 mm) with the same total porosity ($n_t = 0.7$) and the same diffusion coefficient ($D_t = 1.5 \times 10^{-10}$ m^2/s) are considered. The solid lines on Fig. 5 show the theoretical concentration profiles generated with POLLUTE (Rowe and Booker, 1999) through both the thin and relatively thick sample at the end of 20 days. If the effective porosity, n_e, was known for the sample (for example, $n_e = 0.3$) and the effective diffusion coefficient, D_e, was 3.5×10^{-10} m^2/s (i.e., yielding the same value of $D_p = 1.05 \times 10^{-10}$ m^2/s as for the solid line), the profile through the thin sample (shown by the long dashed line) is practically identical to the solid line for the thin sample but quite different for the thicker sample B. To illustrate that the choice of D_p is important even for a thin sample, the calculated profile is also shown at 20 days for $n_e = 0.3$ and $D_e = 1.5 \times 10^{-10}$ m^2/s ($D_p = 0.45 \times 10^{-10}$ m^2/s). It can be seen that this gives a much different curve than that obtained for $D_p = 1.05 \times 10^{-10}$ m^2/s for both the thin and thick sample. Thus for a GCL it is important to establish D_p, but the combination of n and D that give rise to this value of D_p is not critical for practical applications and the range of parameters as indicated by Rowe et al. (2000).

However it should be noted that while the porous media diffusion coefficient approach is suitable for a single GCL, it may not be suitable if multiple layers of GCL are used and the layer ceases to be thin (~ 1 cm or less). More investigation would be required if one was considering the use of multiple layers of GCL, however since this is not usually the case in practice, it is not discussed further here.

280

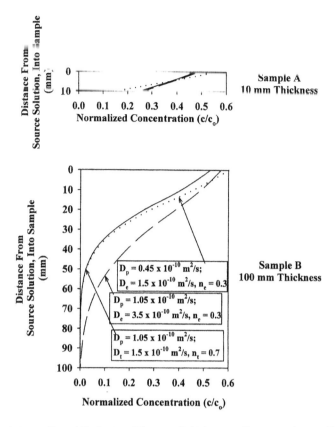

Fig. 5. Difference between D_p and D_e for two different soil thicknesses. Concentration profile with depth for time = 20 days.

3.1. Effect of final bulk GCL void ratio on diffusion coefficient

The final bulk GCL void ratio, defined by Petrov et al. (1997) as the ratio of the volume voids to the volume of solids in the GCL (geotextiles included) at the end of testing (see Appendix A), provides a good basis for evaluating data obtained for both hydraulic conductivity and diffusion tests on different types of geotextile-enclosed GCLs and GCLs of different mass per unit areas. Petrov and Rowe (1997) showed that hydraulic conductivity values for GCLs with different permeants were related to the final bulk GCL void ratio and the nature of the permeating fluid. As will be shown below, sodium and chloride diffusion coefficients deduced for a GCL can also be related to the final bulk GCL void ratio for a given source solution.

The sodium and chloride diffusion coefficients deduced from the various tests are given in Table 3 and are plotted in Figs. 6 and 7 with respect to the final bulk GCL void ratio. The range of final bulk GCL void ratios tested was selected based on a typical range of final bulk GCL void ratios expected to be encountered in field

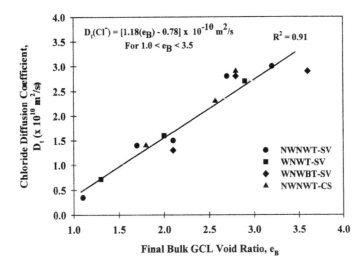

Fig. 6. Variation of chloride diffusion coefficient, D_t, versus final bulk GCL void ratio.

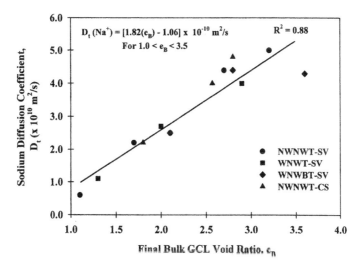

Fig. 7. Variation of sodium diffusion coefficient, D_t, versus final bulk GCL void ratio.

situations. Both sodium and chloride diffusion coefficients vary linearly with the final bulk GCL void ratio with almost an order of magnitude difference in diffusion coefficients from the lowest to the highest final bulk GCL void ratios considered. The R^2 values for a linear regression analysis through the points are 0.91 for chloride (Fig. 6) and 0.88 for sodium (Fig. 7). These R^2 values are considered to be good recognizing that three different types of GCL were tested and two different test methods were employed (CSD and SVD).

Test NWNWTC-CS was performed to simulate a GCL hydrating due to moisture from the underlying soil and then being subsequently consolidated due to the stress exerted during the placement of waste during landfill operations. Even though the sample was initially hydrated to a high bulk GCL void ratio under a 3 kPa stress and subsequently consolidated to a lower bulk GCL void ratio, the bulk GCL void ratio at which the diffusion test was performed ($e_B = 2.6$) controls the diffusion process and the data from this test falls on the relationship between the final bulk GCL void ratio and the sodium and chloride diffusion coefficients in Figs. 6 and 7. As shown by Petrov and Rowe (1997), the main effect of post-confinement stress application (applying a higher stress to the GCL than under which it was initially hydrated) is that it gives a higher final bulk GCL void ratio after consolidation than would have been realized if the full stress had been applied before hydration. This may be due to the fabric of the clay being developed in a more random fashion when hydrated at lower stresses and a higher stress is required to create the more orientated, lower void ratio fabric developed by pre-confinement hydration.

Based on the regression analysis given in Figs. 6 and 7 a simple equation can be written for the diffusion coefficient, D_t, in terms of final bulk GCL void ratio, e_B, viz:

$$D_t(\text{Cl}^-) = (1.18\,e_B - 0.78) \times 10^{-10}\ \text{m}^2/\text{s}, \quad \text{for } 1 < e_B < 3.5, \tag{2}$$

$$D_t(\text{Na}^+) = (1.82\,e_B - 1.06) \times 10^{-10}\ \text{m}^2/\text{s}, \quad \text{for } 1 < e_B < 3.5. \tag{3}$$

However it is important to note that this relationship was only established for source solutions of NaCl of less than 5 g/L (0.08 M). These results will be discussed further in the paper.

3.2. Effect of different contaminant source solutions

The relationships given by Eqs. (2) and (3) and shown in Figs. 6 and 7 were for hydration with distilled water and diffusion with 3 to 5 g/L (0.05 to 0.08 M) NaCl solutions. To investigate if these relationships are dependent on the source solution characteristics, diffusion tests were also performed with higher NaCl salt concentrations and a synthetic municipal solid waste leachate as described below.

3.2.1. Effect of higher NaCl concentrations

To examine the effect of higher NaCl concentrations on the diffusion of sodium and chloride ions through bentonite, a five-phase specified volume diffusion test (test GB5PH) was performed at a 7.1 mm height using a granular bentonite (5500 g/m^2) taken directly from the NWNWT and WNWT GCLs. Bentonite was used alone to allow an examination of the general relationship between bentonite void ratio and increasing NaCl concentration that was independent of the type of GCL.

The first phase of diffusion testing was performed with a 4.6 g/L (0.08 M) NaCl solution. To examine the effect of the chemical composition of the receptor solution after the first phase, the depleted source solution was removed and replaced with another 4.6 g/L NaCl solution (phase II). The granular bentonite sample used for the first phase remained in the apparatus and the receptor reservoir fluid at the end of phase I was used as the receptor fluid for the beginning of phase II. A second diffusion test was performed and diffusion coefficients were again obtained for sodium and chloride. The depleted source was then removed and replaced by a higher NaCl concentration source solution, a diffusion test conducted and diffusion parameters deduced. In total, five tests were conducted with the same sample at the same bentonite void ratio as indicated in Table 4.

In phase I, the initial chloride and sodium concentrations in the receptor were in equilibrium with the initial porewater chemistry of the hydrated GCL and had values of approximately 0 c_o and 0.27 c_o, respectively (with sulphate providing the charge balance). Thus the initial concentration gradient across the sample for Cl^- (0.14 c_o/mm) was 40% greater than for sodium (0.1 c_o/mm) in phase I. In phase II the normalized concentrations in the receptor at the start of the test (relative to the corresponding phase II source concentrations c_o) were approximately 0.27 c_o and 0.33 c_o for chloride and sodium respectively. The initial concentration gradient for chloride (0.1 c_o/mm) was only 6% greater than for sodium (0.094 c_o) and hence the two initial gradients were much closer than in phase I even though the composition of the source solution was the same in both cases. In phase III the normalized initial receptor concentrations were very close at approximately 0.13 c_o and 0.14 c_o for Cl^- and Na^+ respectively while in phase IV and V the normalized initial receptor concentrations of Cl^- and Na^+ were practically identical. As will become evident in later discussions, the initial concentration in the receptor and sample (relative to the source) has an effect on the diffusion coefficient for the GCLs tested. By inference, the porewater chemistry of the soil adjacent to a GCL in a field application may also initially influence diffusion through the GCL. This is consistent with Rowe (1998) who states that D_e is really a mass transfer coefficient depending on things such as complex electrochemical gradients and therefore the contaminant transport model used in this paper will correctly model the tests provided initial parameters are correctly defined for each situation.

As can be seen from the results in Table 4, the repetition of the diffusion test in phases I and II gave slightly different deduced diffusion coefficients for both sodium and chloride. In particular in phase I the deduced diffusion coefficient for chloride is only 68% of that for sodium whereas in phase II the chloride diffusion coefficient has

Table 4
Effect of different NaCl concentrations on sodium and chloride diffusion coefficients (Test GB5PH – bentonite void ratio of 2.4)

Phase	Concentration of NaCl used as c_o (g/L)	Chloride diffusion coefficient, D_t (m^2/s)	Sodium diffusion coefficient, D_t (m^2/s)
I	4.6	1.5×10^{-10}	2.2×10^{-10}
II	4.6	1.9×10^{-10}	2.0×10^{-10}
III	9.1	2.2×10^{-10}	2.3×10^{-10}
IV	35.1	3.8×10^{-10}	3.8×10^{-10}
V	114.3	4.2×10^{-10}	4.2×10^{-10}

increased (relative to phase I) and that for sodium has decreased (relative to phase I) such that for phase II the deduced chloride diffusion coefficient is 95% of that for sodium. This demonstrates the potential influence of electrochemical interactions on deduced diffusion coefficients and will be discussed later in the paper. Phase III represents a doubling of NaCl concentration relative to that used in phases I and II and the diffusion coefficients for chloride and sodium increased by 15% relative to phase II. This small increase may be due to double layer contraction resulting from the increased NaCl concentration of the pore fluid as well as a reduction in the concentration of SO_4^{2-} relative to Cl^- and additional depletion of SO_4^{2-} ions in the receptor fluid due to sampling. Phase IV represents a salt concentration similar to that of seawater. The diffusion coefficient for chloride and sodium subsequently increased by 73% and 65% compared to phase III respectively, probably the result of a combination of double layer contraction and c-axis contraction as shown by Petrov and Rowe (1997). For phase IV the diffusion coefficient for Cl^- and Na^+ were the same. Increasing the salt content similar to that of a concentrated brine solution in phase V resulted in 10% higher deduced diffusion coefficients compared to phase IV for both sodium and chloride and again the two diffusion coefficients were the same.

Test GB5PH chloride diffusion coefficients are plotted on Fig. 8 to illustrate the possible effects of different source solution concentrations on diffusion coefficients and how higher salt contents may affect relationships such as those in Figs. 6 and 7. To allow a direct comparison of the chloride diffusion coefficients given in Table 3 with those obtained for test GB5PH, it is appropriate to plot results in terms of bentonite

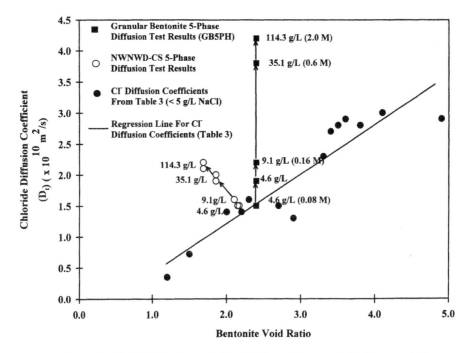

Fig. 8. Change in chloride diffusion coefficient, D_t, due to increases in NaCl concentration.

void ratio (rather than of bulk GCL void ratio) since there are no geotextiles in test GB5PH. The regression line on Fig. 8 is based only on the diffusion coefficients used for Fig. 3 and does not depend on the results of test GB5PH.

If a soil remains saturated and occupies the same volume in a specified volume diffusion test, the bentonite void ratio of GB5PH remains constant. However Fig. 8 shows that the deduced chloride diffusion coefficient increases with increases in NaCl concentration. The changes from phase I to phase II are considered to be due to the relative changes in gradients for chloride and sodium. This is discussed in a later section of the paper. When the NaCl concentration increases from 9.1 g/L (0.16 M) to 114.3 g/L (2 M), the chloride and sodium diffusion coefficients almost double. Thus the relationship developed in Fig. 6 for 3 to 5 g/L (0.05 to 0.08 M) NaCl solutions is not applicable for substantially higher NaCl concentrations suggesting that any relationship between diffusion coefficients and final bulk GCL void ratio or bentonite void ratio is only valid for similar hydration and permeating conditions.

The results given here represent the worst possible situation since the void ratio is constrained to remain constant but the vertical stress between the bentonite and walls of the apparatus decreases with increasing NaCl concentration. For a GCL in a landfill base liner situation, one would expect that the effective stress on the GCL would remain relatively constant but that the final bulk GCL void ratio would decrease due to osmotic consolidation and/or c-axis contraction. For example, Fernandez and Quigley (1991) showed that the effective stress prevented some increase in hydraulic conductivity of a clay sample when permeated with certain hydrocarbon fluids.

To investigate the influence of effective stress and increasing salt contents on GCL diffusion coefficients as the NaCl concentration increases, test NWNWD-CS was performed with an NWNWT GCL at a constant stress of 145 kPa. Details and results of NWNWD-CS are given in Table 5 and results are plotted on Fig. 8 for comparison

Table 5
Effect of different NaCl concentrations on sodium and chloride diffusion coefficients (Test NWNWD-CS)[a]

Phase	Concentration of NaCl used as c_o (g/L)	Chloride diffusion coefficient, D_t (m^2/s)	Sodium diffusion coefficient, D_t (m^2/s)	Final bulk GCL void ratio, e_B (−)	Bentonite total porosity, n_t (−)
I	Spiked*	1.5×10^{-10}	1.5×10^{-10}	1.8	0.68
II	4.6	1.5×10^{-10}	1.6×10^{-10}	1.8	0.68
III	9.1	1.6×10^{-10}	1.7×10^{-10}	1.8	0.68
IV	35.1	(begin) 2.0×10^{-10} (end) 1.9×10^{-10}	(begin) 2.0×10^{-10} (end) 1.9×10^{-10}	(begin) 1.7 (end) 1.5	(begin) 0.68 (end) 0.65
V	114.3	(begin) 2.2×10^{-10} (end) 2.1×10^{-10}	(begin) 2.2×10^{-10} (end) 2.1×10^{-10}	(begin) 1.5 (end) 1.4	(begin) 0.65 (end) 0.63

[a]*Notes*: For phase I, the distilled water in the source reservoir used to hydrate the GCL (0.4 g/L Na$^+$ and 0.8 g/L SO$_4^{-2}$) was spiked with NaCl to establish equal gradients of Na$^+$ and Cl$^-$ through the sample initially (c$_o$ of Cl$^-$ 2.8 g/L, c$_o$ of Na$^+$ 2.2 g/L).

For phases IV and V, the sample deformed during testing and therefore two values of diffusion coefficients are given corresponding to parameters deduced at both the beginning and end of the test.

with GBPH5 (a SVD test). Phase I of this test was performed slightly different than GB5PH to examine the influence of initial gradients on chloride and sodium diffusion coefficients. Instead of placing only 1 cm of deionized, distilled water in the source solution to hydrate the GCL, deionized, distilled water of the same volume as an NaCl solution used to a perform diffusion test was added to the source to hydrate the sample. After hydration of the GCL, the hydration fluid in the source and receptor was sampled and analyzed for major ions to ensure chemical equilibrium. A concentrated NaCl solution was then made to spike this source fluid at the beginning of the test. The dilution of the concentrated NaCl solution into the hydration fluid was such that initial concentrations in the source were 2.2 g/L of Na^+, 0.8 g/L of SO_4^{-2} and 2.8 g/L of Cl^-. Initial receptor concentrations of 0.4 g/L of Na^+, 0.8 g/L of SO_4^{-2} and approximately 0 g/L of Cl^- resulted in gradients of 0.14 c_o/mm for both sodium and chloride. As one can see from Table 5, both sodium and chloride deduced diffusion coefficients are the same for phase I (1.5×10^{-10} m^2/s). For phase II, the source fluid was removed entirely and replaced with a 4.6 g/L (0.08 M) NaCl solution. With some sulphate ions still remaining in the receptor after phase I, the initial concentration of sodium in the receptor (0.32 c_o) was slightly greater than chloride (0.26 c_o), resulting in a slightly higher sodium diffusion coefficient (1.6×10^{-10} m^2/s) compared to that of chloride (1.5×10^{-10} m^2/s). Phase III showed the chloride diffusion coefficient (1.6×10^{-10} m^2/s) to be slightly less than that of sodium (1.7×10^{-10} m^2/s), however both increased slightly relative to phase II results. Increasing NaCl concentrations in Phase IV (35.1 g/L; 0.6 M) and Phase V (114.3 g/L; 2.0 M) resulted in higher sodium and chloride diffusion coefficients compared to the previous three phases as shown in Table 5. For each of these two phases, Na^+ and Cl^- diffusion coefficients were the same for the reasons discussed for test GB5PH.

Comparing Tables 4 and 5 and looking at Fig. 8, diffusion coefficients for NWNWD-CS were much lower compared to the specified volume, GB5PH test at higher NaCl concentrations. This suggests that application of effective stress on a GCL may be beneficial at reducing increases in diffusion coefficients for concentrated NaCl solutions (0.6 M and 2 M). However, for NWNWD-CS, even though the change in diffusion coefficient from phase I (1.4×10^{-10} m^2/s) to phase V (2.2×10^{-10} m^2/s), was not that large, the increasing salt content caused the bentonite void ratio to decrease from 2.2 to 1.7 (see Fig. 8). If Eq. (2) is used to predict the chloride diffusion coefficient for a bulk GCL void ratio of 1.4 (NWNWD-CS, end of phase V), the value predicted (0.88×10^{-10} m^2/s) is approximately 150% lower than what was observed for phase V (2.2×10^{-10} m^2/s). Therefore the change in chloride diffusion coefficient is more than it initially appears if one compares the expected diffusion coefficient at the lower void ratio. However, the ultimate magnitude of diffusion coefficient is still much lower for the 2.0 M NaCl solution for the CSD test compared to the SVD test when both samples started at similar void ratios.

To investigate if this deformation during diffusion testing had any influence on the modelling these tests, the parameters at both the beginning and end of each phase were used to deduce diffusion coefficients. Table 5 shows there is very little difference in the deduced diffusion coefficients for these two possible cases.

The important point to be made here is that the deduced diffusion coefficient depends both on the void ratio and the level of potential interaction of the leachate with the bentonite under the anticipated field conditions and hence tests should be performed using a leachate, receptor solutions and applied stress conditions as close as practical to that anticipated in the field application.

3.2.2. Effect of synthetic leachate

Compared to the NaCl solutions above, the complex electrochemical synthetic leachate solution shown in Table 2 has many different ions that may cause many different levels of ion exchange, double layer contraction, or c-axis contraction to occur during GCL diffusion testing. Synthetic leachate was chosen to simulate a realistic application for a GCL, although it is acknowledged that many other harsher solutions could have been used to examine the effect of factors such as ion exchange. Table 3 shows that for test NWNWT6-SV, the chloride diffusion coefficient of 2.2×10^{-10} m^2/s is substantially higher than the value of 1.5×10^{-10} m^2/s obtained for test NWNWT1-SV or 1.4×10^{-10} m^2/s for test NWNWT5-SV using the same type of GCL and similar final bulk GCL void ratios. It is similar to that in phase III of GB5PH. Thus the different source solution chemistry appears to have increased the diffusion coefficient deduced based on total porosity by almost 50% for chloride (see Fig. 9).

Often in landfill base liner applications an effective stress will be present as leachate migrates through the GCL. Tests NWNWE-CS and NWNWF-CS (Table 3) show chloride diffusion results for the synthetic leachate solution at stresses of 22 kPa and 145 kPa respectively. It should be noted that negligible deformation of NWNWE-CS ($e_B = 2.6$) took place during diffusion testing while sample NWNWF-CS reduced 0.4 mm in height during diffusion testing ($e_{Bstart} = 1.8$; $e_{Bend} = 1.6$). Therefore the

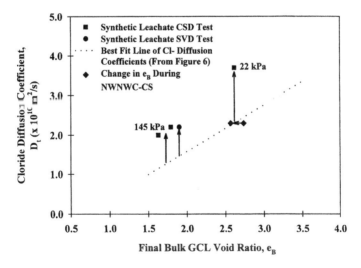

Fig. 9. Effect of synthetic leachate on chloride diffusion coefficient and change in e_B during NWNWC-CS.

range of deduced chloride diffusion coefficients for NWNWF-CS for both the initial and final sample parameters are given on Fig. 9. Again, both results plot above the regression line from Fig. 6 obtained for NaCl concentrations of less than 5 g/L (0.08 M) and the result for the CSD test (145 kPa) is very similar to that obtained from the SVD test. It can be hypothesized that the deviation of the Cl$^-$ results for NWNW6-SV (SVD test) from the 0.08 M NaCl solutions may have been influenced more by the complex interaction of varying gradients for the leachate compared to the 0.08 M NaCl solution than a decrease in effective pore space due to double layer contraction.

It should be noted that no diffusion coefficient is given for sodium in Table 3 (Test NWNWT6-SV). As other cations in the synthetic leachate solution migrate through the clay, they undergo cation exchange. Because Na$^+$ is the dominant ion on the clay exchange sites, some Na$^+$ will be desorbed as the leachate travels through the GCL. No fit to the experimental data can be obtained using a single contaminant transport model if desorption is a predominant mechanism influencing transport.

3.3. Effect of type of GCL manufacture

Three different types of GCL manufacture were tested in this study (Table 1). The NWNWT and WNWT GCLs are similar, except that they employ a different bottom geotextile, while the WNWBT GCL is similar to the WNWT GCL except that it has powdered bentonite in the middle and, more significantly, sprinkled on the top of the GCL. In a companion paper, Lake and Rowe (2000) showed that for confining stresses above about 100 kPa, the three GCLs swelled to similar bulk GCL void ratios for similar confining stresses. The NWNWT and WNWT GCLs exhibited similar swelling behavior at all stresses but below 100 kPa the WNWBT GCL swelled to much higher bulk GCL void ratios. It is interesting to revisit Figs. 6 and 7 to examine if there is any apparent effect of the type of GCL manufacture on deduced GCL diffusion coefficients.

As mentioned earlier, the results in Figs. 6 and 7 show a linear relationship for all the samples tested with 3 to 5 g/L NaCl solutions. This in itself suggests that for the three GCLs and range of bulk GCL void ratios examined, the type of GCL manufacture has little practical effect on deduced GCL diffusion coefficients. If Figs. 6 and 7 are examined closely, there appears to be less scatter of chloride and sodium diffusion coefficients at lower final bulk GCL void ratios than at higher bulk GCL void ratios. However, there is no pronounced trend of one type of GCL having a higher diffusion coefficient than another. Based on the results of Lake and Rowe (2000), one would expect that if any GCL were to deviate from the relationship in Figs. 6 and 7, it would be the WNWBT GCL. However, relative to the scatter that is present at higher bulk GCL void ratios, this does not appear to be the case. The thermally treated needlepunching of the WNWBT GCL can be expected to restrict the bulk void ratio of the bentonite core at higher bulk void ratios, even though the powdered bentonite on the surface of the cover geotextile is allowed to swell unconfined. Thus it can be inferred that the thermally treated needle punching restricted swelling of the core of these GCLs and it may be anticipated that this will control the diffusion through the GCL at the lower confining stresses examined. The powdered surface bentonite may swell and occupy part of the volume but it apparently does not control diffusion.

The fact that the level of restraint in the core can have a significant influence on the results is evident from the results presented by Lake and Rowe (2000) who examined two similar WNW GCLs but where one GCL was thermally treated and needlepunched while the other was only needlepunched. When subjected to a CSD test at 3 kPa with a 4.6 g/L NaCl solution, the non-thermally treated GCL swelled to a higher bulk GCL void ratio (8.2 versus 4.7) and yielded higher sodium and chloride diffusion coefficients than the thermally treated GCL. Apparently the thermal locking of the needlepunched fibres had restricted the bulk GCL void ratio of the thermally treated GCL during hydration at 3 kPa, resulting in lower GCL diffusion coefficients. The effect of thermally treated needlepunching also was apparent from post-test inspections of diffusion test samples listed in Table 3. The samples at the 5.6 mm and 7.1 mm heights were very flat along the bottom and top, with measured heights by a vernier caliper were the same as the ring heights (i.e. 5.6 mm and 7.1 mm). At the 9.1 mm and 11.0 mm heights, even though the measured heights were again the same as the ring height, the bottom of some of the GCLs were "dimpled". In other words the fibres were preventing local areas of the GCL from swelling to the height of the ring. Therefore, even though the GCLs were the desired heights, localized portions of the GCLs were at slightly smaller heights (i.e. smaller void ratios). At the 9 mm height, the effect was not as great as at the 11 mm height but was present nevertheless. Some samples may be affected by this more than others depending on factors such as the mass per unit area of the bentonite in the GCL, the amount of bonding between the fibres and the bottom geotextile, and the density of the needlepunched fibres. This may be a cause of some of the scatter in the deduced diffusion coefficients reported in Figs. 6 and 7 at higher bulk GCL void ratios.

3.4. Constant stress versus specified volume diffusion testing

To examine the effect of the type of diffusion test performed, two different approaches were used: 1) control the final bulk GCL void ratio during diffusion testing (SVD), or 2) control the stress conditions during diffusion testing (CSD). A control over the final bulk GCL void ratio allows a comparison of diffusion coefficients for different types of GCLs regardless of the stresses generated by the bulk GCL void ratio. A control over the confining stress applied during diffusion testing allows comparison of diffusion test results under similar stress conditions (but different void ratios for different GCLs). The CSD test has the added benefit that one can monitor the GCL height during hydration and diffusion testing under the specified confining stress.

Three CSD tests were performed as indicated in Table 3. Tests NWNWTA-CS and NWNWTB-CS were conducted on specimens hydrated and maintained under the noted applied stress. In test NWNWTC-CS, the GCL was allowed to hydrate under a low stress condition (3 kPa) and it was then consolidated under a higher stress (145 kPa) before starting the diffusion test. This was an attempt to simulate a situation that may occur when a GCL becomes hydrated under relatively low stress conditions and then is subsequently consolidated due to the stresses from overlying waste as a landfill is filled.

As previously noted, all the results in Table 3 are consistent with the linear relationship shown in Figs. 6 and 7 regardless of the test method (CSD or SVD) with both tests yielding similar diffusion coefficients at a similar bulk GCL ratio.

Deformation data for test NWNWTC-CS showed that during the course of the diffusion test, the sample decreased in height by almost 0.4 mm with the majority of the deformation occurring after the first day of diffusion testing. This was probably due to a combination of double layer contraction and a relatively high bulk void ratio from being hydrated under 3 kPa stress conditions and subsequently consolidated to 145 kPa. To examine the results of NWNWTC-CS, it is useful to plot the bulk GCL void ratio at the beginning of diffusion testing and the lower bulk GCL void ratio at the end of the test as shown by diamond symbols on Fig. 9. Although the bulk GCL void ratio did change slightly during diffusion testing, the magnitude of the change was small and both bulk GCL void ratios and corresponding diffusion coefficients from this test plot close to the linear relationship established from the other tests.

3.5. Comparison of chloride and sodium diffusion coefficients

Even with the relatively simple NaCl source solutions utilized for the tests in this paper, Na^+ and Cl^- ions do not necessarily move together through the system to maintain charge balance. This is because the hydrated bentonite contains a significant concentration of Na^+ and SO_4^{2-} ions (see Rowe et al., 2000). Thus during hydration, the Na^+ and SO_4^{2-} ions from the bentonite in the GCL diffused into the receptor reservoirs (that initially contained distilled water) until chemical equilibrium was reached between the GCL and the receptor reservoirs. When the source solution of sodium chloride is placed above the hydrated GCL, the concentration gradient of Cl^- is higher than that of Na^+ and there is also some back-diffusion of SO_4^{2-} ions from the soil porewater and receptor solution into the source. This means that the Na^+ and Cl^- ions may not migrate at the same rate through the system. As discussed by Kemper and van Schaik (1966), Yeung and Mitchell (1992) and Rowe et al. (1995), the movement of ions involves several components (i.e. true diffusion, electrical flow, and bulk flow) and in particular, electrical flow may influence the sodium and chloride diffusion coefficients deduced for a particular test.

In a system where the source solution was NaCl and no other diffusion was significant, the change in the normalized source concentrations of Na^+ and Cl^- with time should be identical to maintain a charge balance. However, inspection of Figs. 3 and 4 indicates that this is not the case and the drop in the concentration of Cl^- in the source is greater than the drop in Na^+. Here charge balance is maintained by the upward diffusion of SO_4^{2-} and in order to satisfy charge balance in the source at any given time one must include consideration of the concentration of SO_4^{2-} in the source at that time.

Comparing the initial concentration gradient between the source and receptor (eg. Figs. 3 and 4), it is found that the gradient for chloride is much greater than for sodium (due to initial receptor concentrations of ~ 0 and $0.25\, c_o$, respectively). As a consequence, the initial gradient for chloride is 33% greater than for sodium. However, the need to maintain electrical charge balance means that chloride must, in essence, pull sodium with it and increase the porous media diffusion coefficient ($D_p = n_t D_t$) of sodium while in turn, sodium is retarding the movement of chloride (reducing the porous media diffusion coefficient of chloride). As a consequence, for the tests reported in Table 3, the Na^+ diffusion coefficients are higher than Cl^- diffusion coefficients (see Table 3) even though at infinite dilution in free solution the opposite would be true.

Additional evidence relating to the effect of the electrical interaction of ions on the deduced diffusion coefficients is given by the results of test GB5PH (Table 4). As mentioned previously, phase I is similar to the other tests reported in Table 3 in that initially the gradient for chloride is substantially greater than for sodium. This explains why the deduced diffusion coefficients for sodium and chloride in phase I are consistent with those reported in Table 3. However, after phase I has been completed the mass of SO_4^{2-} in the receptor has been depleted due to sampling and more importantly, the concentration of chloride in the receptor at the end of phase I has increased enough to almost provide a charge balance for Na^+ in the receptor (still some SO_4^{2-}). The gradient of chloride at the start of phase II is only 6% greater than that for sodium and there is a greater need for a similar molar flux to diffuse from the source to the receptor to maintain charge balance. Thus the deduced sodium diffusion coefficient decreases while the deduced chloride diffusion coefficient increases and the ratio of chloride to sodium diffusion coefficients increases from 0.68 in phase I to 0.95 in phase II. Although the difference in diffusion coefficients between phase I and II is not great, it emphasizes the role that electrical interactions play with respect to diffusion through soil.

Since the initial concentration of Na^+ in the receptor is controlled by the bentonite porewater chemistry, one would expect that the difference between Na^+ and Cl^- would become smaller as the source concentration was increased relative to the background concentration. As a consequence, one would expect that as the NaCl source concentration increased, the molar flux and the diffusion coefficient for Cl^- and Na^+ would tend to the same value (to ensure charge balance). This was evident for the results given in Table 4. Sodium and chloride diffusion coefficients were also the same value for phase I of test NWNWD-CS (Table 5) when there was no concentration difference of SO_4^{-2} across the sample due to the hydration procedure and similar initial gradients of Na^+ and Cl^- across the sample as explained in Section 3.2.1. This clearly illustrates the need for caution in using published diffusion coefficient results at infinite dilution in free solution for real systems and emphasizes the fact that the "diffusion" coefficient deduced from typical tests on soil are really mass transfer coefficients rather than true diffusion coefficients since they incorporate the effect of factors in addition to true molecular diffusion (Rowe et al., 1995).

It should be noted again that the diffusion coefficients (D_t) reported in this paper represent mass transfer coefficients deduced using the total porosity. The actual effective diffusion coefficient of both sodium and chloride may also be affected by anion exclusion. It should be recognized that the use of the total porosity n_t and the corresponding value of D_t given in Table 3 represent a convenient way of modeling transport through a GCL and these parameters do not necessarily represent the true effective values. However, they do provide a reasonable estimate of the porous media diffusion coefficient, D_p, and can be used to establish mass transfer across a single GCL.

3.6 Diffusion considerations when designing GCL landfill liner systems

In the design of waste containment systems with GCLs, it would be desirable to have a relationship between the stress levels expected at a waste containment facility and the corresponding diffusion coefficient. However, as shown throughout this paper there are many different variables that interact with the GCL system such as the type of contaminant source solution, GCL hydration conditions, and type of GCL. Although the stress present on a liner system and the diffusion coefficient may be related for a given GCL and contaminant source, this relationship may change if the contaminant source or hydrating conditions change which may cause changes in the void ratio of the bentonite. Barbour and Fredlund (1989) discussed how the results of consolidation tests performed on a Na^+ montmorillonite at different salt concentrations exhibited different void ratios at the same stress level and hence different e-log p curves. This is because when comparing the same stress levels, higher salt contents caused increasing amounts of osmotic consolidation.

The preferred means of obtaining diffusion coefficients is to perform a test under conditions as close as practical to those expected in the field (i.e. under similar hydration conditions, applied stress and source contaminant). However, when no diffusion data is available for initial calculations and the contaminating solution behaves similar to the 3–5 g/L (0.05 M to 0.08 M) NaCl solutions examined in this paper, an initial estimate of chloride diffusion coefficients can be made as described below. Readers are cautioned that the nature of contaminant source solution may alter the diffusion coefficients and give rise to values that are different than those that would be deduced from what is described below.

Chloride is commonly used as a tracer contaminant in landfill design (see Rowe et al., 1995; Ontario Ministry of Environment (MOE), 1998). Thus if one is interested in performing contaminant transport analyses for chloride, an estimate of the final bulk GCL void ratio can be made using swell data. For example if one considers a NWNWT GCL with a vertical stress of approximately 150 kPa (assuming the GCL is hydrated with water under this applied stress), one can use the pre-confinement swell data in Fig. 10a to deduce a bulk GCL void ratio of approximately 1.9. Fig. 11 shows that for a NWNWT final bulk GCL void ratio of 1.9, the chloride porous media diffusion coefficient will be approximately $D_p = 1.0 \times 10^{-10}\ m^2/s$. As explained earlier, for a single GCL layer similar to those tested, the movement of the contaminant through the GCL is controlled by $D_p = nD$. However for most contaminant transport modeling a value of the porosity, n, and diffusion coefficient, D, are needed as inputs. Since it has been shown that transport through a single GCL is insensitive to the precise values of n and D (provided that the product is equivalent to D_p), one can estimate $D_t = D_p/n_t$ where reasonable values of n_t are given in Fig. 11 for a range of GCL bulk void ratios. In this example, for $e_B = 1.9$, a reasonable value of n_t is 0.7 as given in Fig. 11. Alternatively a value of n_t can be deduced for the specific value of

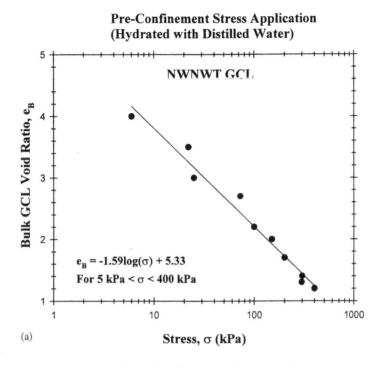

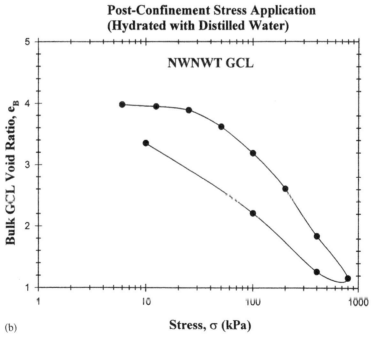

Fig. 10. NWNWT GCL swell data reproduced from Lake and Rowe (2000).

294

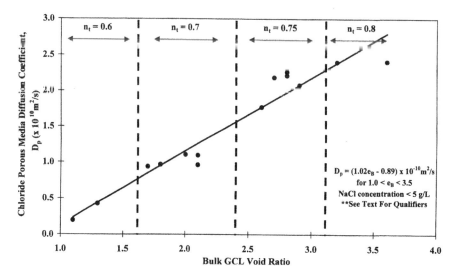

Fig. 11. Typical values of total porosity and chloride porous media diffusion coefficients for bulk GCL void ratios.

e_B and GCL properties as described in Appendix A. For $n_t = 0.7$, the resultant diffusion coefficient, D_t, will be approximately 1.5×10^{-10} m^2/s.

If the situation discussed above was varied such that the NWNWT GCL swelled under low stress conditions (say 6 kPa), and was then consolidated under a 150 kPa stress due to the overlying waste, the final bulk GCL void ratio could be estimated from the post-confinement swell data for the NWNWT GCL in Fig. 10b to be approximately 2.8. Using Fig. 11, a chloride porous media diffusion coefficient of 2.0×10^{-10} m^2/s and bentonite total porosity of 0.75 can be estimated for these conditions. This implies a diffusion coefficient of $D_t = 2.6 \times 10^{-10}$ m^2/s.

Table 3 shows that a chloride diffusion coefficient, D_t, of 2.2×10^{-10} m^2/s was obtained from the testing program for test NWNWT6-SV (synthetic leachate). This test can be used to emphasize the potential dangers of using Fig. 11 data (developed from 0.08 M NaCl solutions) for predicting chloride diffusion coefficients for a synthetic leachate. NWNWT6-SV had a bulk GCL void ratio, e_B, of 1.9, a bentonite total porosity, n_t, of 0.70 and a chloride diffusion coefficient, D_t, of 2.2×10^{-10} m^2/s. However if Fig. 11 is used to try to predict the Cl$^-$ diffusion coefficient for this synthetic leachate test, a higher Cl$^-$ diffusion coefficient is predicted. By taking the bulk GCL void ratio as 1.9, the corresponding porous media diffusion coefficient, D_p, can be estimated from Fig. 11 as 1.2×10^{-10} m^2/s. Since the test was performed at a bentonite total porosity of 0.70, the resultant diffusion coefficient, D_t, is therefore estimated as 1.6×10^{-10} m^2/s (using Fig. 11 and the same procedure as described in the previous paragraph), which is almost a 40% below what was actually experimentally measured (2.2×10^{-10} m^2/s).

If one is interested in estimating sodium diffusion coefficients in a similar manner, Fig. 12 can be used in a similar way to Fig. 11 for chloride to establish GCL diffusion coefficients for sodium.

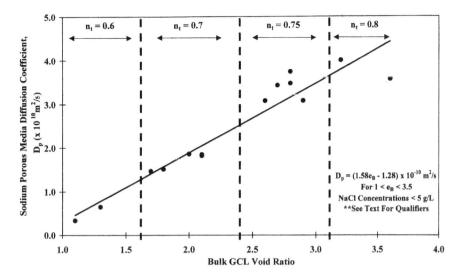

Fig. 12. Typical values of total porosity and sodium porous media diffusion coefficients for bulk GCL void ratios.

The initial clay fabric is known to influence the hydraulic conductivity of a soil when permeated with various fluids. Various researchers (Mesri and Olsen, 1971; Fernandez and Quigley, 1985; Kenney et al., 1992; Petrov and Rowe, 1997) have shown that soils initially hydrated with concentrated salt solutions or pure organic fluids will exhibit a more flocculated fabric and hence larger effective pore space than soils with a dispersed fabric at a given void ratio resulting in a higher hydraulic conductivity than soil initially hydrated with water. Various factors such as effective stress, type of clay mineral, and type of hydrating medium will influence the relative magnitude of difference. Results presented in Table 3 were performed for GCL samples hydrated with distilled water and hence represent a more dispersed clay fabric compared to the possibility of samples being hydrated with a concentrated salt solution or leachate. Therefore there is a possibility that samples hydrated with a concentrated salt or leachate may have a more flocculated clay fabric, larger effective void space and hence higher diffusion coefficient values. As shown by Daniel et al. (1993), whether or not GCLs, when constructed, will remain "dry" before coming in contact with these solutions is dependent on factors such as the suction potential of underlying soils. Daniel et al. (1993) examined the hydration characteristics of the bentonite component of GCLs and found that for a GCL placed in on a soil close to the wilting point (1500 kPa suction) the GCL reached a moisture content of 50%. For a moist soil (suction between 0 and 100 kPa), the moisture content equilibrated to 100% and 140% with a hydration time of approximately one to three weeks. Hydraulic conductivity measurements showed that hydraulic conductivity values of the GCL when permeated with pure organic fluids were less than 10^{-11} m/s when the moisture content of the underlying soil was above 100%. Therefore it is likely that a GCL placed as part of a liner system in a landfill will undergo some hydration after placement of a cover soil and any "first" contact with a leachate will be with a GCL partially or fully hydrated from the contacting soils. If the GCL was to be placed in an arid environment, there may have to be allowances made for pre-hydrating the GCL with water (with an effective stress applied) and preventing the GCL from desiccation until such a time that the leachate is expected to come into contact with the GCL.

4. Conclusions

The diffusion of sodium and chloride through GCLs has been examined for a range of conditions including different final bulk GCL void ratios, different contaminant source solutions, different types of GCL, and different test methods. The following conclusions are applicable to the results presented in this paper but should not be generalized without additional investigation:

1. When considering similar hydration conditions, stress levels, and permeating fluids; the GCLs tested in this paper exhibited a linear relationship between final bulk GCL void ratios and diffusion coefficients. Even when a GCL was hydrated under low stress conditions and subsequently consolidated to a lower final bulk GCL void ratio, it was the bulk GCL void ratio during diffusion testing that controlled the diffusion parameters. Generally, the diffusion coefficient was shown to decrease as the bulk GCL void ratio decreased.
2. The diffusion coefficient may change with changing composition of the source reservoir solution and the linear relationship given in this paper only applies to NaCl solutions with 3 to 5 g/L (0.05 M to 0.08 M) concentrations.
3. At similar bulk GCL void ratios, there was no significant difference in diffusion coefficients obtained for the three GCLs examined in this study. However, there was some scatter of diffusion coefficients at higher bulk GCL void ratios that was attributed to the effect of the thermal treatment at low applied stresses.
4. At infinite dilution, the free solution diffusion coefficient of Na^+ is less than that of Cl^-. However the diffusion coefficient for Na^+ was greater than that for Cl^- for most of the tests reported in this paper, primarily due to back diffusion of SO_4^{-2} from the GCL and the lower gradients of Na^+ compared to Cl^- at the start of the test.
5. Diffusion tests performed with high NaCl concentrations (up to 2 M) were compared by two different test methods, specified volume diffusion and constant stress diffusion tests. Application of an effective stress resulted in much lower absolute diffusion coefficients for 0.6 M and 2 M compared to the specified volume diffusion test. The combination of an effective stress and high NaCl concentrations caused contraction of the sample thickness, reducing the void ratio.
6. The diffusion coefficient of chloride for diffusion tests performed with synthetic leachate was greater than that observed for 0.08 M NaCl solutions. This increase was hypothesized to be due to a combination of increased effective pore space due to double layer contraction as well as the interaction of the many ions in the synthetic leachate as they migrate through the sample.

7. Because of the influences of hydration fluid, source reservoir solution and stress levels during diffusion testing, diffusion tests should attempt to simulate the conditions expected in the field. It is recommended that constant stress diffusion tests be performed to simulate the stress conditions expected in the field. Examples have been provided to illustrate how to obtain an initial estimate of the diffusion coefficient of chloride as well as an estimate of the total bentonite porosity normally required for contaminant transport modeling. A similar approach can be followed for sodium.

Acknowledgements

The research reported in this paper was funded by Terrafix Geosynthetics, the National Research Council of Canada (under the IRAP program), and by the Natural Sciences and Engineering Research Council of Canada. The authors very gratefully acknowledge the value of discussions with Messrs. Don Stewart and Cal Reaume of Terrafix and Mr. Kent von Maubeuge of Naue Fasertechnik during the preparation of this paper. Thanks are also extended to Ms. Leila Hrapovic and Mr. Jamie van Gulck for supplying the synthetic leachate. The information in this paper should not be used without independent examination and verification of its suitability for any particular project.

Appendix A. Definitions

The final bulk GCL void ratio, e_B, is defined by (Petrov et al., 1997):

$$e_B = \frac{H_{GCL} - H_s}{H_s}, \tag{A.1}$$

where H_{GCL} is the GCL height and H_s is the height of solids in the GCL and

$$H_s = H_{sBENT} + H_{sGEO} \tag{A.2}$$

with

$$H_{sBENT} = \frac{M_{BENT}}{\rho_s(1 + w_o)}, \tag{A.3}$$

$$H_{sGEO} = \frac{M_{GEO}}{\rho_{sg}}, \tag{A.4}$$

where H_{sBENT} is the height of bentonite solids; H_{sGEO} is the height of geotextile solids; $M_{GCL} = M_{BENT} + M_{GEO}$ is the mass per unit area of the GCL; M_{BENT} is the mass of bentonite per unit area in the GCL; M_{GEO} is the mass of geosynthetics per unit area in the GCL (Typical values for the three GCLs examined are: NWNWT-620 g/m^2, WNWT-390 g/m^2, WNWBT-500 g/m^2); ρ_s is the density of bentonite solids (typical value of 2.61 mg/m^3); ρ_{sg} is the density of polypropylene geotextile solids (typical value of 0.91 mg/m^3); w_0 is the initial water content of the bentonite (typical value of 0.08).

The void ratio of the bentonite in the GCL is given by

$$e_{BENT} = \frac{H_{BENT} - H_{sBENT}}{H_{sBENT}}, \qquad (A.5)$$

where H_{BENT} is the height of bentonite in the GCL and is given by

$$H_{BENT} = H_{GCL} - H_{GEO}. \qquad (A.6)$$

H_{GEO} is the height of geotextiles in the GCL (this value is difficult to obtain accurately and therefore, based on a number of measurements the geotextile heights were taken as 1.1 mm for the NWNWT and WNWBT GCL and 0.8 mm for the WNWT GCL for all the GCLs examined in this paper).

For a given GCL, H_{GCL}, H_{GEO}, w_0, M_{GCL}, M_{GEO}, ρ_s, and ρ_{sg} are either measured or known and hence e_{BENT} can be deduced from Eqs. (A.1)–(A.6) and the total porosity of the bentonite, n_t, can be calculated from

$$n_t = \frac{e_{BENT}}{(1 + e_{BENT})}. \qquad (A.7)$$

In a consolidation or swell test, e_B is deduced from Eqs. (A.1) and (A.2) based on known values of H_{GCL}, M_{BENT}, M_{GEO}, ρ_s, ρ_{sg} and w_0. If e_B has been estimated from charts such as Fig. 7 and M_{GCL}, ρ_s, ρ_{sg} and w_0 can be measured or estimated for the GCL, then H_{sBENT} can be deduced from Eq. (A.3) and H_{GCL} can be deduced from Eq. (A.1). Thus H_{BENT} can be deduced from Eq. (A.6) and e_{BENT} and n_t deduced from Eqs. (A.5) and (A.7), respectively.

References

Barbour, S.L., Fredlund, D.G., 1989. Mechanisms of osmotic flow and volume change in clay soils. Canadian Geotechnical Journal 26, 551–562.

Daniel, D.E., Shan, H.Y., Anderson, J.D., 1993. Effects of partial wetting on the performance of the bentonite component of a geosynthetic clay liner. Proceedings, Geosynthetics' 93, Vancouver, B.C., Canada IFAI, St. Paul, MN, USA, pp. 1483–1496.

Fernandez, F., Quigley, R.M., 1985. Hydraulic conductivity of natural clays permeated with simple liquid hydrocarbons. Canadian Geotechnical Journal 22, 205–214.

Fernandez, F., Quigley, R.M., 1991. Controlling the destructive effects of clay-organic liquid interactions by application of effective stresses. Canadian Geotechnical Journal 28, 388–398.

Kemper, W.D., van Schaik, J.C., 1966. Diffusion of salts in clay-water systems. Soil Science Soc. Amer. Proc. 30, 534–540.

Kenney, T.C., van Veen, W.A., Swallow, M.A., Sungaila, M.A., 1992. Hydraulic conductivity of compacted bentonite-sand mixtures. Canadian Geotechnical Journal 29, 364–374.

Lake, C.B., Rowe, R.K., 2000. Swelling characteristics of needlepunched, thermally treated GCLs. Geotextiles and Geomembranes 18 (2–4), 77–101.

Lo, M.C., 1992. Development and evaluation of clay-liner materials for hazardous waste sites. Ph.D. Dissertation, The University of Texas at Austin.

Mesri, G., Olsen, R.E., 1971. Mechanisms controlling the permeability of clays. Clays and Clay Minerals 19, 151–158.

Ontario Ministry of Environment (MOE), 1998. Landfill Standards: A guideline on the regulatory and approval requirement for new expanding landfill sites.

Petrov, R.J., Rowe, R.K., 1997. GCL - chemical compatibility by hydraulic conductivity testing and factors impacting its performance. Canadian Geotechnical Journal 34 (6), 863–885.

Petrov, R.J., Rowe, R.K., Quigley, R.M., 1997. Selected factors influencing GCL hydraulic conductivity. Journal of Geotechnical and Geo-Environmental Engineering, ASCE 123, 683–695.

Rowe, R.K., 1998. Geosynthetics and the minimization of contaminant migration through barrier systems beneath solid waste – Keynote Lecture. Proceedings, 6th International Conference on Geosynthetics, Atlanta, Vol. 1, pp. 27–102.

Rowe, R.K., Booker, J.R., 1999. POLLUTE v.6.4 – 1-D pollutant migration through a non-homogeneous soil©. Distributed by GAEA Environmental Engineering Ltd., 44 Canadian Oakes Dr., Whitby, Ontario, L1N 6W8, fax (905) 725–9657.

Rowe, R.K., Lake, C.B., Petrov, R.J., 2000. Apparatus and procedures for assessing inorganic diffusion coefficients through geosynthetic clay liners. ASTM Geotechnical Testing Journal, (in press).

Rowe, R.K., Quigley, R.M., Booker, J.R., 1995. Clayey Barrier Systems for Waste Disposal Facilities E and FN Spon, 390pp.

Yeung, A. T-C and Mitchell, J.K., 1992. Coupled fluid, chemical, and electrical flows in soil, Geotechnique 42 (4).

Papers from

Journal of Terramechanics

Simulation of soil deformation and resistance at bar penetration by the Distinct Element Method☆

Hiroaki Tanaka [a],*, Masatoshi Momozu [b], Akira Oida [b], Minoru Yamazaki [b]

[a]*Laboratory of Agricultural Machinery, Department of Hilly Land Agriculture, Shikoku National Agricultural Experiment Station, 2575, Ikano, Zentsuji, Kagawa 765, Japan*
[b]*Agricultural Systems Engineering Section, Graduate School of Agriculture, Kyoto University, Kyoto 606-01, Japan*

Abstract

A mechanical model of soil is constructed using the Distinct Element Method (DEM) which makes it possible to analyze the discontinuous property of soil. To discuss the applicability of the soil model by the DEM, a bar penetration test was conducted and the result was compared with the simulation results. From the results of the behavior of elements, it could be said that the mechanical model by the DEM could well simulate the discontinuous behavior of soil and the parameters used in the simulation play important roles to make the soil model useful. As for the penetrating resistance, some problems which lie in the present DEM model are discussed and the key to solving these problems is indicated. Moreover, the method to determine the time interval used in the DEM simulations is mentioned in terms of the stability of the solution in the calculation. © 2000 ISTVS. All rights reserved.

1. Introduction

There are many interactive phenomena between soil and farm machines. To investigate mechanical interactions between soil and machines is very important for designing farm machines or evaluating tillage methods. In terms of computational

---

☆ This report was awarded "best paper" at the 12th International Conference of the ISTVS (Beijing, 1996).
* Corresponding author. Tel.: +81-877-62-0800; fax: +81-877-62-1130.
E-mail address: hirtana@skko.affrc.go.jp (H. Tanaka).

Reprinted from *Journal of Terramechanics* **37 (1)**, 41-56 (2000)

Nomenclature

$[d_n]_t, [d_s]_t$	normal and tangential damping forces on an element at time t
$[e_n]_t, [e_n]_{t-\Delta t}$	normal spring forces on an element at time t and $t - \Delta t$
$[e_s]_t, [e_s]_{t-\Delta t}$	tangential spring forces on an element at time t and $t - \Delta t$
$[E]_t, [E]_{t-\Delta t}$	energies of a system at time t and $t - \Delta t$
$[f_n]_t, [f_s]_t$	normal and tangential contact forces on an element at time t
g	gravitative acceleration
I_i	moment of inertia of the element i
k_{np}, k_{sp}	normal and tangential spring constants between elements
k_{nw}, k_{sw}	normal and tangential spring constants between element and wall
$[M_i]_t$	resultant moment on the element i at time t
m_i	mass of the element i
r_i, r_j	radii of the element i and j
r_{ij}	distance between the centers of the elements i and j
Δt	time interval
$\Delta u_i, \Delta u_j$	horizontal displacements of the elements i and j during Δt
$\Delta u_n, \Delta u_s$	normal and tangential relative displacements of the element i to the element j during Δt
$[\Delta u_i]_t, [\Delta u_i]_{t-\Delta t}$	horizontal displacements of the element i at time t and $t - \Delta t$
$[\dot u_i]_t, [\dot u_i]_{t-\Delta t}$	horizontal velocities of the element i at time t and $t - \Delta t$
$[\ddot u_i]_t$	horizontal acceleration of the element i at time t
$\Delta v_i, \Delta v_j$	vertical displacements of the elements i and j during Δt
$[\Delta v_i]_t, [\Delta v_i]_{t-\Delta t}$	vertical displacements of the element i at time t and $t - \Delta t$
$[\dot v_i]_t, [\dot v_i]_{t-\Delta t}$	vertical velocities of the element i at time t and $t - \Delta t$
$[\ddot v_i]_t$	vertical acceleration of the element i at time t
$[X_i]_t$	horizontal resultant force on the element i at time t
$[x_i]_t, [x_i]_{t-\Delta t}$	horizontal positions of the element i at time t and $t - \Delta t$
$[Y_i]_t$	vertical resultant force on the element i at time t
$[y_i]_t, [y_i]_{t-\Delta t}$	vertical positions of the element i at time t and $t - \Delta t$
α_{ij}	angle that the line connecting the centers of the elements i and j makes with the X axis
η_{np}, η_{sp}	normal and tangential coefficients of viscous damping between elements
η_{nw}, η_{sw}	normal and tangential coefficients of viscous damping between element and wall
$\Delta \varphi_i, \Delta \varphi_j$	rotational displacements of the elements i and j during Δt
$[\varphi_i]_t, [\varphi_i]_{t-\Delta t}$	rotational positions of the element i at time t and $t - \Delta t$
$[\Delta \varphi_i]_t, [\Delta \varphi_i]_{t-\Delta t}$	rotational displacements of the element i at time t and $t - \Delta t$
$[\dot \varphi_i]_t, [\dot \varphi_i]_{t-\Delta t}$	rotational velocities of the element i at time t and $t - \Delta t$
$[\ddot \varphi_i]_t$	rotational acceleration of the element i at time t
μ_p, μ_w	coefficients of friction between elements and between element and wall

mechanics, there have been many reports which have dealt with soil–machine mechanical interactions but almost all of them treated soil analyzed as a continuum. Soil originally consists of granular assemblies which can be cut and separated by the action of machine parts. In such a case, the continuum model is hard to apply.

In 1979, Cundall and Strack [1] developed a discrete model to analyze behaviors of granular assemblies. It is called the Distinct Element Method (DEM). The DEM is applied in several fields such as powder technology, civil engineering [2] and soil mechanics. The mechanical model by the DEM has the capability to analyze the discontinuous property of the materials which is usually difficult to be analyzed by the continuum model. However, at the same time, the DEM also has some problems that the concrete method to determine the parameter has not been established yet, much time should be taken for calculation and so on. This report will begin with the explanation of the theory for construction of the soil model by the DEM and lead to the discussion about the constructed DEM model of soil by comparing with the result of the experiment. If the soil model by the DEM is constructed with high accuracy, it could be applied to many mechanical and dynamic problems between soil and machines in the field of Terramechanics.

2. Mechanical model of soil using the DEM

2.1. Expression of soil by circular elements

The most characteristic feature in the DEM is to assume that the object analyzed is an assembly of elements. In this study, the soil is represented as the assembly of two dimensional circular elements as shown in Fig. 1. It becomes possible to analyze the behavior of the soil by tracing the motion of each element which is derived from the mechanical interaction between elements or between the element and wall of the container or penetrating bar.

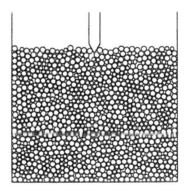

Fig. 1. The soil model by circular elements.

2.2. Mechanical interaction between elements

Each element receives contact forces from contacting elements and/or sides of the walls. The magnitude of the contact force is determined by the relative displacement and relative velocity of the element. In the DEM simulation, a spring with the spring constant k, a dashpot with the coefficient of viscous damping η and a slider with the coefficient of friction μ are supposed to be between elements. The same assumption is put between the element and the wall. The mechanical relationships are shown in Fig. 2.

2.3. Judgement of contact

The judgement of contact of elements is based on the geometrical relationship between the element and the element/wall. In other words, if the distance between the centers of two elements i and j is smaller than the sum of the radii of these two elements, it follows that the element i contacts with the element j. Also in the case of the contact between element i and the wall of the container or the bar, if the distance between the center of the element i and the side of the wall or bar is smaller than the radius of the element i, it follows that the element i contacts with the wall/bar. In this study, the side of the metal bar is treated as the "wall" in the calculation.

2.4. Principle of motion of elements

In the DEM, the positions of each element are determined step-by-step at intervals of Δt s. The position and contact force on each element determine the motion of the element. Simulation is conducted by repeating the calculations of these two values. The concept of this algorithm is shown in Fig. 3. The details of these calculations are mentioned in later sections.

2.5. Calculation of the contact force on each element

Contact forces which act on a contacting element are proportional to the relative displacement and velocity of the element to the other contacting element. In this section, the calculation of contact force is formulated. As shown in Fig. 4, two elements

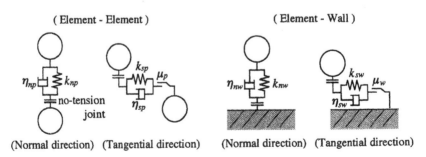

Fig. 2. Mechanical relationships between the element and the element/wall.

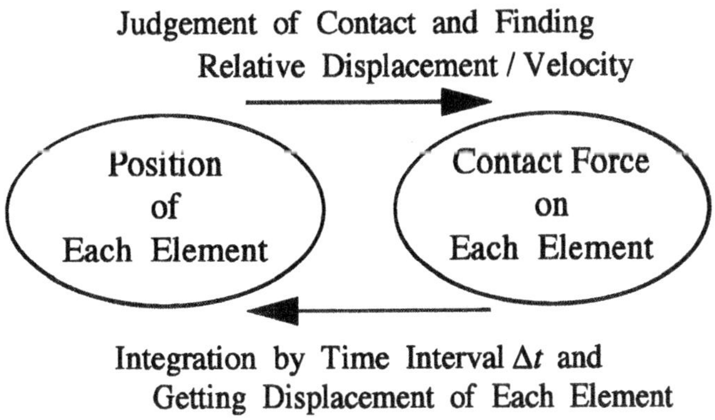

Judgement of Contact and Finding
Relative Displacement / Velocity

Position
of
Each Element

Contact Force
on
Each Element

Integration by Time Interval Δt and
Getting Displacement of Each Element

Fig. 3. The concept of the simulation algorithm.

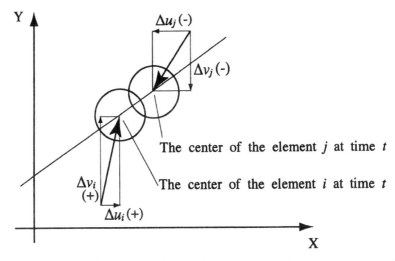

Fig. 4. The displacements of the contact elements (Δu_i, Δv_i are positive, Δu_j, Δv_j are negative in this case).

i and j which contact with each other are taken as an example. This figure shows the positions of the elements i and j at time t, respectively. The tracks of motion of each element during Δt are represented by arrows. The horizontal (the X direction) and vertical (the Y direction) displacements of the elements i and j during Δt are expressed by the notations Δu_i, Δv_i, Δu_j, Δv_j, respectively.

Here the angle and directions are defined, which concern the formulation in this section. As shown in Fig. 5(a), considering a line which connects the centers of the elements i, j, and taking the center of the element i as the vertex, the angle that the line makes with the X axis is defined as α_{ij}. The angle α_{ij} is taken as positive in the counterclockwise direction. The sign rule for relative displacement of the element i to the element j is shown in Fig. 5(b). In the normal direction, Δu_n is taken as positive for compression. In the tangential direction, Δu_s is taken as positive in the

308

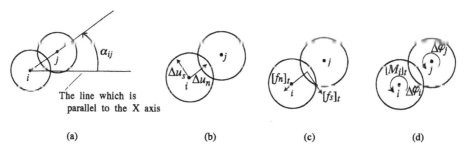

Fig. 5. Definition of the angle α_{ij} [(a)] and the sign rule [(b), (c) and (d)].

clockwise direction around the element j. In Fig. 5(c), the sign rule for contacting force on the element i given by the element j is indicated. $[f_n]_t$ and $[f_s]_t$ are taken as positive in the directions opposite to Δu_n and Δu_s. As for the rotational displacement $\Delta\phi_i$ and the moment $[M_i]_t$, each value is taken as positive in the counterclockwise direction as shown in Fig. 5(d).

In accordance with these definitions, the relative displacement of the element i to the element j in the normal direction is expressed by Eq. (1), and in the tangential direction by Eq. (2).

$$\Delta u_n = \left(\Delta u_i - \Delta u_j\right) \cos\alpha_{ij} + \left(\Delta v_i - \Delta v_j\right) \sin\alpha_{ij} \tag{1}$$

$$\Delta u_s = -\left(\Delta u_i - \Delta u_j\right) \sin\alpha_{ij} + \left(\Delta v_i - \Delta v_j\right) \cos\alpha_{ij} + r_i\Delta\varphi_i + r_j\Delta\varphi_j \tag{2}$$

If these two elements contact for the first time at time t, relative displacements in each direction are expressed by Eqs. (3) and (4), respectively, which are equal to the overlap quantities of the two elements.

$$\Delta u'_n = \left(r_i + r_j\right) - r_{ij} \tag{3}$$

$$\Delta u'_s = \Delta u_s \cdot \frac{\Delta u'_n}{\Delta u_n} \tag{4}$$

From the relative displacements during Δt, the contact forces on the element i at time t are obtained based on the mechanical relationships in Fig. 2. In the normal direction for example, the spring force at time t, which is proportional to the overlap quantities, is obtained by adding the increment of the spring force $k_{np}\Delta u_n$ to the spring force at the previous step $[e_n]_{t-\Delta t}$ as follows:

$$[e_n]_t = [e_n]_{t-\Delta t} + k_{np}\Delta u_n \tag{5}$$

Therefore, the normal contact force on the element i given by the element j is obtained by adding the force by the dashpot $[d_n]_t$, which is proportional to the relative velocity of the element i to the element j during Δt, to the force by the spring $[e_n]_t$ as follows:

$$[f_n]_t = [e_n]_t + [d_n]_t = \left([e_n]_{t-\Delta t} + k_{np}\Delta u_n\right) + \frac{\eta_{np}\Delta u_n}{\Delta t} \tag{6}$$

In the same way, the contact force in the tangential direction is calculated as follows:

$$[f_s]_t = [e_s]_t + [d_s]_t = \left([e_s]_{t-\Delta t} + k_{sp}\Delta u_s\right) + \frac{\eta_{sp}\Delta u_s}{\Delta t} \tag{7}$$

As shown by the no-tension joint in Fig. 2, the tensile force between contacting elements is not considered in this model, therefore:

$$\text{If } [e_n]_t < 0, \quad [d_n]_t = [e_s]_t = [d_s]_t = 0 \tag{8}$$

The magnitude of contact force by the spring in the tangential direction can not exceed the maximum force caused by friction, therefore:

$$\text{If}$$
$$|[e_s]_t| > \mu_p[e_n]_t, \quad [f_s]_t = [e_s]_t = \mu_p[e_n]_t \, \text{sig}\left[[e_s]_t\right] \text{ and } [d_s]_t = 0 \tag{9}$$

where, $\text{sig}\left[[e_s]_t\right]$ is the sign of $[e_s]_t$.

Transforming the contact forces in the normal and tangential directions into the forces in the horizontal, vertical and rotational directions and summing up the contact forces caused by all of the elements which contact with the element i, resultant forces which act on the element i in the horizontal, vertical and rotational directions are obtained as follows:

$$[X_i]_t = \sum_j \left(-[f_n]_t \cos\alpha_{ij} + [f_s]_t \sin\alpha_{ij}\right) \tag{10}$$

$$[Y_i]_t = \sum_j \left(-[f_n]_t \sin\alpha_{ij} - [f_s]_t \cos\alpha_{ij}\right) \tag{11}$$

$$[M_i]_t = -r_i \sum_j \left([f_s]_t\right) \tag{12}$$

The same procedure is operated in the calculation between the element and wall.

2.6. Calculation of the position of each element

The accelerations of the element i in the horizontal, vertical and rotational directions are obtained by dividing the contact forces in each direction [Eqs. (10)–(12)] by the mass or moment of inertia of the element i as follows:

$$[\ddot{u}_i]_t = \frac{[X_i]_t}{m_i} \tag{13}$$

$$[\ddot{v}_i]_t = \frac{[Y_i]_t}{m_i} \tag{14}$$

$$[\ddot{\varphi}_i]_t = \frac{[M_i]_t}{I_i} \tag{15}$$

Integrating the accelerations over the time interval Δt and adding the values to the previous velocities, the velocities of the element i in each direction at time t are obtained as follows:

$$[\dot{u}_i]_t = [\dot{u}_i]_{t-\Delta t} + [\ddot{u}_i]_t \Delta t \tag{16}$$

$$[\dot{v}_i]_t = [\dot{v}_i]_{t-\Delta t} + ([\ddot{v}_i]_t - g)\Delta t \tag{17}$$

$$[\dot{\varphi}_i]_t = [\dot{\varphi}_i]_{t-\Delta t} + [\ddot{\varphi}_i]_t \Delta t \tag{18}$$

The displacements of the element i in each direction at time t are given by integrating the velocities over the time interval Δt. However, the average of the displacements at time t and $t - \Delta t$ was adopted as the displacement at time t in actual calculation in order to get the stability of solution [2] as follows:

$$[\Delta u_i]_t = \frac{1}{2}\left([\Delta u_i]_{t-\Delta t} + [\dot{u}_i]_t \Delta t\right) \tag{19}$$

$$[\Delta v_i]_t = \frac{1}{2}\left([\Delta v_i]_{t-\Delta t} + [\dot{v}_i]_t \Delta t\right) \tag{20}$$

$$[\Delta \varphi_i]_t = \frac{1}{2}\left([\Delta \varphi_i]_{t-\Delta t} + [\dot{\varphi}_i]_t \Delta t\right) \tag{21}$$

Finally, adding the displacements of the element i at time t to the positions of the element i at time $t - \Delta t$, the renewed positions of the element i at time t is obtained as follows:

$$[x_i]_t = [x_i]_{t-\Delta t} + [\Delta u_i]_t \tag{22}$$

$$[y_i]_t = [y_i]_{t-\Delta t} + [\Delta v_i]_t \tag{23}$$

$$[\varphi_i]_t = [\varphi_i]_{t-\Delta t} + [\Delta \varphi_i]_t \tag{24}$$

Adopting these procedures to all of the elements in turn, the contacting forces, the displacements and the positions of all elements at time t are calculated. The values of the displacements $[\Delta u_i]_t$, $[\Delta v_i]_t$, $[\Delta \varphi_i]_t$ calculated by Eqs. (19)–(21) are used as Δu_i, Δv_i, $\Delta \varphi_i$ in Eqs. (1) and (2) in the next step of the calculation to progress the simulation.

2.7. Stability of solution in the DEM simulation

It is generally said that one of the DEM parameters effects so complicatedly the simulation result that the method to determine the DEM parameters is not clear. It is also said that the stability of solution is affected especially by the normal spring constant and the mass of element. In this study, the spring constant was set prior to the time interval and the value of the time interval was decided subsequently in order to retain the stability of solution. The coefficient of viscous damping was set as $1/\sqrt{2}$ times of the critical damping coefficient $2\sqrt{mk}$ [3]. In this section, the interrelationship between normal spring constant k, mass of element m and time interval Δt is discussed and the limitation of the time interval is proposed.

To simplify the problem, a certain system is assumed as shown in Fig. 6. This figure means that an element on the boundary of the wall at time $t - \Delta t$ moves perpendicularly to the wall during the time interval Δt, and overlaps with the wall at time t. The limitation of the time interval is to be considered in terms of the energy of the system at these two points in time $t - \Delta t$ and t.

Assuming the displacement of element during Δt as Δu_n, the velocity of the element at $t - \Delta t$ is:

$$[\dot{u}]_{t-\Delta t} = \frac{\Delta u_n}{\Delta t} \tag{25}$$

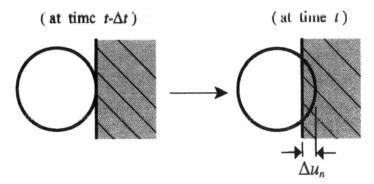

Fig. 6. Collision of an element to the wall.

At the point in time $t - \Delta t$, as the element does not contact with the wall, the energy of the system at $t - \Delta t$ equals the kinetic energy of the element as follows.

$$[E]_{t-\Delta t} = \frac{1}{2}m[\dot{u}]_t^2 \quad \Delta t \tag{26}$$

After the time interval Δt, the element contacts with the wall and contact force generates. From the Eqs. (6),(10),(13) and (16), the velocity of the element at time t is:

$$[\dot{u}]_t = [\dot{u}]_{t-\Delta t} - \left(k\Delta u_n + \frac{\eta \Delta u_n}{\Delta t}\right) \cdot \frac{\Delta t}{m} \tag{27}$$

At this point in time, the energy of the system equals the sum of the kinetic energy of the element and the potential energy of the spring as follows:

$$[E]_t = \frac{1}{2}k(\Delta u_n)^2 + \frac{1}{2}m[\dot{u}]_t^2 = \frac{1}{2}k(\Delta u_n)^2 + \frac{1}{2}m\left\{\frac{\Delta u_n}{\Delta t} - \left(k\Delta u_n + \frac{\eta \Delta u_n}{\Delta t}\right) \cdot \frac{\Delta t}{m}\right\}^2 \tag{28}$$

Here, it is plain that the energy of the system is consumed by the dashpot during the time interval Δt. Accordingly, a certain condition is given to the energy of these two points in time as follows:

$$[E]_t < [E]_{t-\Delta t} \tag{29}$$

namely,

$$\frac{1}{2}k(\Delta u_n)^2 + \frac{1}{2}m\left\{\frac{\Delta u_n}{\Delta t} - \left(k\Delta u_n + \frac{\eta \Delta u_n}{\Delta t}\right) \cdot \frac{\Delta t}{m}\right\}^2 < \frac{1}{2}m\left(\frac{\Delta u_n}{\Delta t}\right)^2 \tag{30}$$

In Eq. (30), the coefficient of viscous damping η can be replaced by $\sqrt{2mk}$ as mentioned above. And Eq. (30) leads to a cubic inequality of Δt as follows:

$$\Delta t^3 + 2\sqrt{2}\sqrt{\frac{m}{k}}\Delta t^2 + \frac{m}{k}\Delta t - 2\sqrt{2}\left(\frac{m}{k}\right)^{\frac{3}{2}} < 0 \tag{31}$$

The solution of this inequality within the range of $0 < \Delta t$ is:

$$\Delta t < 0.75\sqrt{\frac{m}{k}} \tag{32}$$

In this way, the limitation of the time interval becomes clear as shown in Eq. (32). In actual calculation, some kinds of values are used as k and m because two kinds of normal spring constant (k_{np} and k_{nw}) are assumed in Fig. 2 and two kinds of diameters of the elements are assumed in the simulation. Therefore, the least value of the limitations calculated by Eq. (32) should be taken as the limitation of the time interval in the calculation.

3. Experiment and simulation

3.1. Experimental procedure

A laboratory experiment was conducted to confirm the soil model by the DEM. As shown in Fig. 7, the soil was put into a container with the volume of $15 \times 15 \times 1$ cm^3. The metal bar was penetrated vertically into the soil at the speed of 1 cm/s, and the resistant force which acts on the bar and the penetration depth of the bar were measured by the sensors. The behavior of the soil was photographed continuously. The properties of the soil are shown in Table 1.

3.2. Simulation procedure

In the simulation of this experiment, the sizes of the container and the bar were set up the same as those in the above experiment. The diameters of the elements were determined to be 0.4 and 0.5 cm. The density of the element was determined to be 1.53 g/cm^3 which is equal to that of soil in the experiment. These elements were located at random as the initial condition of the simulation. The number of elements

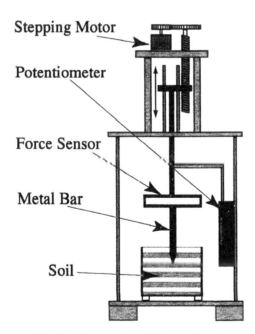

Fig. 7. The apparatus of the experiment.

was counted as 944. According to the parameters used in this simulation as shown in Table 2, the time interval Δt was set to 1×10^{-4} s considering the stability of solution mentioned above. The outputs of calculation (the positions of elements and resistant force) were obtained every 1000 cycles of the calculation which equals 0.1 s of the simulation time.

4. Results and discussion

4.1. Result of the experiment (behavior of soil)

The soil behavior at several points in time is shown in Fig. 8 as an example. As the soil was marked by powdered chalk every 3 cm from the bottom of the

Table 1
Physical properties of soil used in the experiment

Composition	Moisture content	Liquid limit	Plastic limit
Sand 57% Silt 31% Clay 12%	8.82%	37%	19.2%

Table 2
Parameters used in the simulation (cf. Fig. 2)

Element–element			Element–wall		
k_{np}	1000	[N/m]	k_{nw}	1000	[N/m]
k_{sp}	250	[N/m]	k_{sw}	250	[N/m]
η_{np}	0.774	[N·s/m]	η_{nw}	0.774	[N·s/m]
η_{sp}	0.387	[N·s/m]	η_{sw}	0.387	[N·s/m]
μ_p	0.0, 0.25, 0.5, 0.75	[–]	μ_w	0.0, 0.25, 0.50, 0.75	[–]

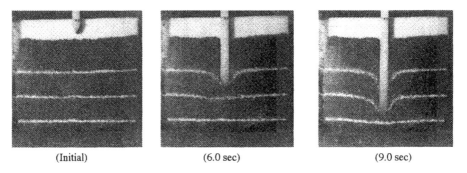

(Initial) (6.0 sec) (9.0 sec)

Fig. 8. The soil behavior in the experiment.

container, it can be seen that the soil near the bar moved downward following the movement of the bar.

4.2. Result of the simulation (behavior of elements)

Some results of the simulation are shown in Fig. 9. The elements near the bar moved downward following the movement of the bar. It can be said that the tendency of the behavior of elements was the same as that of the soil in the experiment.

4.3. The effect of the coefficient of friction between the element and wall

In this study, as the first step to build up the concrete method to determine the parameters, the effect of one of the parameters (i.e. the coefficient of friction) was examined. Fig. 10 shows some results of the experiment. These results show the positions of the elements at the same elapsed time of 9.0 s when the coefficient of friction between element–element was assumed to be 0.75. But the coefficient of friction between the element and wall was varied as shown in the figure.

The effect of the coefficient of friction clearly appeared in the behavior of elements. It can be seen that as the coefficient of friction becomes higher, the movement

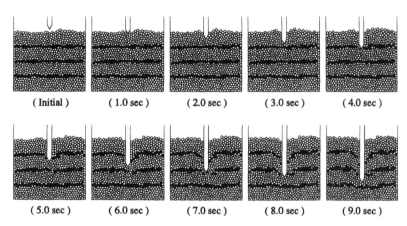

| (Initial) | (1.0 sec) | (2.0 sec) | (3.0 sec) | (4.0 sec) |

| (5.0 sec) | (6.0 sec) | (7.0 sec) | (8.0 sec) | (9.0 sec) |

Fig. 9. Some results of the simulation.

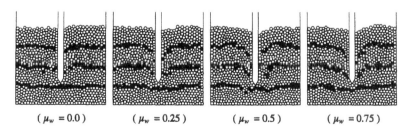

| ($\mu_w = 0.0$) | ($\mu_w = 0.25$) | ($\mu_w = 0.5$) | ($\mu_w = 0.75$) |

Fig. 10. The effect of the coefficient of friction between the element and wall.

316

of the elements near the bar comes larger. It also can be seen that as the coefficient of friction becomes higher, a surface bump appeared clearer. Comparing these results with those of the experiment, the simulation with the coefficient of friction between the element and wall of 0.25 can best simulate the behavior of the soil in the experiment. Therefore, it could be confirmed that the value of parameters have essential effects on the behavior of the elements and the accuracy of the simulation becomes higher if the value of each parameter is selected appropriately.

4.4. The result of penetration resistance (soil and simulation)

The penetration resistance in the experiment and the simulation are shown in Fig. 11. Each graph shows the relationship between the penetration depth and the resistance. The tendency that the penetration resistance becomes larger according to the penetration depth is shown in both results of the experiment and simulation, but the difference between experiment and simulation appeared.

The penetration resistance in the simulation considerably fluctuated. One reason could be that the diameter of the elements is large, and the other reason could be that the model is lacking the effect of cohesion because of the no-tension joint (cf. Fig. 2) which is put in the mechanical model, so that the frequent change of the resistance given by one element would cause a big change of the penetration resistance.

It also can be seen that the value of penetration resistance in the simulation is apparently smaller than that of the experiment. One reason could be that the value of normal spring constant was not appropriate, and the other reason is possibly that the interlocking between elements is difficult to be achieved by using the circular elements which are unstable and are apt to be stirred by the penetrating bar through the calculation.

These characteristic behaviors are the problem with establishing the accurate soil model by the DEM. These problems are to be discussed more in the next section.

4.5. Additional experiment using rigid balls and discussion

As mentioned in the previous section, there is a large difference between the result of penetration resistance of soil and that of simulation. Therefore, it is necessary to discuss the reason for such a difference. In order to confirm the mechanical property of the DEM model used in this study, an additional experiment was conducted using rigid balls which are made of alumina (Fig. 12). Because the alumina ball is perfectly discontinuous, such a mechanical condition seems to be similar to that of the simulation.

The penetrating resistance in the case of the rigid ball is shown in Fig. 13. The tendency of fluctuation in the rigid ball qualitatively agreed with that obtained by the simulation. Therefore, it follows that the discontinuous property of the present DEM model is very strong and this result would justify the supposition in the previous section that the fluctuation is caused by the large diameter and no-tension joint. The elements with adequate diameter and cohesive effect would be essential to settle the fluctuation of force in order to predict the force between soil and machine parts.

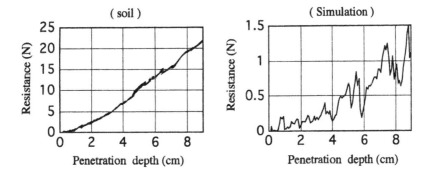

Fig. 11. The penetration resistance of the experiment and the simulation.

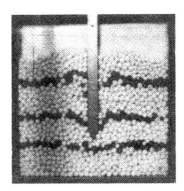

Fig. 12. The additional experiment using the rigid balls.

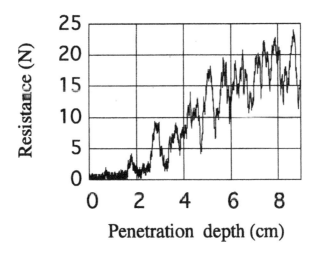

Fig. 13. The result of penetration resistance in the case of rigid balls.

The magnitude of resistance in the simulation is still smaller than that in the rigid ball. Therefore, it follows that the representation of the effect of interlocking in the simulation is still insufficient even in simulating the behavior of rigid balls. In a further study, this problem should be discussed in order to improve the mechanical model in the DEM.

5. Conclusions

The mechanical model of soil was constructed by the Distinct Element Method, and the bar penetration test was simulated using the soil model. As mentioned above, the behavior of the soil was able to be well simulated by the movement of the elements. This paper also shows the effect of one of the parameters (the coefficient of friction between the element and wall) by comparing some results of the behavior of elements. This discussion could contribute to the establishment of the concrete method to determine the values of the DEM parameters based on the soil properties.

As shown in the result of the penetrating resistance, there lie some problems for the DEM to predict the force which acts on machine parts. This paper confirmed the characteristic property of the present DEM model by way of the experiment using the rigid balls. The discussion offers a key to further improvement of the soil model by the DEM.

Furthermore, this paper also contributes to the efficiency of simulation works using the DEM since the limitation of the value of the time interval to get the stability of solution was derived as the practical problem in the calculation. The DEM would improve its ability to analyze the mechanical interaction between soil and machine if this method is discussed more and becomes sophisticated.

Acknowledgements

We wish to express our warmest gratitude to A. Professor Dr. Eiichiro Sakaguchi (Tokyo University of Agriculture) for his assistance in developing the DEM program used in this study.

References

[1] Cundall PA, Strack ODL. A discrete numerical method for granular assemblies. Geotechnique 1979;29(1):47–65.
[2] Kiyama H, Fujimura H. Application of Cundall's discrete block method to gravity flow analysis of rock-like granular materials (in Japanese). Proc JSCE 1983;333:137–45.
[3] Sakaguchi, E. et al. Simulation on flowing phenomena of grains by distinct element method. Proc CIGR Ag Eng Milano '94, 1994, Rep. No. 94-G-025.

Bearing capacity of forest access roads built on peat soils

M.J. O'Mahony, A. Ueberschaer, P.M.O. Owende *, S.M. Ward

Forest Engineering Unit, Department of Agricultural and Food Engineering, University College Dublin, Earlsfort Terrace, Dublin 2, Ireland

Abstract

Significance of the thickness of peat substratum on the bearing capacity of forest access roads laid on peat soils in Ireland is evaluated. Bearing capacity of an experimental pavement was assessed on the basis of its surface deflection measured using a Benkelman beam. The mean deflections for winter, spring and summer seasons were 2.7, 5.1 and 5.4 mm, respectively, and the lower value for winter was attributed to the frozen pavement and lower soil moisture conditions. Pavement response in winter was a function of the interaction term of linear components of thickness of pavement layers and the peat substratum ($R^2 = 0.67$), while in spring ($R^2 = 0.70$) and summer ($R^2 = 0.72$), it also included a moderating quadratic term of thickness of the peat substratum. Deflection generally increased with thickness of pavement and the peat substratum, and effect of pavement thickness was pronounced under peat layer greater than 1000 mm which was attributed to inherent weakness of the pavements over such areas. It is suggested that thickness of the peat substratum may be a basis for developing specifications for timber haulage vehicles, or routeing of such traffic for minimal environmental impact. © 2000 ISTVS. Published by Elsevier Science Ltd. All rights reserved.

Keywords: Timber haulage; Ireland; Subgrade; Benkelman beam

1. Introduction

Over one-seventh of the land-surface in the Republic of Ireland is covered by peat, a non-homogeneous deposit of partially decomposed vegetative matter saturated with water [1]. The organic nature and high degree of dispersion characterise peat as having elastic–viscous–plastic properties. As a system with highly mobile components of water and dispersed solids, a peat bed deforms easily under mechanical pressure, with a reduction in the water potential [2]. Its deformation modulus decreases with

---

* Corresponding author. Tel.: +353-1-706-7418; fax: +353-1-475-2119.
E-mail address: philip.owende@ucd.ie (P.M.O. Owende).

Reprinted from *Journal of Terramechanics* **37 (3)**, 127-138 (2000)

Nomenclature

σ_z	Vertical compressive stress (kPa)
σ_x	Horizontal compressive stress (kPa)
ϵ_z	Vertical compressive strain
ϵ_c	Horizontal compressive strain
ϵ_t	Horizontal tensile strain
a	Radius of circular tyre contact area (m)
P	Wheel contact pressure (kPa)
E	Modulus of deformation (kPa)
D	Pavement deflection (m)
z	Depth from pavement surface (m)
d	Linear component of thickness of peat substratum
d^2	Quadratic component of thickness of peat substratum
r	Linear component of thickness of pavement
A	Interaction term for linear components of thickness of peat substratum and thickness of pavement
B	Interaction term for quadratic component of thickness of peat substratum and linear component of thickness of pavement
β_i	Regression coefficients ($i = 1\ldots.5$)
α	Regression constant
R^2	Coefficient of determination
p	Statistical level of significance

---

water content, and increases with the degree of decomposition. Consequently, there are variations in deformability, bearing capacity and stability of peat soil foundations under varying weather conditions and time frames. These present a considerable challenge to the maintenance and serviceability of flexible road surfaces and other similar structures constructed on peat soils.

There are about 92 000 km of public roads in the Republic of Ireland, and 12% is underlain by peat subgrade. With respect to strength, a strong subgrade has a California Bearing Ratio (CBR) of 15–30% [3]. Roads with peat subgrade have a CBR of 2 to 4%, which indicates their inherent weakness. About 1000 mm of overlaying pavement is needed to limit field deflection, when exposed to say 80 kN axle loads, to ensure that such roads remain in serviceable condition [4]. However, the load imposed by such a pavement may over-stress the underlying peat giving rise to significant settlement [5]. Controlled pre-consolidation at the initial stage of construction is effective in forestalling this phenomenon, but it requires long construction periods and also leads to loss of road elevation, hence, is unpopular. Excavation to more stable materials such as rock, gravel or clay is therefore recommended when the peat layer is deeper than 1500 mm [6], which is only economical when constructing major trunk roads. Minor roads often are laid directly upon peat, an option that is recompensed with extremely poor performance. The ride comfort on such roads deteriorates rapidly, and with cumulative damage, they are eventually unable to bear loads commensurate with their design capacity [7].

Forests in the Republic of Ireland are predominantly established on peat soils, which are unsuitable for the production of agricultural crops. Access roads with peat subgrades therefore, have to be used in general forest management and during logging operations. Haulage of large loads of timber and the movement of other heavy machinery peculiar to mechanised forestry operations accelerate their deterioration. This imposes expensive repair and maintenance costs, hence, makes transportation a costly factor in the overall timber production process [8], and other road users are also aggravated. Consequently, a simple and effective assessment of their future serviceability or bearing capacity facilitates a rational approach to minimising the damage and related haulage costs.

Bearing capacity of a pavement system refers to the number of wheel passages of a specified type that it can support before it reaches an unacceptable level of functional or structural distress [9]. A road should therefore support the applied traffic loading within acceptable limits of ride quality and deterioration over its design life. Primarily, the pavement system must reduce its surface strains and stresses on the subgrade, such that its surface does not crack or deform excessively under peak and cumulative traffic loads [10]. Vehicle operational characteristics may also be adapted to attenuate the load regimes due to traffic on the road structure [11].

Pavement maintenance policies are based on the relationship between the bearing capacity (as indicated by some measure or rating of its structural condition) and applied loads [12]. An accepted method of assessing pavement response [9,13,14] is the measurement of its transient deflections under a standard load. The technique allows prediction of serviceability using empirical relationships between surface deflection and bearing capacity. It is also used to identify weak sections of road networks to enable their restoration, or for the purposes of imposing suitable traffic regulations and restrictions before the structural integrity is seriously impaired. Correlation between trafficability and soil strength is most consistent when the critical soil layer is considered [15]. The objective of this study was to investigate the significance of the peat substratum as the determinant of bearing capacity of roads with thin asphalt surfacing layers and peat subgrade.

2. Loading of flexible pavements

A flexible pavement consists of one or more layers of bituminous material (wearing course) laid on a base of granular material, and a subgrade of natural soil (Fig. 1). Failure of such roads refers to the degradation of the structural integrity or surface profile when trafficked by vehicles over their service life. The most important pavement defects that may be attributed to vehicular traffic are fatigue cracking, longitudinal rutting, plucking and potholing, and polishing or reduced skid resistance [16,17]. Plucking, potholing and polishing are related to the wearing course, while fatigues cracking and rutting are associated with bearing capacity issues (insufficient thickness of pavement layers, and weak subgrade or inadequate drainage) [18].

322

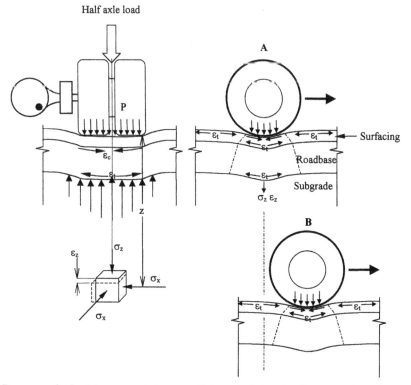

Fig. 1. Response of a flexible pavement during trafficking depicting the phenomena from front and side views of a dual wheel. The successive wheel positions A and B depict the cyclic loading which results in tensile and compressive strains in its strata, and compressive stress and strain on the subgrade.

Under service condition, the top and bottom of the pavement strata move rapidly from compression to tension as they return to their original profiles as illustrated by the successive wheel positions in Fig. 1. Fatigue cracking results from the repetitive tensile strains. From Boussinesq's theory [9], the normal stressing of an infinitely deep elastic, isotropic and homogeneous layer along the centre line of the wheel load that is uniformly distributed over a circular area, is expressed as:

$$\sigma_z = P\left[1 - \frac{z^3}{(a^2 + z^2)^{3/2}}\right] \tag{1}$$

For elastic deformation with no volumetric change, and assuming Poisson's ratio of 0.5 for soil [10], the resultant vertical strain is expressed as:

$$\varepsilon_z = \frac{3P}{2E}\left[\frac{a^2}{(a^2 + z^2)^{1/2}}\right] \tag{2}$$

Under practical conditions, the quantitative values of strain and displacement deviate from Boussinesq's equations due to the difference in stiffness of the paving layers and the subgrade. However, they display similar trends [19].

3. Materials and methods

3.1. Description of the test site

A 4.2 km long experimental road, was located in County Mayo on the West of the Republic of Ireland, running between 9° 35′ W Longitude and 9° 39′ W Longitude and adjacent to 53° 45′ N Latitude. The section considered was flat over a distance of 700 m on the 110 m elevation contour, gradually ascending to 220 m at 1600 m distance, then following the 220 and 240 m contours to the 1900 m distance. The road was constructed in the second half of the 19th century and was considered to be representative of an Irish forest access road laid over peat. Its structure consisted of a crushed limestone base of thickness varying between 150 and 220 mm and sealed (topped) with 5 mm thick bitumen bound layer, a sub-base of sandy gravel of thickness between 220 and 300 mm, and a subgrade of blanket peat (850–2750 mm) and clay. Based on the constituents, it was classified as a flexible pavement [20]. It provided the only access to the adjoining forest, hence, it was assumed that the entire stretch had been exposed to similar traffic. The most recent repairs undertaken were in 1989 and 1996, after damage had been incurred from haulage of timber, and included strengthening of weak sections using coarse aggregate followed by compaction and surface sealing.

Thickness of the peat substratum, visual pavement characteristics and road drainage conditions including depth of ditches and level of flow in them varied considerably over the experimental section. Six separate blocks (Table 1), considered as uniform on the basis of the existing drainage, but including sections with and without evidence of previous repair or restoration, were selected and marked out. The experimental section was marked at 100 m intervals, and at every 10 m within each

Table 1
Description of the experimental road[a]

| Block | MC (%d.b.) | Shear strength (kPa) | Cone index (kPa) | Depth of road layers (mm) | | |
				Base	Sub-base	Peat subgrade
A	590	7–30	280	220	300	1570–2100
B	625	17–38	280	200	300	1200–2750
C	340	22–29	400	160	220	0–1250
D	550	17	200	150	250	1000 1900
E	550	17	200	150	250	1100–1700
F	340	21–29	420	160	220	0–1000

[a] The moisture content (MC), shear strength, and penetration resistance (Cone index) values are for the subgrade and these were measured in summer (July) 1996.

324

block for data points. Points with visible pavement defects were avoided, resulting in eight data points selected in each section.

3.2. Measurement procedures

Pavement construction was verified by digging trenches at four different locations across the experimental road. Depth of peat was further measured at every data point by probing the adjacent road embankment. The extent of cracking as a pavement condition was assessed by visual inspection, and classified according to Lister [21]. Rut depth was measured in the summer (July) with a vertical scale using a 2 m straight bar placed transversely to the wheel-track. The width of pavement and the embankment, at respective blocks were also noted.

Soil parameters of moisture content, shear strength and cone index were used to characterise the subgrade at the initial conditions of test (Table 1). Moisture content was determined by gravimetric method. Six soil samples from each block were collected in plastic bags and transferred to the laboratory. The samples were then dried in an oven at 105°C [22] for 24 h. Moisture content was then evaluated as the ratio of the mass of water extracted to the mass of dry soil.

Shear strength of the subgrade was measured in situ at four locations in the subgrade of each block using the vane shear apparatus [23]. The apparatus was fitted with a 75 by 150 mm vane and its axis was rotated at a rate of 0.1°/s during measurement. Cone index was measured using a recording penetrometer with a 60° cone angle and 500 mm^2 base area. The cone was pushed through the top 300 mm of the subgrade at a rate of approximately 0.1 m s^{-1}. The values in Table 1 are averages from four profile records.

Pavement deflection was measured using a Benkelman beam [24]. The working set-up of the beam is depicted in Fig. 2. During measurement, the measuring beam is passed between the rear dual wheels of the experimental truck and its contact point

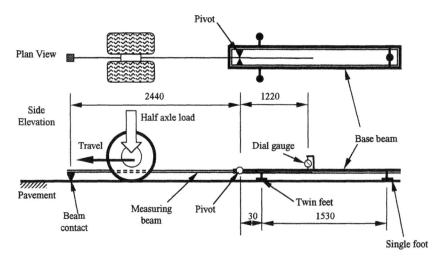

Fig. 2. Salient features of the Benkelman deflection beam.

is rested at the point under test. The base beam is levelled laterally using the adjustable feet, and the dial gauge is zeroed. The test truck then travels at creep speed (< 2 km/h), and the maximum deflection is read on the dial gauge as the wheel passes over the point under test and moves from its zone of influence. The equipment measures pavement deflection to an accuracy of 10 µm. Characteristics of the experimental truck are presented in Table 2. Deflection measurements were taken in three different occasions to include the effects due to season variations. Initial measurements were taken in winter (January) of 1996, and subsequent measurements in spring (March) and summer (July). Weather data (Table 3), including amount of rainfall and temperatures over 14 days prior to and during the days of measurement, was obtained from the meteorological service. The temperature at the surface of the experimental pavement and at 100 mm depth was also recorded. It is known that pore water pressure influences deformability, bearing capacity and stability of saturated soils such as peat [25]. However, since the measurements were carried out with transient loads and in single wheel passes, its effects were ignored.

3.3. Data analysis

Possible association between measured deflection and depth of peat in the subgrade for the three series of experiments was evaluated by multiple non-linear regression and correlation analyses. Based on the scatter plots of the deflection

Table 2
Specifications of the vehicle used for loading the pavement for deflection measurement

Specification	Recommended [24]	This study
Vehicle weight		
Rear axle load (kN)	62.3	62.2
(kg)	(6350)	(6340)
Tyre		
Size	7.50R20 or 8.25R20 or 9.00R20	9.00R20
Inflation pressure (kPa)	590	590

Table 3
Weather conditions prior to and during the experiments in 1996

	Pre-test average	Winter (January)	Pre-test average	Spring (March)	Pre-test average	Summer (July)
Temperature (°C)						
Air	2	1	10	11	12	14
Soil (top 100mm)	2	2	10	11	11	16
Surface		−2		11		21
Rainfall (mm)						
Days with	19	1	25	4	47	1
< 0.05 mm	10		2		4	

measurements and theory of elastic deformation of flexible pavement, the association (functional forms) between surface deflection, and thickness of pavement and peat substrata for the three series of experiments was determined. The base and subbase were considered as a single unit since, for practical purposes, their function is to attenuate the stress on the subgrade by spreading the load as indicated by the dotted envelop in the road structure illustrated in Fig. 1. The bituminous surfacing layer is usually employed as protection from ingress of water, and it does not contribute to the structural rigidity of these pavements. The relationship with all possible terms for variation of deflection with linear component of thickness of pavement, and linear and quadratic components of thickness of the peat substratum, including all possible interaction terms was prescribed as:

$$D = \alpha + \beta_1 d + \beta_2 d^2 + \beta_3 r + \beta_4 A + \beta_5 B \tag{3}$$

where $A = dr$ and $B = d^2 r$

Possible association between rut depth and class of surface cracking, with thickness of pavement and depth of peat in the subgrade was evaluated by multiple linear regression and correlation analyses. Stepwise multiple regression technique [26] was used in each case to eliminate the insignificant regression terms, by including each term successively based on their relative contributions to the variation in pavement deflection, rutting and cracking [27].

4. Results and discussion

The results of the variation of pavement deflection with depth of peat in the subgrade and thickness of the pavement are presented in 3D contour surfaces in Fig. 3. The respective multiple regression models are presented in Table 4. Deflection of the pavement surface almost doubled in spring (March) compared to the winter season (January). The differences ranged from 1.78 to 3.36 mm, corresponding to increases of between 48 and 98%. The differences between the deflection data of March and July were statistically insignificant, ranging from 50 to 590 μm, equivalent to increase of between 1.4 to 8.3%. The data indicated higher deflections under wetter and warmer weather conditions. For example, the mean pavement deflection for experiments in winter (2.7 mm), was lower that for spring (5.1 mm) and summer (5.4 mm). The corresponding average precipitation over 14 days prior to measurements was 19, 25 and 47 mm, with pavement temperatures of −2, 11 and 21°C, respectively (Table 3). For subgrades with no peat deposits ($d = 0$), the deflection approximately tripled for spring and summer compared to the corresponding winter values.

Variation in deflection values for winter (January) was best described by the interaction of linear components of pavement thickness and thickness of peat substratum ($R^2 = 0.67$), i.e. higher deflections were recorded in sections with thicker peat substrata, and also were pronounced in areas with thicker pavement layers (Fig. 3). Models for spring (March) and summer (July) included the interaction of the linear components of the independent variables (d, r), and the quadratic components of

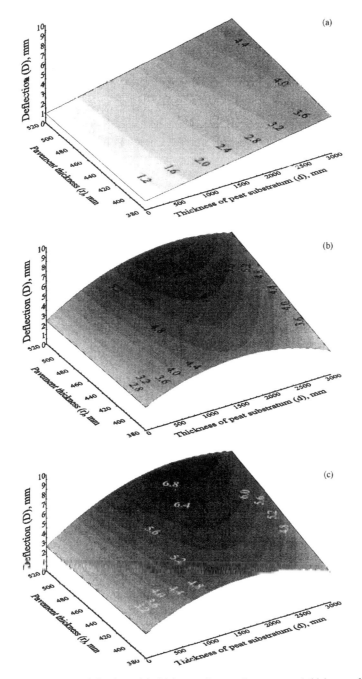

Fig. 3. Variation of pavement deflection with thickness of peat substratum and thickness of experimental pavement for measurements in (a) winter, (b) spring and (c) summer of 1996. The corresponding pavement temperatures were −2, 11 and 16°C, respectively. The colour scale levels are deflection values in mm, and range from 0.89 (white) to 7.2 (black).

328

Table 4
Multiple regression models and correlation for measured surface deflection against thickness of pavement and depth of peat in the subgrade [Eq. (3)]

Season	α	β_2	β_1	p	R^2
Winter (January)	0.94		2.48×10^{-6}	< 0.01	0.67
S.D.	0.196		2.60×10^{-7}		
Spring (March)	2.41	-1×10^{-6}	7.91×10^{-6}	< 0.01	0.70
S.D.	0.289	2.7×10^{-7}	1.29×10^{-6}		
Summer (July)	2.52	-1×10^{-6}	8.42×10^{-6}	< 0.01	0.72
S.D.	0.293	2.7×10^{-7}	1.31×10^{-6}		

depth of peat in the subgrade (d^2) with R^2 values of 0.70 and 0.72, respectively. For constant pavement thickness, surface deflection was observed to increase with thickness of peat substratum, at a reducing rate (Fig. 3), which is in agreement with the elasticity theory of pavement loading by wheeled traffic [9]. Up to 1000 mm thickness of the peat substratum, deflections for spring and summer were mainly dependent on the inherent size of the peat-bed. From 1000 mm thickness of peat substratum, the deflection also increased with the pavement thickness.

The linear variation and lower values of pavement deflection in winter (January) could be attributed to the frozen surface (temperature−2°C) and the prevailing moisture conditions (Table 3). The expected higher pavement stiffness at freezing temperatures may have reduced surface displacement, and also improved the pavement's load spreading ability, to limit the stress influence on the subgrade. The higher surface deflection with thicker pavement layer at constant thickness of peat substrata (Fig. 3), was an interesting observation in that, a pavement essentially should limit strains at the bottom of the surface layer and stresses on the subgrade. Considering elastic solutions for a three-layer pavement system using Method of Equivalent Thickness to allow for application of Boussinesq's equations [9], increasing the thickness of pavement layers is expected to lower the stress on a subgrade, to limit surface deflection. Thicker pavements therefore effectively enhance serviceability of roads laid on peat soils [4].

For the range of pavement thickness (380 to 520 mm) and depth of peat substratum (0 to 2750 mm) studied, the effectiveness of the greater pavement thickness seems to be negated in areas with peat substrata of greater than 1000 mm thickness. This could have been due to inherent structural weakness that was exemplified by discernible layer colours in the profiles of the excavated road bases. The layer colours could be attributed to successive repairs on the road, hence, the pavement stiffness in the sections may not be commensurate with the measured thickness. Unpublished results with the authors, from pavement layer assessment using Ground Penetrating Radar, also suggests that such repairs are intended mainly to regulate the road surface for maintenance of a good the ride quality, hence, are confined to the damaged areas such as rutting in the wheel tracks. However, the significance such repairs (mainly overlaying with aggregate) on the overall structural performance of the pavement systems is still unclear at this stage and may require further investigations.

The observations suggest that thickness of the peat substratum is a constraint on the bearing capacity of such roads, and could also determine the effective techniques for their management. Correlation of rut depth with thickness of peat substratum and pavement was also attempted, and only the coefficient for the peat layer was significant and suggesting that ruts were deeper in areas with thicker peat layers in a linear relationship. Forty three percent of the variation in rut depth could be attributed to the underlying peat, while there was no correlation between the observed pavement cracking and thickness of pavement or peat substrata.

Magnitude of transient deflection has been used with empirical relationships (deflection charts), to estimate bearing capacity of pavements for scheduling maintenance of road networks and the design of overlays [13]. The charts are based on deflections of less than 2 mm, but values ranging between 0.5 and 9 mm were recorded in this study, which suggest that the relationships may not be rational for pavements with soft soil subgrade such as peat. The data for winter (Fig. 3a) suggests a critical depth of peat that may limit field deflection to below 2 mm, but it is affected by the prevailing temperature and subgrade moisture conditions. Hence, for all-season timber transportation in Ireland, the roads are under-designed for the loads for which they may be exposed. However, due to their low volume of traffic, it is unlikely that they will be re-laid or strengthened in the short term to enable them to withstand the load regimes associated with forestry operations. Therefore, limitation of pavement damage by control of vehicle operation parameters such as axle load characteristics [17], adoption of permanently reduced and variable tyre pressure technologies [11], and routeing based on seasonal road strength, provide some alternatives for forestry operations including timber haulage. Economics of individual alternatives or different combinations should determine the practical option.

5. Conclusion

Surface deflection of flexible pavements laid over peat soils is mainly affected by thickness of the peat substratum. Size of the peat substratum is therefore important to the bearing capacity of forest access roads with thin asphalt surfacing layers. It could therefore be a basis for prescribing specifications for environmentally sensitive timber haulage vehicles, or haulage and routeing restrictions for logging traffic in areas with such roads.

Acknowledgements

This study was supported by COFORD (Project 3-7-1995) and Coillte Teoranta.

References

[1] Galvin LF. Physical properties of Irish peats. Irish J of Agric Res 1976;15(2):207–21.

[2] Solopov SG, Volarovich MP, Korchunov SS, Tsuprov SA, Mogilevskii II, Abakumor ON. Physical and mechanical properties of peat. In: 3rd Int. Peat Congress 1968. p.155–6.

[3] Anon. Road development strategy 1000–2000, Dublin County and City Engineers Association, 1988.

[4] MacFarlane IC. Muskeg engineering handbook. Toronto: University of Toronto Press, 1969.

[5] Davitt S, Killeen RC. Maintenance techniques for bog roads [National Roads Authority (NRA) report RC.375]. Dublin: NRA, 1996.

[6] Anon. Specifications for roadworks. Dublin: Department of Environment Stationary Office, 1978.

[7] Hampson IDM. Constructing low cost un-surfaced roads. Forestry Engng Specialist Group. Edinburgh: Heriot Watt University, 1993.

[8] Anon. Pathway to progress: harvesting and transport. Dublin: National Council for Forest Research and Development, 1994. p. 57–75.

[9] Ulliditz P. Pavement analysis. Amsterdam Elsevier, 1987. p.8,29–33, 44–52.

[10] Lay MG. Handbook of road technology, vol. 1. London Gordon and Breach Science Publishers, 1990. p.193–233.

[11] Bradley A. The effect of reduced tire inflation pressure on road damage: a literature review(Report SR-123). Quebec: FERIC, 1996.

[12] De Pont J, Pidwerbesky B. Vehicle dynamics and pavement performance models. Proc 17th ARRB Conf 1994;17(2):17–31.

[13] Kennedy CK, Lister NW. Prediction of pavement performance and the design of overlays (Report LR 833). TRRL, 1978.

[14] Hunter RN, editor. Bituminous mixtures in road construction. London: Thomas Telford Publishers, 1994.

[15] Knight SJ, Meyer MP. Soil trafficability classification scheme. In: Proc. 1st int. conf. mech. soil–vehicle systems, Torino, 1961. p. 567–74.

[16] Atkinson K. Highway maintenance handbook. London. Thomas Telford Ltd. p. 93–8.

[17] Cebon D. Vehicle-generated road damage: a review. Vehicle System Dynamics 1989;18:107–50.

[18] Anon. Catalogue of pavement defects. Dublin: National Inst for Physical Planning and Construction Research, 1985.

[19] Yoder EJ. Principles of pavement design. London. John Wiley. p. 20–51.

[20] Croney D, Croney P. The design and performance of road pavements. London McGraw-Hill, 1991. p. 15–21.

[21] Lister NW. Deflection criteria for flexible pavements (Report 375). TRRL, 1972.

[22] Day JH, Rennie PJ, Stanek W, Raymond GP. Peat testing manual (technical memo 125). National Research Council of Canada, 1979.

[23] Anon. B.S.1377: methods of tests for soils for civil engineering purposes. London: British Standards Institution, 1975.

[24] Kennedy CK, Fevre P, Clarke CS. Pavement deflection: equipment for measure in the United Kingdom (Report 834). TRRL, 1978.

[25] Amaryan LS. Soft soil properties and testing methods. Rotterdam: Balkema, 1993. p. 51–4.

[26] Sen A, Srivstava M. Regression analysis: theory, methods and applications. New York: Springer Verlag1990. p. 233–52.

[27] Anon. Minitab statistical software release 8 Pennsylvania: Minitab Inc, 1991

AUTHOR INDEX

Printed and bound by CPI Group (UK) Ltd, Croydon, CR0 4YY

03/10/2024

01040317-0019